Smart Materials and Structures

Smart Materials and Structures

M.V. Gandhi and B.S. Thompson

Intelligent Materials and Structures Laboratory,
Department of Mechanical Engineering,
Michigan State University, East Lansing, USA

CHAPMAN & HALL
London · Glasgow · New York · Tokyo · Melbourne · Madras

Published by Chapman & Hall, 2–6 Boundary Row, London SE1 8HN

Chapman & Hall, 2–6 Boundary Row, London SE1 8HN, UK

Blackie Academic & Professional, Wester Cleddens Road, Bishopbriggs, Glasgow G64 2NZ, UK

Van Nostrand Reinhold Inc., 115 5th Avenue, New York NY10003, USA

Chapman & Hall Japan, Thomson Publishing Japan, Hirakawacho Nemoto Building, 6F, 1–7–11 Hirakawa-cho, Chiyoda-ku, Tokyo 102, Japan

Chapman & Hall Australia, Thomas Nelson Australia, 102 Dodds Street, South Melbourne, Victoria 3205, Australia

Chapman & Hall India, R. Seshadri, 32 Second Main Road, CIT East, Madras 600 035, India

First edition 1992

Typeset in Times 10/12pt by Columns Design & Production Services Ltd., Reading
Printed in Great Britain by St Edmundsbury Press Ltd., Bury St Edmunds, Suffolk

ISBN 0 412 37010 7 0 442 30876 0 (USA)

A catalogue record for this book is available from the British Library

Library of Congress Cataloging-in-Publication data available

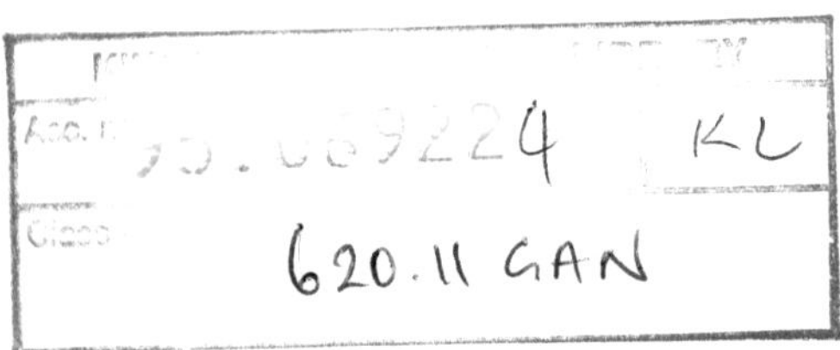

Contents

Preface

Human civilization has been so profoundly influenced by materials technologies that historians have defined distinct time periods by the materials that were dominant during these eras. Thus as humankind embarked upon the continual quest for superior products and weaponry fabricated from superior materials, terms such as the *Stone Age*, the *Bronze Age*, and the *Iron Age* have entered the vocabulary. The current *Synthetic Materials Age* featuring plastics and fiberous composites is providing a viable precursor to the dawn of a new era, the *Smart Materials Age*, which will capitalize on these synthetic materials in order to exploit several eclectic emerging technologies for the synthesis of smart materials exhibiting nervous systems, brains, and muscular capabilities. The degree of sophistication displayed by this new generation of materials will depend mostly on the individual application, however, it is anticipated that these innovative material methodologies and technologies will eventually be utilized in several diverse fields of science, such as nanotechnology, biomimetics, neural networking, artificial intelligence, materials science, and molecular electronics, for example.

This new generation of smart materials will significantly impact civilization. For example, some classes of materials will be able to select and execute specific functions autonomously in response to changing environmental stimuli, others will only feature embedded sensory capabilities in order that a structural member is manufactured to comply with the quality control specifications. Self-repair, self-diagnosis, self-multiplication and self-degradation are also some of the characteristics anticipated to be a feature of the supreme classes of smart materials. It is clearly evident upon reviewing the capabilities of these smart, or intelligent, materials in an engineering context that all aspects of civilization will be influenced by these new generations of innovative materials as designers capitalize on their unique capabilities in industries as diverse as aerospace, manufacturing, automotive, sporting goods, medicine, and civil engineering.

The objective of this book is to develop an introduction to the embryonic fields of smart materials and structures and also intelligent materials in one volume, although the precise nomenclature and definitions in this relatively immature field have not been fully researched and formally documented in the scientific literature. The book is targeted at the undergraduate and graduate engineering population, applied scientists, and professional engineers in industry and government. Nevertheless, we believe that the contents of most chapters are largely intelligible to a more general, broader lay audience, since the book is devoid of mathematical derivations. The reader who thirsts for this challenging facet of the field is directed to the comprehensive bibliography at the end of the book which contains a rich fountain of ideas to satisfy a variety of palates. The objective of compiling this bibliography is to develop a comprehensive data-base of scientific publications and patents which, at this time, provide the nucleus of this diverse discipline. Naturally, there may be some omissions from this list, for which the authors assume full responsibility, and they would welcome correspondence acknowledging their oversights.

This book is formatted to provide an introduction to the field of smart materials and structures and also intelligent materials prior to highlighting several facets of the field such as actuator systems, sensory systems, and future research issues. Chapters 1–3 introduce the reader to intelligent materials, smart structural systems, and potential applications. Chapters 4–7 then present a more comprehensive treatment of some of the technologies inherent in the current generation of smart structural systems, while Chapter 8 presents projections for the maturation of the field and also a variety of research issues, that must be addressed for fully exploring the potential of smart materials.

The development of this book could not have been undertaken without the authors receiving education from academically and intellectually gifted individuals and also assistance from several key support staff, graduate students, and post-doctoral fellows.

Mukesh V. Gandhi received his undergraduate education in Mechanical Engineering at the University of Bombay. Professor Gandhi was awarded an M.S. degree in Mechanical Engineering at Wayne State University, Detroit, where his thinking on mechanics and materials was greatly influenced and molded by Professor C.N. DeSilva, who was a fountain of knowledge and confidence. He received his Ph.D. degree in Applied Mechanics from the University of Michigan, Ann Arbor, where his dissertation work in nonlinear mechanics and materials was guided and supervised by Professor A.S. Wineman, who was always full of creative ideas, enthusiasm, and encouragement. The first author thanks these two unique individuals, and also Professors K.A. Kline, C.S. Yih, and W.H. Yang for making such a positive contribution to his education and career. On a more personal note,

Mukesh V. Gandhi thanks his family for providing relentless support and encouragement, and also Mr Bhupinder Bedi for molding the philosophical and spiritual basis of his life.

Brian S. Thompson resided in the English Lake District during his formative years. He was educated at Workington Secondary Technical School and Workington Grammar School where he benefitted immensely from the traditional, somewhat Victorian, educational environment imposed upon him by the masters in these institutions. Upon successfully completing this phase of his education, Professor Thompson was privileged to receive a university education in both England and Scotland where he studied Mechanical Engineering. He was awarded B.Sc. (Hons) and M.Sc. degrees at the University of Newcastle-upon-Tyne where his interest in engineering design and machines was stimulated by the lecturing of Mr A.A. Fogarasy and Professor L. Maunder. Subsequently, he undertook doctoral studies on nonlinear vibration phenomena in linkage mechanisms at the University of Dundee under the direction of Professor A.D.S. Barr. Dr Thompson is extremely grateful to Professor Barr for his visionary thinking, and also for the stimulating and innovative research environment he generated. This research training was augmented by numerous stimulating discussions with Dr J.M.B. Brown of the Department of Mechanical Engineering. Dr Thompson acknowledges with deep gratitude the influence of these gifted academics on his education, and the continued support of his family during these years of formal education.

This book would not have reached fruition without the stimulus of Mr Mark Hammond from *Chapman & Hall* who was a continual source of guidance during this undertaking. The burden of typing the manuscript was most efficiently handled by Miss Anita VanHall, a student from the Michigan State University Business School who is employed in the Intelligent Materials and Structures Laboratory, while the diagrams and book-jacket were created and skillfully prepared by one of our undergraduate students, Mr Dave Downes. We are extremely grateful to these talented individuals for their help in this endeavor. In addition we wish to acknowledge the editorial help provided by Mr Mark Wolschon and Dr Kasiviswanathan Sethupathi, and also the constructive comments and suggestions of the anonymous reviewers of the draft manuscript of the book which we have subsequently incorporated in the text. Last, but certainly not least, the authors are indebted to their wives for their encouragement and continual support.

Many of the experimental results and methodologies presented in the book are consequences of the research endeavors of the authors who were funded by several federal agencies. The authors wish to acknowledge the financial support provided by the Defense Advanced Research Projects Agency (DARPA), program monitor Dr R.L. Rosenfield; the US Army Research Office (ARO), program monitor Dr G.L. Anderson; the Mechanical Systems Program of the National Science Foundation (NSF), program monitor Dr E. L. Marsh; and the Manufacturing Systems Program of the National Science Foundation (NSF), program monitor Dr M.E. DeVries; and the Michigan Department of Commerce Research Excellence Fund.

We are conscious of human frailty and limitations, consequently we realize that some errors must have inevitably escaped our scrutiny as we toiled to compile this, the first book exclusively dedicated to the embryonic eclectic field of smart materials. The authors would, therefore, be obliged if upon detection, readers would advise them of any omissions and oversights that they may have discovered. We trust that this book will stimulate the readership to cogitate upon the significant role of materials in the shaping of society at the dawn of the twenty-first century which will certainly witness the emergence of an era in which scientists and engineers will learn from biological systems in order to synthesize smart materials. The authors have the clarion conviction that smart materials will profoundly influence human civilization and the maturation of various embryonic advanced technologies and that the integration of such technologies will trumpet humanity's quest for synthesizing new generations of materials whose capabilities may indeed surpass those bestowed by Mother Nature. The odyssey through these uncharted terrains promises to be an exciting and memorable one.

M.V. Gandhi
B.S. Thompson
East Lansing, Michigan

1

Historical prologue

Materials technology has had such a profound impact on the evolution of human civilization that historians have characterized periods in that evolution by such terms as the Stone Age, the Bronze Age, and the Iron Age. Each new era was brought about by the continuing quest for even better products, a quest that is very much in evidence today. The current 'Synthetic Materials Age' has been precipitated by humankind's demand for materials with superior performance characteristics, inspired primarily by the quest to conquer the last frontier of space. The dawn of the 21st century will witness the emergence of the 'Smart Materials Age'. It will be catalyzed by a technological revolution that will exploit several emerging technologies, such as materials science, biotechnology, biomimetics (the development of synthetic systems by the use of information obtained from biological systems), nanotechnology (the use of molecular-scale objects as components of molecular machines), molecular electronics, neural networks (the electronic simulation of the neural composition of the human brain), and artificial intelligence. These technologies will provide the nervous system, the brains, and the muscles for a new generation of advanced materials and structures that are at present a mere skeleton compared with the anatomy perceived in the not-too-distant future. This quantum jump in materials technology seems certain to revolutionize the future in ways far more dramatic than the way the electronic chip has catalyzed the evolution of our life-styles.

The relentless pursuit of excellence in all technologies during the dusk of the 20th century and also the dawn of the next century will be dominated by advanced materials and the impending revolution in smart materials. The principal ingredients of the exceedingly diverse classes of engineered materials upon which these eras are based are the synthetic plastics, and mundane natural materials such as clay and sand. By exploiting state-of-the-art technologies in diverse fields of science, groups of engineers, chemists, physicists and materials scientists have devised innovative inter-disciplinary techniques for synthesizing, analyzing, and manufacturing new generations of engineered materials.

These new classes of engineered synthetic materials comprising ceramics, plastics and composites have had a major impact upon our lifestyles in areas as diverse as those utilizing the more traditional naturally-occurring materials. Figure 1.1 provides a flavor of this diversity. Within the sporting goods industry, for example, fiberous composite materials mimicking biological materials have revolutionized the design and manufacture of fishing rods, golf clubs, archery bows, skis and tennis rackets. Synthetic biomaterials have had a major role to play in the numerous prosthetic devices available to *Homo sapiens sapiens* for prolonging life and relieving human suffering. Advanced lightweight composites and high performance plastics have revolutionized concepts in the aerospace and automotive industries. Super-conductivity, where electricity flows through material without energy loss, promises faster computers and cheap electro-magnetic devices which could be exploited in numerous applications such as high-speed bullet trains,

Classes	**Examples**	**Principal properties or characteristics**	**Examples of usage**
Metals and alloys	Steels, superalloys, non-ferrous	Strength, toughness	Automobiles, aircraft, pressure vessels
Ceramics	Alumina, silicon nitride, metal carbides	Temperature and corrosion resistance, high hardness	Furnace refractories, cutting tools, engine components
Plastics	Polymers, rubbers, polyurethanes	Strength, low density, corrosion resistance,	Pipes, panels, process plant
Composites	Fiber-reinforced plastics, metals or ceramics	High toughness high-strength, low-weight	Aircraft and other transport components
Construction materials	Building stone, cement	Durable, plentiful supply, cheap	Buildings, roads, bridges

Fig. 1.1 Classes of materials and their usage.

for example. A new generation of integrated circuits featuring gallium arsenide are much more powerful than conventional silicon chips. They promise a new class of superior computers which could positively impact numerous military systems, robotic devices, and instrumentation. Photonics is transforming data networks and telecommunication by permitting light pulses to transmit high-volume information incredibly quickly through optical fibers. These diverse examples provide glimpses of some of the

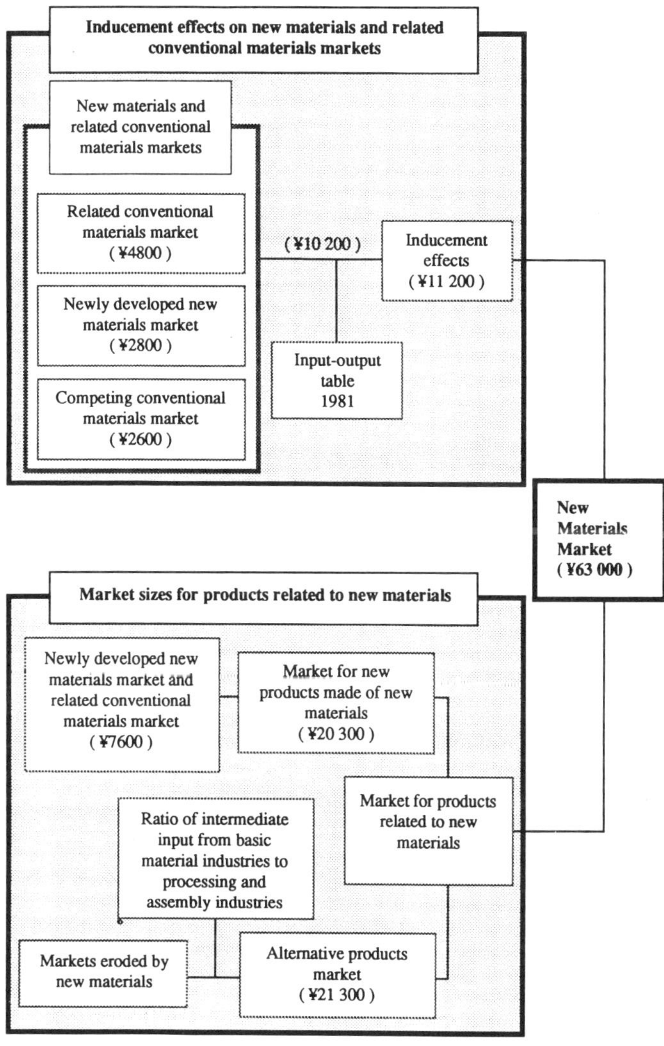

Fig. 1.2 Estimated total value of new materials markets.

revolutions currently taking place in different marketplaces as a consequence of the revolutionary developments in the field of advanced materials.

The innovative developments in advanced materials technologies are fuelled by global competition between the Western European nations, the Pacific Rim nations lead by the Japanese, and the United States. Generally, these advanced materials not only create a conventional market for the materials themselves, but they are a necessary precursor for the evolution of new products and the enhancement of existing products by exploiting these advanced materials as shown in Figure 1.2. Indeed it has been estimated that between two and three workers in other sectors of the economy depend on the livelihoods of individuals employed in the primary materials industries. Thus materials are not just end-products in themselves. Rather, they permit end-products to attain their design specifications by providing the vital framework for the diverse mechanical, biomedical, building, chemical, and transportation segments of the economy, for example. Material-specific innovations should, therefore, be considered in the context of the final engineered product.

1.1 An historical perspective

It is clearly evident therefore, that progress in materials science stimulates the economy and largely dictates the standard of living enjoyed by society. This observation is not new. The dominant role played by materials in the evolution of human civilization has been clearly recognized by historians, and has been responsible for terms such as the 'Stone Age', the 'Bronze Age', and the 'Iron Age', for example. In these time-periods the standard of living was largely dictated by the most advanced material of that era. While the impact of the material dominating these three time-periods was significant, the rate of progress was relatively slow on the time-line of human civilization, being somewhat evolutionary in nature. This is in sharp contrast to the very rapid progress made in materials science and technology during the last century which has been revolutionary and somewhat explosive in nature. This explosion is attributed to a number of factors including the uninhibited dissemination of information, global competition, and advanced communication technologies.

At the beginning of the Paleolithic period, some one million years ago, the earliest humans, *Homo habilis*, selected tools from whatever natural objects were readily available to this primitive species, rather than creatively manufacture appropriate tools. These natural objects included fibers, skins, and bones and horn, but the hardest and densest material was stone. While various types of stones and minerals were available, the embryonic philosophies of materials science are clearly evident in this time-period because these primitive people judiciously selected flint as the most appropriate material for both tools and weapons. Thus, while flint is a somewhat brittle material it was nevertheless not too fragile for serving as a knife, scraper, or spearhead for example. With the availability of high-quality flint in certain regions, centers trading in flint were established and these evolved into settlements. Civilization as we now know it, was about to begin and this pattern of interaction between technology and human settlement has been a continual feature of mankind's odyssey through time.

At the beginning of the Old Stone Age, *Homo habilis* could generate fire by striking pieces of iron pyrites and flint against one another, in order to develop sparks which would ignite adjacent flammable materials. Alternatively he would twirl a stick between the palms of his hands with one end located in a hole in a second wooden member in order to start a fire. The exploitation and control of fire had tremendous cultural consequences. Fire provided heat to warm these prehistoric human beings and furthermore it provided a cooking resource. The art of cooking triggered the evolution of utensils, pots and pans which facilitated the boiling, baking, frying and stewing of foods. However, in order to create the utensils, pots and pans, these early human beings had to further develop the embryonic field of materials science by molding and subsequently fire-hardening clay vessels. Clay was, therefore, the first organic structural material to have its physical properties dramatically modified intentionally by advanced hominids. This discovery triggered the establishment of one of the first industries in civilization which was focused upon pottery.

The transformation of the properties of soft pliable clay into a product featuring a hard stone-like material which could serve a useful purpose further stimulated the search for new materials. This curiosity subsequently resulted in experiments with fire and numerous minerals. An outcome of this activity was the glazing of pottery, which was partially responsible for the development of glass, and also the smelting of metals from ores at the dusk of the Paleolithic Period.

During the Mesolithic and Neolithic Periods, man began to utilize different materials to decorate pottery and also to create cave paintings of his life-style, such as those in Lascaux, France, and Altamira in Spain. He also began to develop the field of agriculture. This involved the development of various tools in wood and stone to till the ground, and thresh grain, and to finally grind it into flour. These developments required man to select appropriate materials for the implements based upon his primitive knowledge of materials science. At the same time a textile industry also began to be established, based upon natural fibers such as flax, reeds, and animal hairs. The fabrics manufactured in this embryonic industrial environment were principally used for clothing, and the industry also spawned the development of primitive machinery and devices.

Towards the end of the Neolithic Period of the Stone Age the beginnings of writing began to be established in Egypt and also in Iraq. This cultural development was probably motivated by commercial growth which stimulated the necessity for recording business transactions. Two different types of writing emerged from these early civilizations, and both forms were dictated and motivated by the different types of materials that were available to each group of peoples. Egyptians living in the Nile delta developed cursive hieroglyphic writing using brushes and ink to record their thoughts on papyrus, a flexible planar medium developed from the interweaving of a reed abundantly available in the delta region. In sharp contrast to this philosophy of documenting information, the Sumerians in the Euphrates-Tigris valley developed cuneiform writing by recording their thoughts on clay tablets. Clay was abundant in the Sumerian alluvial region, and inscriptions were made in the soft fine-grained mineral using sharp needle-like instruments, prior to drying the tablets by utilizing the radiated heat from the sun.

In the Middle and the New Stone Ages, man domesticated animals and in addition to inventing writing, man invented the plough, the sailing ship, and also developed a high temperature kiln which further stimulated the smelting and deployment of metals. This latter capability precipitated the demise of stone as the premier material, and facilitated the development and dominance of bronze and ultimately iron.

The Bronze Age is typically characterized by the period 3500 BC to 1000 BC. During this time tools and weapons were fabricated in this alloy of copper and tin, thus bronze is the oldest alloy utilized by mankind. This metallurgical feat is further evidence of *Homo sapiens sapiens* increasing his skills in materials science and technology. This achievement was motivated by the ability of castings to function as tools and weapons. Typical artefacts from this era were cups, urns, ornaments, helmets, swords and shields.

The Iron Age, which began in approximately 1500 BC in Asia Minor and continued through to the 20th century AD, is the historical period in which technological innovations facilitated the utilization of a cheap and abundant ferrous material resulting in broad ramifications. The existence of iron ore had been known for some time before the Greco-Roman period but the ore could not be refined because of the inability to develop furnaces which could initiate the chemical reactions at 2800° F. This temperature compared with the 1900° F required for the refinement of copper ores. Once the technology had been established, the manufacturing techniques spread rapidly since iron ores were abundant and cheap. This economic factor was really significant in the provisioning of an army, where a leader could readily furnish all his troops with iron swords rather than have a small number armed with bronze swords, while the remainder entered battle with more primitive weapons such as spears, or bows and arrows.

Furthermore, iron products which were wrought rather than cast, were harder and stronger than their bronze counterparts. These characteristics profoundly changed the resources and capabilities of society. During the dawn of the Iron Age, mechanical geniuses such as Archimedes, Ctesibius, and Hero exploited these characteristics in numerous mechanical devices such as the pulley, the screw and the lever, and also compressed-air engines, hydraulic devices, and screw-cutting machines. Iron plough shares also permitted deeper plowing techniques to be developed which resulted in the cultivation of inferior-quality soils, and human settlement in regions where these new technologies could now be applied successively.

While iron was undoubtedly the dominant material during the Roman Empire, the discovery of hydraulic cement at that time had an important role to play in the growth of the Roman influence throughout Europe. This new structural material comprising silica, limestone, and alumina was exploited in the construction of civil engineering monuments such as roads, buildings, bridges, and aqueducts.

Natural magnets, based upon iron ore, Fe_3O_4, were employed for compasses by the Chinese between approximately 500 BC and AD 1000. However, the deployment of these naturally-occurring materials for navigation purposes was not developed in Europe until the 13th century, whereupon there followed increased exploration by sea which resulted in the transportation and trading of goods.

Iron continued to be the premier material exploited by civilization until the emergence of steel in the 18th century. These two materials were instrumental in motivating the English Industrial Revolution which resulted in both significant socio-economical and cultural changes, as well as technological changes. This process of dramatic change from an agrarian economy to an industrial economy was repeated in other parts of the world during the 19th century and is still being repeated in the 20th century.

The Industrial Revolution was fuelled by the synergistic interactions of new materials, an increasing utilization of new materials, an increasing utilization of scientific data, new machines and energy sources, new organizational structures for the workforce, and improved transportation and communication technologies. While these interactions precipitated mass production environments and the increased consumption of raw materials, there were numerous non-technical consequences of this revolution. These included an increase in international trade; workers were required to develop machine-oriented skills and be subjected to industrial discipline; the modification of laws in response to industrialization; the growth of suburban areas; the more uniform distribution of wealth; the mechanization of agriculture enabled the industrial communities to be fed, and humankind received a psychological boost from the ability to further exploit natural resources in a controlled manner.

Thus naturally-occurring materials were a principal ingredient of the Industrial Revolution which was responsible for major changes in society. The complex interactions between materials, technological innovations, and civilizations has subsequently become more inextricably intertwined at the dusk of the 20th century.

1.2 Plastics

At the dawn of the 20th Century, natural materials were still the dominant resources for the industrial, commercial, and military segments of the economy. However, this dominance was to be subjected to an onslaught, and subsequently overwhelmed by an embryonic technology focused on the development of synthetic materials, namely, plastics. From beginnings in laboratories in the 1860s, synthetic plastic materials, such as celluloid and xylonite, were invented prior to the synthesis of cellophane in 1908 and bakelite in 1909. This material synthesis philosophy represented a major departure from the traditional approach of utilizing natural materials with their known defects and limitations.

From these early beginnings at the dawn of the twentieth century, the plastics industry has now expanded to such an extent that there are now tens of thousands of different plastics available in the marketplace, and in 1979 the volume of plastics manufactured in the United States exceeded that of steel for the first time. The principal elements from which these advanced materials are fabricated are carbon, hydrogen, oxygen, and nitrogen, and these elements are typically obtained from coal, limestone, petroleum, salt, and water. Thus the technologically advanced western world is not as dependent upon raw materials from the third world nations as it was in the past. These plastic materials are engineered by carefully synthesizing the appropriate micro-structure in order to develop the desired material characteristics. The properties of the resulting synthetic materials can be tailored to comply with the required design specification to yield a diverse range of properties which are hard or soft, wear resistant, transparent, degradable, and heat resistant, for example.

This diverse range of tailorable material properties has resulted in plastics replacing traditional materials in many industries. In the packaging industry, for example, the outstanding durable properties offered by many plastics has resulted in these materials replacing metal, glass and paper parts, and this trend is also clearly evident, for example, in both the household appliance industry and the electronics industry, as revealed by any historical review of the materials employed to fabricate coffee makers, blenders, vacuum cleaners, hi-fi equipment, and televisions.

The automobile industry has also witnessed a materials revolution with the replacement of numerous metallic parts by a variety of high-grade plastics. This revolution-engine has been fuelled by legislation demanding greater safety, ecological concerns associated with exhaust emissions, and mandates for reduced fuel consumption. In the USA, automobiles are subjected to bumper impact tests in which a car must be capable of withstanding a 5 mph impact without damaging the bumper at the front or rear of the vehicle. The introduction of this legislation in the early 1970s triggered the demise of the traditional steel structural members that previously fulfilled this protective role, and subsequently thermoplastics and polyurethanes captured the market.

In the same decade, the western world was also confronted with several oil crises which further fuelled the notion of more effectively utilizing one of the earth's principal natural resources. Since the weight of an automobile is responsible for almost 50% of the fuel consumed, this troubled economic climate further stimulated the efforts of engineers and scientists to develop and utilize plastics rather than steel in the construction of automobiles. The payoffs associated with this philosophy can be gleaned from the rule-of-thumb that a car will save about half a liter of fuel for every 100 kilometers traveled if the weight of the vehicle is reduced by 100 kilograms. By implementing this philosophy, not only can engineers fully exploit a diverse group of synthetic materials that never rust and which can be manufactured more easily than their steel counterparts, but furthermore, the significant reduction in gasoline consumption associated with the implementation of this philosophy would result in a considerable reduction in air pollutants.

The exploitation of synthetic plastic materials in the automobile, packaging, and household appliance industries is due to the diverse range of physical properties offered by such materials. These diverse physical properties are obtained by synthesizing or engineering, the macromechanical structure of polymeric materials in order to precisely comply with the design specification of each structural member. While in the past, materials development was primarily the responsibility of metallurgists, now it is undertaken by diverse teams of polymer scientists, ceramists, chemists, and other specialists using sophisticated equipment. For example, computer graphics now permits the materials scientist to develop a desired molecular structure on the computer screen prior to manufacturing that material. Indeed in 1990 it was estimated that there were over 60 000 different plastics available in the marketplace. This situation contrasts sharply with the fact that there are only two groups of substances from which almost all skeletal tissues of animals and plants are formed. These are the amino-acid-based proteins and the polysaccharides, which generally occur together in different proportions in the vast majority of biological materials. Thus while the chemistry of these naturally-occurring structural materials is apparently limited, this is compensated for by tremendous diversity in their microstructure.

Nevertheless, this inability of synthetic materials to compete with the limited chemistry and diverse microstructure of biological materials has not prevented the scientific community from utilizing the diverse properties of plastics and other classes of state-of-the-art materials, for example, to develop medical tools for the treatment of diseases and for the relief of human suffering. Consider for example the disease, diabetes mellitus, a slow savage killer which afflicts an estimated 100 million people worldwide, and which currently costs the US $20 billion a year in health care services. It is the seventh leading cause of death in the US, and furthermore, it more than doubles the risk of a diabetic suffering a stroke or heart-attack; it is the leading cause of blindness in adults; it is the leading cause of limb amputations after traumatic injuries; it accounts for one third of all kidney failures; it causes nerve damage; it is responsible for wounds healing more slowly; it makes infections difficult to control, and it is responsible for pregnant women bearing children with congenital defects. The disease results from an inability of the pancreas to produce insulin, which is a hormone essential for the nourishment of cells by glucose in fat, muscle and the liver. Furthermore, with rising obesity in the human population, and also the reduction in physical exercise, this terrible disease is expected to flourish not only in the affluent western cultures but also in the developing nations. The science of advanced materials is being exploited by inter-disciplinary teams of physicians, engineers, and materials scientists fighting the disease as they develop superior kidney dialysis machines; artificial pancreases; open-loop miniature portable insulin pumps for patients; prototype blood-glucose sensors for the closed-loop autonomous control of insulin-dependent diabetics; portable blood-glucose monitors; disposable plastic syringes, and also sophisticated drug delivery systems such as the porous synthetic membrane utilized in the exploratory micro-capsule technique for treating insulin-dependant diabetics.

1.3 Advanced composite materials

The study of biological systems has resulted in the emulation of these systems by engineers and materials scientists. Two principal fields have emerged from these studies. The first is the field of biotechnology and biomimetics where typically artificial body organs are developed to enhance the quality and span of life, and to also relieve human suffering. A variety of artificial tissues and organs can be synthesized by employing plastics, composites, ceramics, and glasses which can harmoniously interface with or replace body tissues and parts as indicated in the cartoon sketch of Figure 1.3. Polyurethanes, carbons and silicones are currently being employed for bone implants and replacement hip joints. Tendons and ligaments can be fabricated from polymers and carbon fibers, as are artificial lens implantations. Attempts are currently underway to develop viable synthetic human skin which can provide relief to burn victims. Other applications in the biomaterials revolution include *in vivo* artificial healing systems, dental structures, artificial larynxes, and inter-aortic balloons.

The second discipline to be stimulated by the study of biological materials is the field of composite materials where these synthetic engineered materials are synthesized with two distinct phases comprising a load-bearing material housed in a relatively weak protective matrix to mimic the

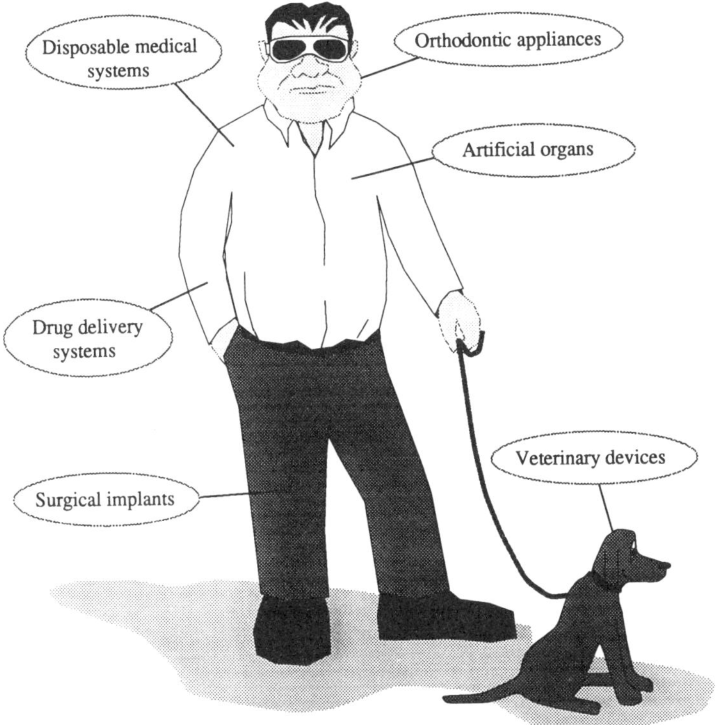

Fig. 1.3 Sample biomaterials applications.

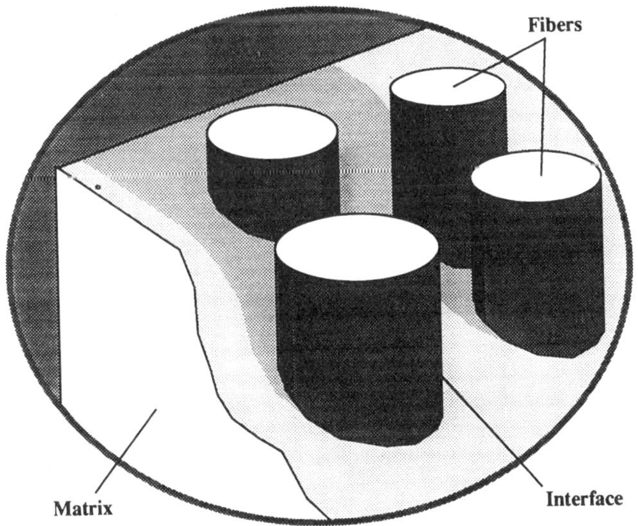

Fig. 1.4 The principal features of a fiberous composite material.

microstructure of biological structures. Figure 1.4 schematically presents the principal features of a fiberous composite material; namely, the fibers, the matrix material, and the interface region between these two dissimilar materials. This class of new structural material can be classified as metallic, ceramic, or polymeric, depending upon the load-bearing or reinforcing material employed. The reinforcement may be particulates, whiskers, laminated fibers or a woven fabric. These reinforcements are bonded together by the matrix which distributes the loading between them. Generally the reinforcement is a fiberous or particulate material, with the latter category of reinforcement permitting far superior structural properties to be achieved at the expense of more challenging fabricating technologies and higher costs. A characteristic of these advanced materials is that the combination of two or more constituent materials yields a composite material with engineering properties superior to those of the constituents, as depicted in Figure 1.5 by the shaded areas where the reinforcements domain overlaps the polymeric, ceramic and metallic domains. The associated materials are termed polymer-matrix composites, ceramic-matrix composites and metal-matrix composites.

There are numerous examples in nature where this type of microstructure, comprising a load-bearing structural phase housed in a protective matrix, is present in biological materials. Consider the humble tree, where the trunk and branches comprise flexible cellulose fibers in a rigid lignin matrix. The class of synthetic materials such as fiber-glass composites and graphite-epoxy laminates, which emulate biological systems, have in recent decades revolutionized the automotive, aerospace and sporting goods industries. However, it should be noted that composite materials have been initialized for centuries by mankind, as evidenced by one of the first recordings of this

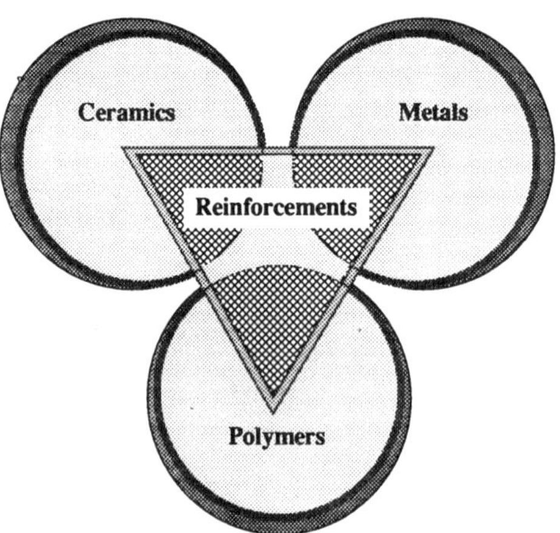

Fig. 1.5 Structural composite materials.

technology in *Exodus Ch. 5, Vs. 5* of the Bible in 450 BC. This early citation concerned the time spent by the Israelites in Egypt where they were compelled to manufacture bricks for the Pharaoh featuring a composite material of clay and straw.

The current generation of advanced composite materials have yielded some of the lightest, strongest, stiffest, and corrosion-resistant materials available to the engineering community. Consider, for example, the magnitudes of the stiffness-to-weight ratio or the strength-to-weight ratio of the commercial metals relative to those of the advanced composite materials. A graphical presentation of the evaluation process is depicted in Figure 1.6. While the specific stiffness of aluminum can be increased three-fold by the addition of silicon carbide fibers to create a metal-matrix composite, the specific stiffness of graphite-epoxy, fiber-reinforced, polymeric materials can be over four times greater than the specific strength of

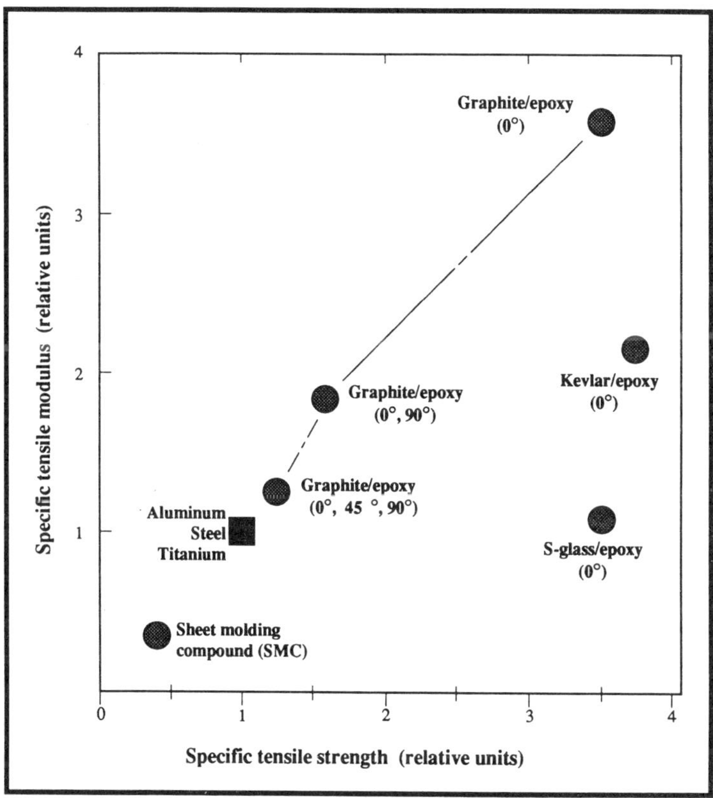

Fig. 1.6 A comparison of the specific strength and stiffness properties of composites and metals.

steel. The ramifications of this comparison are immense because lightweight, high-strength, high-stiffness structures can be fabricated in these advanced polymeric composite materials with a weight saving of approximately 50%. This class of designs translates into superior performance for diverse products in the aerospace, defense, automotive, biomedical and sporting goods industries for example.

While the high specific stiffness and the high specific strength properties of polymer matrix composites is an impressive set of credentials relative to the commercially available monolithic materials, they only represent a couple of data points associated with this class of engineered materials where the material properties are specifically tailored to comply with a design specification. Indeed, the degree of sophistication of a broad range of materials is increasing dramatically while the weight of these engineered materials per product is decreasing as indicated in Figure 1.7. By appropriately designing the macrostructure of composite materials, the engineer can develop materials with different properties in different

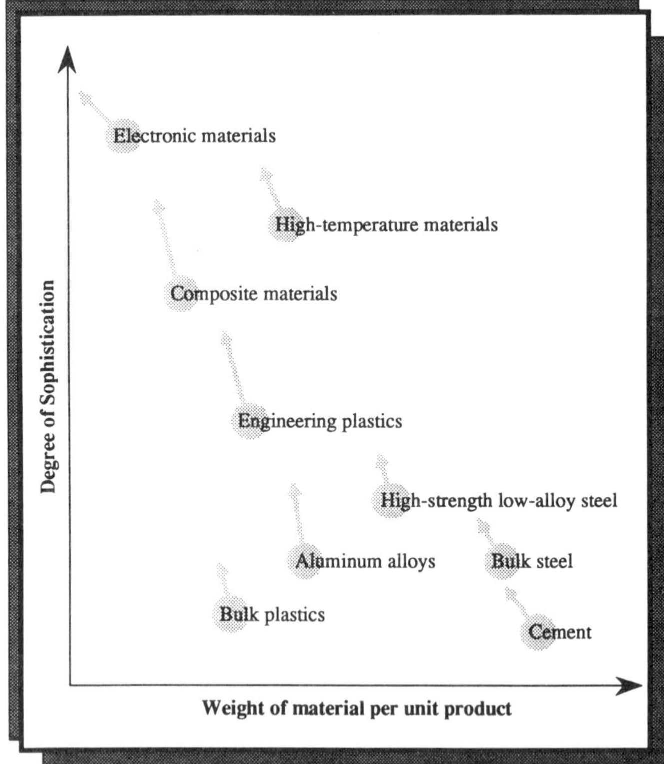

Fig. 1.7 A product design trend concerning the increased degree of sophistication in materials employed in practice and their reduced weight.

directions or alternatively different properties at different local regions in a structural member. Thus, for example, by judiciously selecting and placing appropriate reinforcements in a composite material, the strength properties can be enhanced only in those locations where it is required in order to achieve design and cost efficiencies. An appreciation of these efficiencies is presented in Figure 1.8 where the relative importance of both cost and performance for fabricated composite components is illustrated in the context of the diverse industrial sectors that utilize these engineered materials. Thus, for example, the performance of a biomedical composite material is the dominant design parameter, heavily outweighing the importance of the cost of the composite material. Thus, while the cost of a biomedical composite material may be several orders of magnitude greater than the cost of a composite material for the construction industry, the cost will only be of secondary importance to the performance in biomedical applications. On the other hand, the performance of a composite material for the automotive industry will not dominate the selection process because

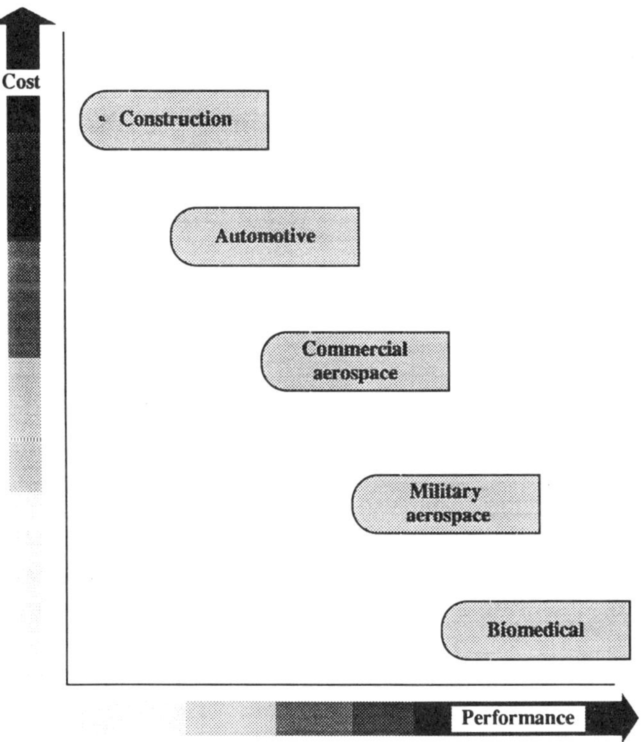

Fig. 1.8 The relative importance of cost and performance on structural composite materials employed in different industries.

this high-volume industry is extremely cost conscious. Consequently, the cost of the material will probably be more important than the performance of the material.

1.3.1 The aerospace industry

The primary fiber reinforcements in the aerospace industry are carbon or graphite fibers, high-strength glass fibers, and ceramic fibers. The primary matrices for high strength, high stiffness applications are generally epoxy resin systems. This industry has pioneered the field of designing and manufacturing with composite materials. The motivation for this pioneering work is the significant payoffs associated with these materials. For example, an aluminum wing rib has approximately 20 parts with as many as 500 mechanical fasteners. In sharp contrast to this, a thermoplastic wing rib is generally manufactured as a single part which permits significant savings in manufacturing costs to be accrued. Similarly, a conventional aluminum helicopter fuselage typically comprises 10 000 parts, but a comparable fuselage can be fabricated in advanced polymeric composite materials by using only 1500 parts, which again results in considerable savings.

The principal advantage of advanced polymer matrix composites in aerospace applications, is their superior strength-to-weight ratios and their superior stiffness-to-weight ratios compared with the traditional, commercially available monolithic materials. These superior characteristics are responsible for weight savings of typically 30%, but they can be as high as 60% depending on the application. The consequences of these weight savings can be measured in terms of an increased range, speed and payload, or else reduced fuel consumption. Examples of this technological revolution include the *Voyager* aircraft which flew nonstop round the world without refueling, the 'all plastic' corporate passenger airplane *Starship* manufactured by Beech Aircraft Corporation, and the recent versions of the British Aircraft Corporation Harrier jump-jet which features 30% composite materials in the wings and airframe structure. It has been estimated that composite parts will represent 65% of the structural weight of commercial aircraft by the year 2000. This translates into a $1.5 billion market for the material alone, if the advanced composite materials retail for US $60 a pound.

1.3.2 The automotive industry

The task of transferring composite materials technologies from the aerospace industry to the automotive industry is currently being undertaken in an effort to not only attain improved fuel economy but also to reduce manufacturing costs. Indeed it is estimated that the automotive industry will consume the greatest volume of polymer matrix composites in the future.

This technology-transfer task is not a simple undertaking as clearly evidenced in Figure 1.9. The aerospace industry is characterized by high-cost, high-quality structures which are manufactured at low production rates in high-strength, high-stiffness, high-toughness, and high-temperature thermoplastics. In sharp contrast to this situation, parts for the automotive industry are generally characterized by low-cost, and they must be manufactured at high-production rates in order to satisfy the less demanding performance specifications of the industry. This dramatically different set of constraints has triggered a revolution in design-and-manufacturing technologies for the production of fast-curing medium-toughness thermoplastic parts featuring hybrid micro-structures characterized by different types of fibers or combinations of continuous and chopped fiber systems.

The greatest volume of polymeric matrix composite materials currently being employed in the automotive industry is sheet molding compound (SMC) which is used for the exterior body panels of vehicles. The General Motors (GM) Pontiac *Fiero* sports car is an example of this technology

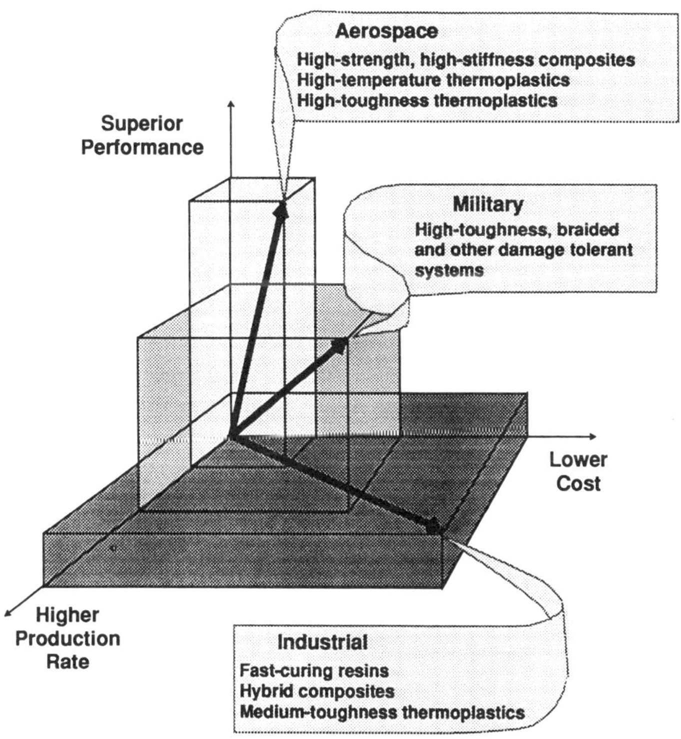

Fig. 1.9 The diverse costs, production rates, and performance characteristics of composite components for the aerospace, military, and industrial sectors.

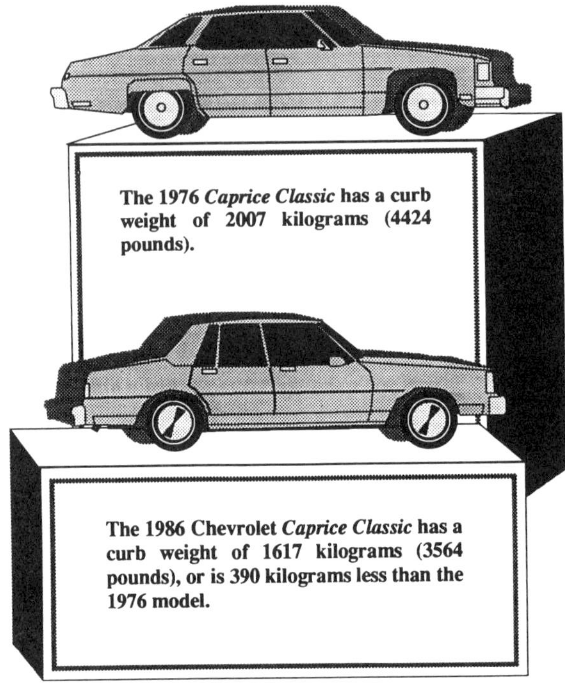

Fig. 1.10 Weight reduction trends in US automobiles.

which utilized a steel frame to which the SMC panels were attached. Automobile structural components are currently being developed using continuous fiber composites. An example of this weight-reduction strategy is illustrated with the utilization of composite drive shafts on Ford *Econoline* vans. These members are manufactured by filament winding a hybrid fiber configuration using both graphite and E-glass fibers in a polyester resin. Glass fiber reinforced leaf springs are also in production in the GM Chevrolet *Corvette* sports car and several other GM vehicles such as Chevrolet's *Lumina* and *M* vans.

While the drive in the automotive industry towards the deployment of more advanced composite materials in their product lines has been fuelled by cost savings in manufacture, the US government's mandate for vehicle fleets with superior fuel consumption, has triggered an increased focus on the weight saving features offered by composite parts, as succinctly illustrated in Figure 1.10. Consider, for example, the 1976 GM *Caprice Classic* automobile which had a curb weight of 4424 pounds. The 1986 model weighed 661 pounds less than this because engineers substituted lightweight materials throughout the vehicle resulting in savings of 119 pounds in the sheet metal parts and grille, 99 pounds in the suspension and brakes, 66 pounds in the frame, and 55 pounds in the steering gear and wheels. Thus it is clearly evident that considerable benefits accrue from the judicious selection of materials in the design of this class of product.

The desirable mechanical properties offered by fiberous composite materials, can be exploited in the design of the reciprocating members of automobile engines to yield a product with superior performance characteristics. Consider for example the design and fabrication of a connecting-rod in an advanced composite material, rather than a traditional steel member which is typically manufactured as either a forging or a casting. A fiberous polymeric design has several distinct advantages over the traditional monolithic design and these include lower weight, increased specific stiffnesses and strengths, and superior damping characteristics. These distinct advantages are responsible for engines with superior vibrational response characteristics and also reduced levels of acoustical radiation. Furthermore, there are several design cascading effects accruing from the utilization of advanced materials in the moving elements of an internal combustion engine, which frequently enable balance-shafts to be eliminated from the engine configuration. Engine mounts are subjected to smaller dynamical loads and consequently can be simplified, transmission shafts can be designed with fewer bearings, and the dynamical loading imposed on the connecting-rod and crank shaft bearings is less severe. Prototype engines have been installed in vehicles and test results indicate improved fuel consumption, reduced noise levels in the passenger compartment, and a more responsive power plant. Thus there are significant payoffs associated with the judicious selection of materials for critical components in articulating mechanical systems.

1.3.3 The machinery products industry

The design of robot arms and also component members of high-speed machinery in polymeric matrix composite materials would greatly enhance the performance of these machine systems. This philosophy would, for example, positively impact both machine systems processing light media, where the operating speed is limited by the inertia of the articulating members, and also robot arms, which typically have arm to pay-load weight ratios of approximately 10:1. These applications can exploit the benefits offered by composites, such as, anisotropic properties; superior vibration damping characteristics compared to the commercial metals; superior specific stiffnesses and strengths, and also superior fatigue characteristics.

The potential advantages of utilizing composite laminates in the design of machine systems where the productivity is governed by the stiffness-to-weight ratio of one or more machine elements, has been clearly documented in the scientific literature through a variety of industrial case-studies in different fields. Typically, these advantages have been demonstrated in machines with fast-moving components whose speeds are limited by their own inertia rather than external surface tractions. Examples include production machinery for various industries such as printing, packaging, and textiles, and also robotic systems handling small parts for the pharmaceutical, electronics, and food-processing industries.

The manufacture of flexible conductors for electric cables requires the twisting of a number of individual fine wires with a regular pitch of twist. This is typically achieved in a production environment by employing a double-twist bunching machine which features a bow that performs the twisting operation. The production rate of the machine is directly proportional to the bow speed, and one revolution of the bow imparts two twists in the wire. Since the bow material must withstand stresses generated by centrifugal loading, the principal design task in the development of this class of machine is the synthesis of a bow with a high strength-to-weight ratio. This design task can be accomplished by fabricating the bow in a long-staple-graphite-epoxy material which permits the speed of the machine to increase from 500 rpm, to almost 3000 rpm relative to a machine featuring a bow fabricated in a commercial metal. The cost of the laminated bow is an order of magnitude higher than the original bow design but the payoff is a significant increase in productivity which provides a viable return on the initial higher investment.

The warp thread in a category of narrow fabric textile looms is raised and lowered by an H-shaped heddle frame approximately 15 inches long. The productivity of this textile machine has been limited by the performance of this reciprocating member, which is traditionally fabricated in an aluminum-alloy that repeatedly failed because of the fatigue environment in which it operated. The inertia and stiffness of the heddle frame were the crucial factors in the component design. Subsequently a graphite-epoxy composite member was designed, engineered and fabricated, and the operating speed of 2250 weft insertions a minute was increased by 50%, resulting in a similar increase in machine productivity. It is clear, therefore, that advanced composite materials have a significant role to play in the design of new generations of high-speed machine systems.

1.3.4 The sporting goods industry

The current generation of advanced composite materials have so dramatically influenced the sporting goods industry that the athlete must carefully scrutinize the materials from which equipment is fabricated prior to adopting it for an event because this decision can significantly influence the difference between victory and defeat. Thus Olympic pole-vaulters employ fiberous polymeric poles rather than the traditional aluminum poles to establish records. On the golf course, a golfer on the tee can take a wood from his bag with a head that is no longer a natural biological material.

Subsequently on the fairway he can use an iron which does not contain any ferrous material in the traditional metallic head. Both woods and irons are now fabricated from composite materials and yield superior performance.

Carbon-fiber composites have revolutionized the sport of archery, transforming the fabrication of the bow and the arrows from wooden or metallic members to fiberous polymeric members. In this sport, the distance and accuracy of the shot are both dependent upon the speed of the arrow from the bow. With an advanced composite bow, for example, the shaft of a carbon-fiber arrow can be 20 feet/s faster out of the bow compared with an aluminum arrow. Furthermore, the smaller cross-sectional dimensions of the arrow fabricated in a composite material reduce aerodynamic drag and minimize the inaccuracies caused by cross-winds.

Bicycle designers, like the designers of archery equipment, are also confronted with the challenge of reducing aerodynamic drag because it minimizes the maximum speed and endurance of the rider. The bicycle frames employed by the US Olympic team in 1988 were fabricated from carbon fibers and each machine weighed only 3.3 pounds. Each tube of the frame was manufactured as an airfoil section by the braiding process in order to reduce the drag coefficient. Furthermore, the wind resistance of the standard conventional spoked wheel was reduced by 50% by designing disk wheels which were fabricated in Kevlar, an aramid fiber developed by DuPont. Similarly, lightweight Kevlar-reinforced tubular tires were designed to reduce rolling resistance. Thus the bicycle designer is able to synthesize a superior product by exploiting the versatility of composite materials in design and manufacture.

Boats and kayaks have also benefitted from the exploitation of composite materials technologies. State-of-the-art racing boats now typically feature a honeycomb core sandwiched between skins of glass and graphite-epoxy prepreg materials. This structural design philosophy permits a stiffer and lighter hull to be fabricated, which reduces hull flexing and hull vibration during racing conditions which in turn permits a superior performance to be achieved through the reduction of turbulence and drag.

World-class athletes attain speeds of 120 mph in downhill ski races because the designers of ski equipment have successively addressed the challenges of designing equipment for this sport. In order to attain these high velocities and perform sharp fast turns in slaloms it is essential for the equipment to be manufactured utilizing lightweight, high-strength, high-stiffness, tough, abrasion resistant materials that dissipate energy. Thus skis are typically fabricated in various combinations of graphite fibers, aramid fibers and glass fibers in order to synthesize hybrid materials with the desired mechanical properties.

Ski poles must be lightweight and stiff in order to perform their function and minimize the energy-consumption of the skier as the poles are swung to-and-fro. Low-cost aluminum and pultruded glass-fiber poles are being superseded by lighter and stiffer, hybrid graphite poles featuring combinations of woven and unidirectional continuous fibers. These poles are typically fabricated with epoxy resin systems to reduce impact failures, which are common with the previous generations of this class of equipment. Furthermore, these poles, which only weigh 100 grams are articulated approximately 5000 times in a 15 kilometer cross-country event. Consequently this state-of-the-art design can save the competitor 2500 ft/lb of energy during the race. This is not an insignificant saving in energy for the athlete in an event where only a fraction of this amount could determine the victor.

1.4 Design-for-manufacture algorithm for fiberous composite materials

The development of advanced engineered materials has been responsible for new integrated approaches to the design and manufacture of fiberous composite materials. The development of these new methodologies has been necessitated by the creation of materials with distinct micro-structural properties during each manufacturing process. Engineers traditionally fabricate parts in monolithic materials, such as the aluminum alloys or the low-carbon steels, and the material is first selected from a design handbook based upon the criteria associated with the specific application, prior to selecting the discrete manufacturing processes with which to fabricate the final structure. This classical philosophy is not viable in the field of advanced composite materials.

With engineered composite materials, the designer first focuses upon the performance requirements prior to *simultaneously* synthesizing the necessary combination of constituent materials and also their micro-structure in an *integrated* design and manufacture algorithm. Thus with advanced tailored materials, the traditional notions of the distinct phases of material selection, design, and manufacturing processes all merge into a continuum philosophy embodying both design and manufacture in an integrated fashion.

While the current methodologies for designing composite-based components have been relatively well developed, largely on an ad-hoc basis, such methodologies do not explicitly incorporate the constraints and characteristics imposed by the specific manufacturing processes employed to fabricate composite parts. Each manufacturing process imparts significantly distinct mechanical properties to the part, due to the unique macromechanical characteristics associated with each process. Thus, the state-of-the-art in the area of designing composite-based structural elements is very similar to the traditional design of components in the classical, monolithic, homogeneous isotropic materials, where the choice of the appropriate manufacturing processes is made *a posteriori*. This philosophy has some serious limitations since the materials selection process, the part geometry, and the constraints

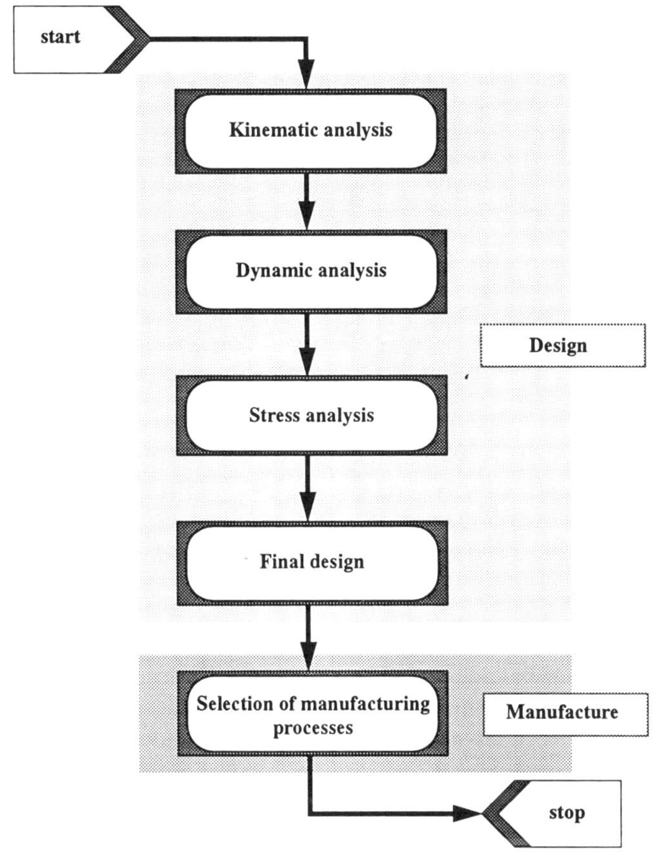

Fig. 1.11 Traditional sequential algorithm for the design of articulating machinery.

and characteristics of the manufacturing processes are inextricably intertwined.

Consider, for example, the problem of designing the connecting-rod of an internal combustion engine in a monolithic material. Given the design specifications, the traditional design process conveniently decomposes into the distinct phases of analysis comprising kinematics, machine dynamics, and strength of materials, as shown in Figure 1.11, which precedes the selection of an appropriate manufacturing sequence. The kinematic analysis and machine dynamic phases provide information on the velocity and acceleration characteristics in addition to the forces in the member and at the bearings. The subsequent stress analysis phase is employed to determine the principal stresses, the characteristics of the time-dependent deformation field, and the fatigue life of the part prior to determining the final details of

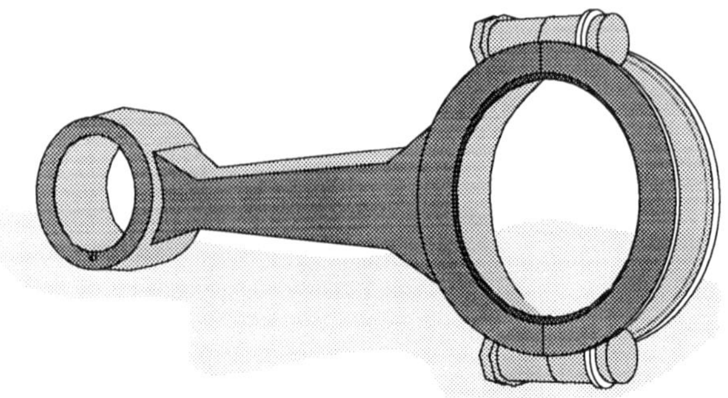

Fig. 1.12 A typical monolithic connecting-rod for an internal-combustion engine.

the form-design of the part. Generally, the decision on the selection of appropriate manufacturing processes to be employed in the fabrication of the connecting-rod, is subsequently undertaken discretely in isolation from the design process. Clearly, the choice of the appropriate manufacturing processes, such as forging and casting, for example, affects the Young's moduli and Poisson's ratios of the material because of the distinct grain structure imparted by each process. These two classes of material characteristics which emerge from the consideration of the manufacturing processes, are subsequently incorporated in the design process in order to finalize the part geometry, and yield the design shown in Figure 1.12.

The design of a connecting-rod fabricated in fiberous composite materials contrasts sharply with this philosophy, because design decisions cannot be conveniently divorced from the consideration of the appropriate manufacturing processes. If the connecting-rod is pultruded, filament wound, or braided, then the geometry of the part will be constrained by each manufacturing process, and each process will, in turn, impart distinct stiffnesses, damping and mass characteristics to the member due to the distinct fiber-volume fractions, spatial distribution of the fibers, fiber orientations and the lay-ups associated with each process. A composite connecting-rod which permits the fibers to have preferential orientation in the directions of the maximum loads, can be developed in which compression molding is employed to manufacture the bearing housing of the small-end of the articulating member, while the compressive loading imposed upon the connecting-rod is carried by a central pultruded part between the two bearing housings, illustrated in Figure 1.13. The tensile loading is carried by a filament wound part, in which the continuous fibers laid down by this process are wrapped round the two end bearings and the pultruded part in order to unite the principal elements of the design. Clearly, innovative designs of this kind can significantly capitalize on coherent design-for-manufacture methodologies, in order to satisfy the

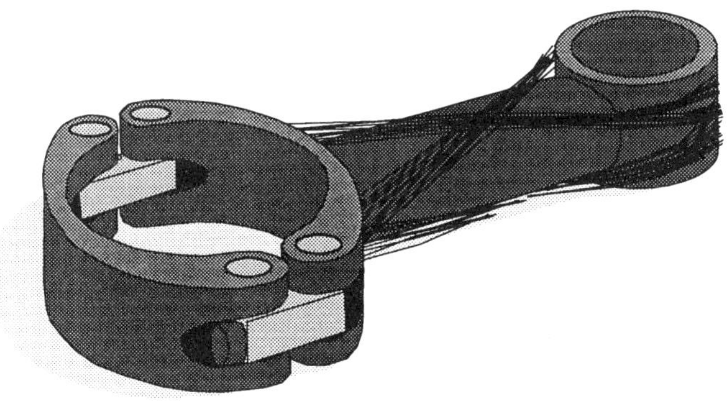

Fig. 1.13 Fiberous composite connecting-rod for an internal-combustion engine.

insatiable demand for high-performance low-cost components in the international marketplace.

In sharp contrast to the classical design-for-manufacture (DFM) approaches for monolithic materials, the DFM strategies for composite materials should not only incorporate the manufacturing processes, but they are in fact exclusively driven by the manufacturing processes themselves. A DFM strategy will manifest itself at the macro-mechanical level in the synthesis of the desired material properties and part shapes, while at the subassembly level, for example, this methodology would render obsolete some conventional design-for-assembly (DFA) methodologies for monolithic materials. Parts which are currently fabricated as sub-assemblies comprising a multitude of individual parts manufactured in monolithic materials, could be fabricated as a single part in composite materials, which can permit significant savings in manufacturing costs to be accrued.

A rigorous DFM strategy for composite parts and subassemblies requires an interactive algorithm which encompasses the materials selection process, and also issues pertaining to part geometries and manufacturing considerations, for example, as presented in Figure 1.14. Such a DFM strategy would be initiated by the part/subassembly specification which would include diverse constraints imposed on the static and dynamic deflections, mass and inertial characteristics, payload requirements, impact resistance, fatigue life, environmental factors such as temperature and humidity, and of course, cost, quantity, and production-rate considerations. The part subassembly specification provides the basis for initiating an interactive DFM algorithm, which integrates the various facets of the DFM strategy by simultaneously considering part geometry, manufacturability, and material selection and synthesis.

The fibers employed in the manufacture of a composite part may be either continuous or discontinuous. Continuous reinforcing fibers are commercially

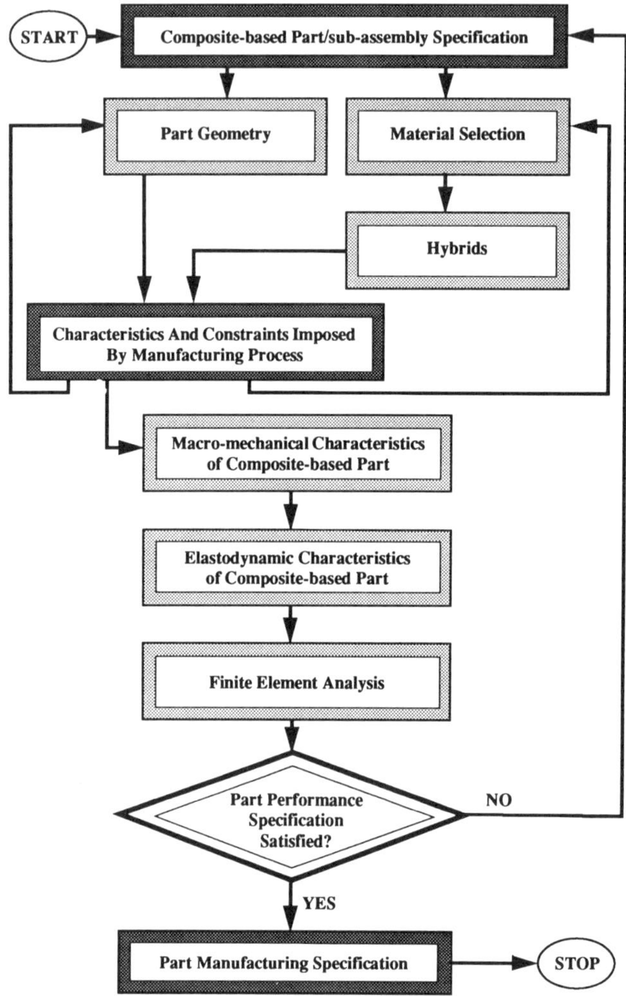

Fig. 1.14 DFM algorithm for members fabricated in composite materials.

available in many product forms, ranging from monofilaments to filament fiber bundles, and from unidirectional ribbons to single layer fabrics and multi-layer fabric mats. The reinforcing fibers and matrix resins may be combined into many different intermediary product forms that are designed for subsequent use by specific fabrication processes. Combinations of unidirectional fiber ribbons or woven fabrics with the resin system are called prepregs, which have very precisely controlled fiber/resin ratios, highly controlled tack and drape for thermoset matrices, controlled resin flow

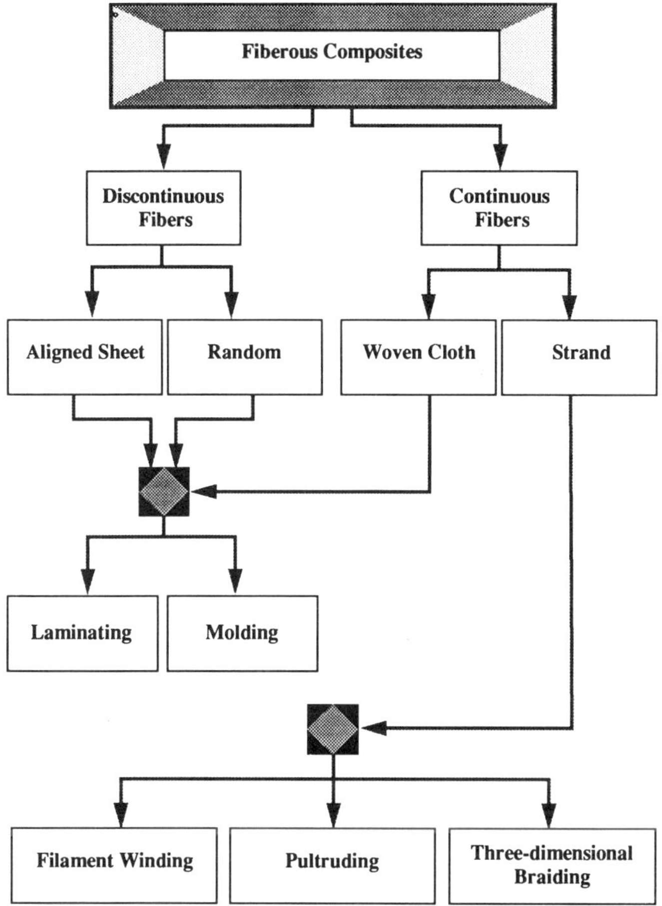

Fig. 1.15 The relationship between the selection of continuous or discontinuous fibers for a part, and the selection of appropriate manufacturing processes.

during the cure process, and in some fabrication processes prepregs allow better control of fiber angle and placement, which facilitates the fabrication of composite-based parts to stringent tolerances.

Thus, for example, the choice of manufacturing a part using discontinuous fibers would clearly constrain the manufacturing process to molding or laminating; the choice of continuous fibers would be a necessary condition for fabricating a part by employing filament-winding, pultrusion, or three-dimensional weaving and layup techniques as shown in Figure 1.15. In a continuous fiber reinforced composite part the fibers provide virtually all of the load-carrying capacity of the structure, and the multiple fibers in such a structure permit a redistribution of the loading to occur, in the event of several fibers failing due to the part being exposed to a fatigue environment for example. This situation is much more desirable than a catastrophic failure of the structure. Of course the structure fabricated with discontinuous fibers would generally be much weaker than an equivalent structure fabricated with continuous fibers and hence it is more prone to catastrophic failure.

It is clearly evident that the choice of continuous fibers versus discontinuous fibers in the material-selection phase of the DFM strategy has a significant impact on the choice of appropriate manufacturing processes, which in turn constrains the choice of part geometry. Furthermore, each manufacturing process results in a distinct fiber volume fraction, and also a distinct fiber orientation and spatial distribution of the fibers in the part. This results in distinct anisotropies, such as aleotropies, and orthotropies, for example, which impart distinct global mechanical characteristics such as stiffness, damping and mass, to the part under consideration. For example in the case of filament winding techniques alone, different combinations of hoop, longitudinal and helical winding patterns will invariably result in filament wound parts with different stiffnesses, damping capacities and mass characteristics, which in turn govern the overall elastodynamic performance of the part under the prescribed service conditions. Thus, while fiberous composite materials clearly offer the designer significant freedom and also economic advantages, the complex coupling of the design, analysis, and fabrication processes mandates that they be considered simultaneously during the development of a part.

1.5 Ceramics

Ceramics occupies the number one position on the list of industries with the greatest potential for commercial development, according to data that has been compiled by Japan's Ministry of International Trade and Industry. The markets for structural ceramics are projected to total between $1 billion and $5 billion in the year 2000 and the principal ingredients of these markets are presented in Figures 1.16 and 1.17. This new generation of fine ceramics bears little resemblance to the sand and clay combinations of terrestrial materials utilized by *Homo habilis* in the Mesolithic and Neolithic periods for the fabrication of earthenware utensils. These new materials feature minerals such as aluminum oxide, silicon nitride and silicon carbide which are manufactured as extremely pure, fine powders prior to consolidation at high temperatures to yield a durable dense structure. The control of these new ceramics provides the design engineer with a class of ceramics which are stronger, lighter, and harder than most metallic competitors. Furthermore, the ceramics can operate at much higher temperatures and they do not

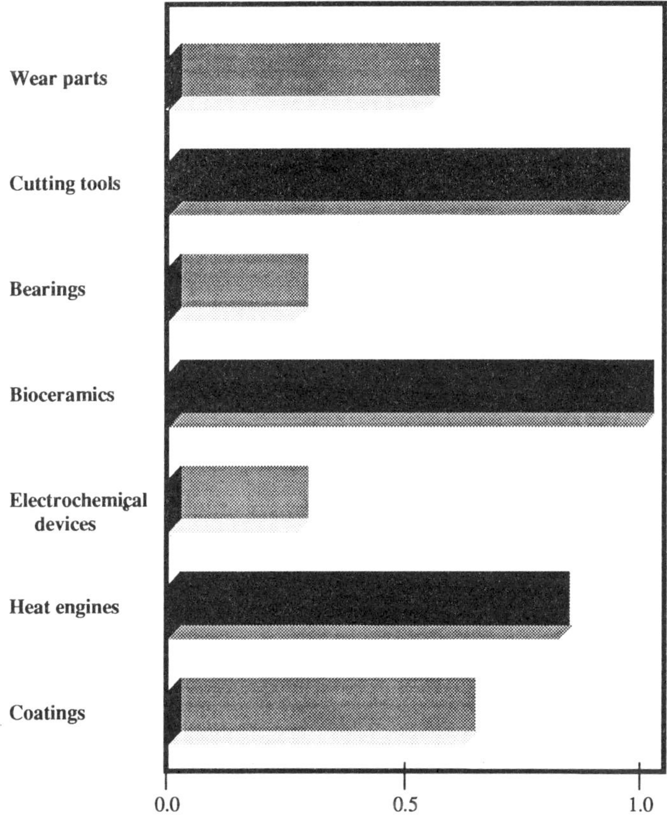

Fig. 1.16 Projected US markets for structural ceramics in the year 2000.

degrade significantly due to oxidation as do commercial metals such as the irons and steels.

These materials do, however, have an *Achilles heel*: brittleness. This brittleness is associated with the number of extremely small flaws such as cracks, voids, and impurities present in the material, and also the amount of energy required to fracture the material in the presence of these flaws. Thus the manufacturing processes must be extremely carefully controlled, since flaws as small as 10 to 50 micrometers can reduce the strength of a ceramic part to only a few percent of its ideal strength thereby rendering the part useless. While material toughness is a major limitation of the current generation of advanced structural ceramics, the many significant advantages of these materials have not deterred further research on these compounds while other research efforts are focused on this limiting material characteristic. These latter efforts are typically focused on synthesizing tougher

Application	Performance advantages	Examples
Wear parts seals bearings valves nozzles	High hardness, low friction	Silicon carbide, alumina
Cutting tools lathe tools milling cutter	Hot hardness, high strength	Silicon nitride
Heat engines diesel components gas turbines	Thermal insulation, high temperature strength, fuel economy	Silicon carbide, silicon nitride, zirconia
Medical implants teeth joints	Surface bonds to tissue, corrosion resistance, biocompatibility	Bioglass, alumina, zirconia hydroxylapatite
Construction highways bridges buildings	Improved durability, lower overall cost	Advanced cements and concretes

Fig. 1.17 Future applications of structural ceramics.

materials through the prosecution of research on microstructural design, transformation toughening, and ceramic composites where the incorporation of ceramic particulates, whiskers, or continuous fibers in a ceramic matrix can yield a composite material that absorbs more energy during fracture, than a geometrically identical part fabricated in the matrix material alone.

The highly desirable characteristics offered by these new ceramics will impact a broad segment of the industrial sector such as the automotive, electronics, aerospace, medical and telecommunications segments. Ceramics technologies have developed a high profile in the advanced ceramic consumer goods industries through the marketing of pliers, scissors and ball-point pen tips, for example. Structural ceramic parts featuring alumina, silicon carbide and silicon nitride are commercially available for wear-resistant parts, turbo-chargers, rocker arms, and cutting tools. Coatings and bearings for orthopedic or dental implants are also available. Some of these materials function as resorbable materials to provide temporary support until the human body can gradually replace it, while others are surface-active materials which form a bond with the surrounding tissue in order to stimulate growth.

One of the principle foci of ceramic materials research and development is concerned with the automobile engine. The Japanese, who are clearly the leaders in this field of ceramic technology, are projecting the development of engines which are 30% to 40% more fuel efficient then the current generation of engines, and furthermore, these engines run at higher temperatures without lubrication or cooling. The electronics industry currently consumes the vast majority of technical ceramic production at this time, but it too will benefit from this new generation of ceramics though the development of superior computer chips, photonic devices, capacitors, and the like, which will trigger the evolution of the optical computer where information-pulses travel at the speed of light. This new generation of computers will be able to process information much more quickly than the current generation of electron-based machines, and they will also be capable of accommodating a greater volume of data. Clearly, yet again, the revolution in materials science research is the kernel of significant new developments in many eclectic technological fields.

1.6 Pure glass

Technological progress in materials science has catalyzed a revolution in data transmission and hence information technology. This revolution exploits the ability of material scientists to not only manufacture new classes of semi-conducting materials but also to manufacture thin flexible strands of plastic or purest glass, approximately the size of a human hair, through which laser light can be transmitted at velocities comparable to the speed of light. The contribution made by the materials science community was the ability to develop pure materials in order to address the critical issues of scattering and absorption, which are the principal sources of inefficiency that previously limited the transmission of information through transparent media. Scattering losses are typically associated with changes in a light wave caused by changes in the density of the medium through which the wave is being transmitted. Techniques were developed to meticulously control the density of the thin fibers. The second type of losses are associated with absorption caused by impurities in the glass caused by various types of ions such as copper, vanadium, and iron. The development of vapor deposition techniques has facilitated the manufacture of extremely pure glass without these impurities.

These thin strands of optical material serve as the wave-guides for the light beams or pulses, and they have been responsible for the photons replacing electrons as the preferred mode of rapidly transmitting large volumes of information and the spawning of the embryonic field of photonics, or opto-electronics. This field not only encompasses telecommunications but optical computing and also numerous sensing systems. The most common use of optical fibers is as a transmission medium connecting two electronic circuits comprising a transmitter and a receiver. These latter devices interface the photonic and electric signals. The optical fiber connecting these interface units permits a light beam to be transmitted by the principle of internal reflection, since the transparent core material has a higher refractive index than the surrounding cladding. This basic electromagnetic phenomenon is largely responsible for the recent revolution in telephone and data transmission systems.

1.7 Gallium arsenide

The field of photonics, while significantly benefitting from the development of optical fibers, has also benefitted immensely from developments in the semi-conductor industry associated with a new generation of synthetic materials. One of these strategic developments is the synthesis of the man-made material gallium arsenide. Integrated circuits, or chips, featuring this material are beginning to revolutionize segments of the semiconductor industry and also the diverse range of products contingent upon high-speed computational facilities. In addition, they have utility in many classes of electronic systems and devices including both optical and electronic communication systems.

The success of this class of synthetic materials is largely based upon the superior performance of gallium arsenide chips relative to conventional silicon chips. These advantages include reduced power consumption; the ability to process both photonic and electronic data; the ability to process data at extremely high frequencies; the ability to process electronic information at an order of magnitude faster than silicon systems, and the ability to operate at higher temperatures.

The advantage of this new generation of chips must be weighed against the higher cost of manufacturing them in this synthetic material. These higher costs are associated with the difficulties of removing impurities from both the gallium and the arsenic which are responsible for unacceptable heterogeneities in the gallium arsenide structure. Furthermore, the task of synthesizing the atomic structure of these materials remains a challenging frontier.

However, while a gallium arsenide wafer for manufacturing chips costs about twenty times more than the comparable silicon wafer, this financial circumstance is offset by the higher processing speeds offered by integrated circuits that combine minute lasers with signal-processing capabilities on a single chip. This revolution in opto-electronics again offers hope for the development of ultra-high speed computational facilities based upon light pulses rather than the moving of electrons.

1.8 Superconductors

Research on superconductivity in the last decade has resulted in significant breakthroughs which have received extensive coverage in the mass media so that the term is now a household word. The kernel of the research is to synthesize materials which conduct electricity without energy loss. This phenomenon only occurs when the conductor is subjected to intense cold, typically round −452° F or 4° K. Thus the operation of systems featuring superconducting materials has been an expensive undertaking because the conducting material has been cooled in liquid helium which is an expensive undertaking. Research has, therefore, focused on increasing the temperature at which a material becomes superconductive in order to employ liquid nitrogen as the cooling agent. This material is abundant and very cheap, and it assumes a gaseous state at −320° F, or 77° K, which is readily attainable in practice. By synthesizing materials featuring rare-earths, such as yttrium, barium and thallium, the temperature at which these materials become superconducting has been raised beyond the critical economic temperature for liquid nitrogen cooling.

The commercial consequences of these achievements all exploit the fact that a superconductive material loses all electrical resistance and hence does not waste energy when conducting an electrical current. Potential applications for superconducting materials include the development of powerful electromagnets which would propel high-speed trains for example, more powerful computers, electric automobiles, more efficient national grids for electricity transmission, superior magnetic-resonance machines and imaging machines in medicine.

The field of superconductivity research has fuelled a global battle for superiority in this emerging technology with the numerous potential spin-offs in diverse segments of the economy. The key players are, quite naturally, the US, the principal technology-oriented Western European nations, and the Japanese. The Japanese program is substantial, as evidenced by the fact that the Japanese have filed more patent applications for superconductivity materials and the associated systems than the whole of the remainder of the world combined. This situation is duplicated in the other fields of cutting-edge materials research where a global battle for commercial and economic supremacy is being waged by the principal industrialized nations. Figure 1.18 presents projections for markets of new materials by the year 2000.

Function	New Materials 1981	New Materials	Competing Conventional Materials	Materials Total
Highly-functional high polymer materials	0.2	1.5	0.5	2.0
Fine ceramics	0.2	1.9	1.9	3.8
New metals	0.1	1.5	2.3	3.8
Composite materials	--	0.4	--	0.4
Total	**0.5**	**5.4**	**4.8**	**10.2**

Fig. 1.18 Market sizes for new materials by the year 2000.

1.9 Closure

The insatiable demand for new generations of industrial, military, commercial, medical, automotive and aerospace products have fuelled research and development activities focused on advanced materials. This situation has been further stimulated by the intellectual curiosity of *Homo sapiens* in synthesizing new classes of biomimetic materials, and of course global competition by the principal industrial nations is also a parameter in the equation governing the rate of technological progress. A fundamental axiom of this field of advanced materials is that the ultimate materials are the biological materials which are assumed to be somewhat optimal in performance, design and manufacture; consequently the principal focus of this field is to replicate these characteristics and properties in synthetic materials which can be employed in diverse scientific and technological applications.

Thus by integrating the knowledge-bases associated with advanced materials, information technology and biotechnology, these three megatechnologies are facilitating the creation of a new generation of biomimetic materials and structures with inherent brains, nervous systems and actuation systems which are currently a mere skeleton compared with the anatomy perceived in the not-too-distant future. This quantum jump in materials technology will revolutionize the future in ways far more dramatic than the way the electronic chip has impacted our lifestyles. These new materials are termed *smart materials* or *intelligent materials* and they will typically feature fiberous polymeric composite materials, embedded powerful computer chips of gallium arsenide which will be interfaced with both embedded sensors and embedded actuators by networks of embedded optical-fiber waveguides through which large volumes of data will be transmitted at high speeds.

Today's material revolution is the cornerstone of the triumvirate of megatechnologies which comprise the essential ingredients of this embryonic field. These technologies will have a mutually symbiotic relationship and will significantly impact one another resulting in synergistic technological advances which cannot be foreseen today. However, a natural consequence of advancing on these technological disciplines, will be the impending revolution in smart materials and structures.

The classes of smart materials and intelligent structures are diverse and the application of them is largely unknown. However, what is known is that this new generation of materials will certainly revolutionize our quality of life as dramatically as the state-of-the-art materials did in the past, with stone implements triggering the Stone Age, alloys of copper and tin triggering the Bronze Age, and the smelting of iron ore triggering the Iron Age. The time-line of humankind is located at the dawn of a new age, The Smart Materials Age, which is the topic of this treatise.

2

Intelligent materials

The history of the science of materials from its conception in Paleolithic times through the Stone Age, the Bronze Age and the Iron Age to the current Synthetic Materials Age and beyond to the Smart Materials Age is pictorially chronicled in Figure 2.1. A review of the historical evolution of this science is presented in Figure 2.2 which highlights the distinct transition from structural materials to functional materials; and now smart materials;

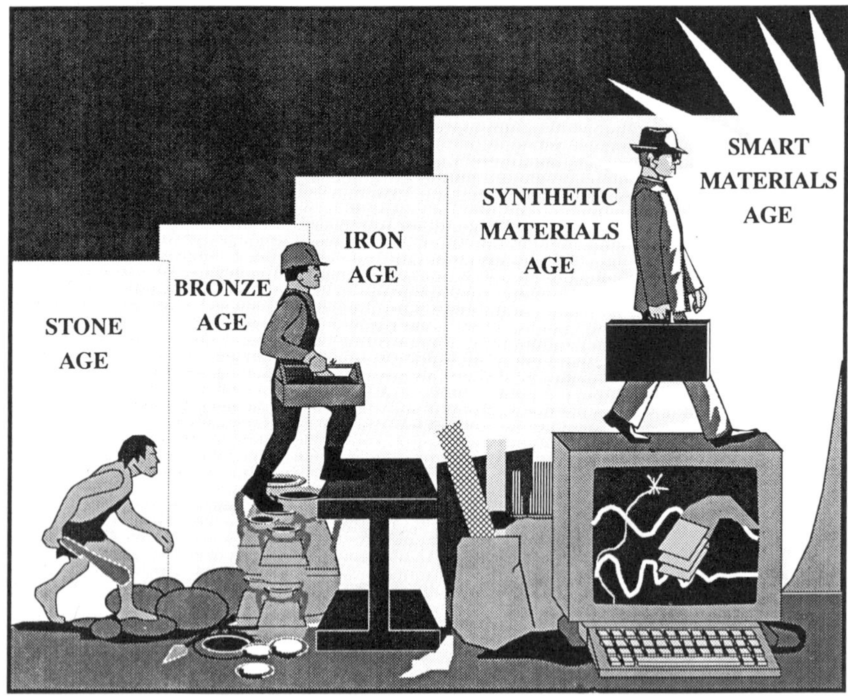

Fig. 2.1 The eras of materials science.

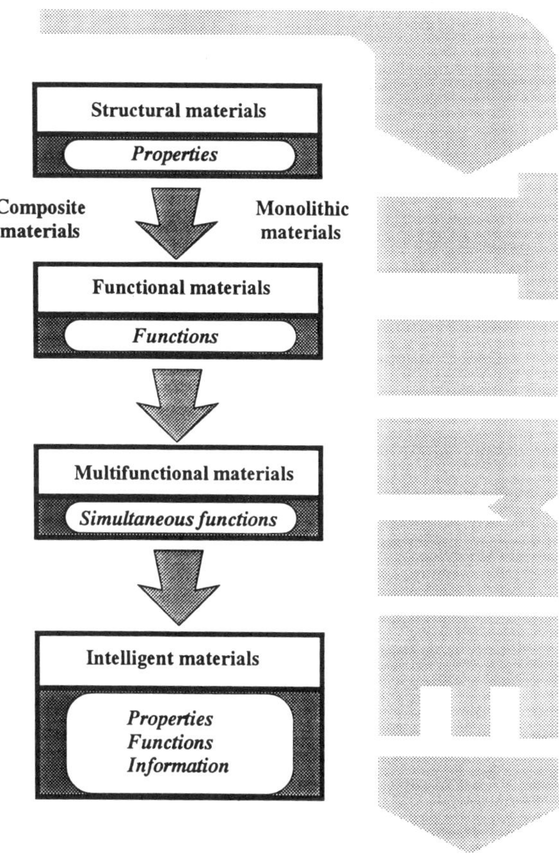

Fig. 2.2 Evolution of materials science.

as humankind's scientific and technological prowess has matured.

Structural materials are those materials that are principally characterized by their mechanical strength and are generally employed in load-bearing situations. Consequently some one million years ago *Homo habilis* selected flint as the most appropriate material for tools and weapons because it was structurally superior to the other natural materials available such as bone and wood. Similarly aeronautical engineers in the 1990s will design the load-bearing members of advanced fighter aircraft in polymeric fiberous composite materials, because they possess structural properties which are substantially superior to those of the monolithic structural materials. Figure 2.3 presents a list of some common structural materials deployed in practice.

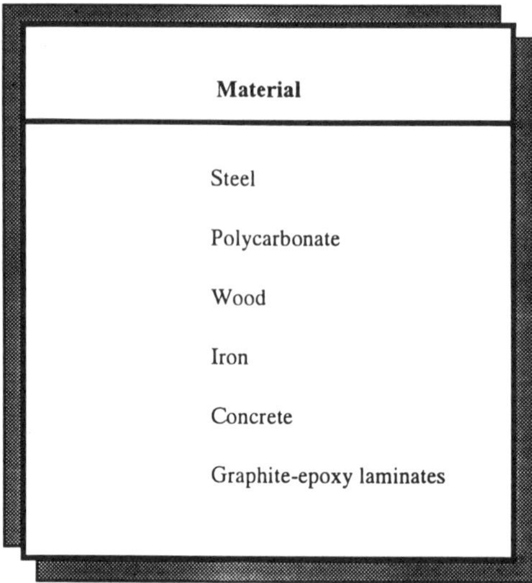

Fig. 2.3 Structural materials.

During the past two centuries, materials science has witnessed the emergence of a second wave of research and development thrusts which has evolved in parallel with those activities focused on the development of new generations of structural materials. This second thrust has focused on the development of *functional* materials; namely, materials whose principal functional characteristic is exploited in the fields of science and technology rather than the inherent structural properties of the material. Figure 2.4 lists a variety of functional materials.

Thus, for example, vibration transducers typically feature a piezoceramic element, which, when fixed to a vibrating structure, is subjected to dynamical deformations and thereby develops an appropriate time-dependent electrical signal for measuring the vibrational response of the structure. The mechanical strength of the piezoelectric element is not exploited in this application, instead the functional piezoelectric characteristics of the material, which couple the mechanical and electrical behavior, dominate the selection process.

Another example of a functional material is germanium which is a photoconductive material. When this class of functional materials is exposed to a light source it generates an electric current as the photons of light are absorbed at the surface of the material and electrons are subsequently released from the atomic structure. Since the associated current is generally small, an amplifying circuit must be employed to utilize this phenomenon for commercial applications. An analogous phenomenon occurs with

Material	Properties
nickel-titanium	shape-memory
cadmium sulphide	piezoelectric
terbium iron	magnetostrictive
quartz	pyroelectric
electro-rheological fluids	viscoplastic
aluminum soap solution	viscoelastic
barium titanate	ferroelectric
copper oxide	photoelectric
potassium dihydrogen phosphate	electro-optic
selenium	photoelectric
germanium	photoconductive

Fig. 2.4 Functional materials.

photoelectric materials. When radiant energy in the visible, X-ray, ultraviolet or gamma-ray segments of the spectrum infringes on these materials, charged atoms or molecules are subsequently released.

The rapid evolution of hardware in computer science has been made possible by integrated circuits featuring silicon chips. Again, these circuits have not benefitted from the mechanical strength of the material. Instead electronics engineers have exploited the functional properties of silicon, namely its electrical and physical properties. The dependence of the microprocessor industry on electrical signals and silicon chips may soon be

Material	Properties
Lead Zirconate Titanates	Piezoelectric
Bone	Mineral Homeostasis Structural Homeostasis Piezoelectric Pyroelectric
Lead-Magnesium Niobate	Thermo-electrostrictive
Terbium-Iron-Dysprosium	Thermo-magnetostrictive

Fig. 2.5 Poly–functional materials.

superseded by a new technology involving gallium arsenide chips, photonics and *multifunctional* materials. Multifunctional materials are materials characterized by several functional properties which are utilized in practice rather than their structural properties. Figure 2.5 presents a table documenting some of these materials. The molecular units in biological materials are generally multifunctional in nature because this configuration facilitates efficiency and economy. This research and development trend towards poly-functional materials exploits notions of biomimetics, and it represents a necessary first step towards establishing the essential ingredients for the evolution of smart materials and intelligent materials.

Synthetic poly-functional materials have been developed by modifying structures at the bulk and molecular levels in order to synthesize molecular and polymeric systems with the desired physical and chemical properties. These systems exploit organic materials with multifunctional properties which provide a crucial ingredient for the evolution of the next generation of ultra-high-speed computational facilities based upon photonics technology.

Thus *Homo sapiens sapiens* currently has the ability to develop structural materials, functional materials, and also poly-functional materials. These capabilities are essential for the synthesis of smart materials. The most sophisticated class of smart materials and structures are currently based upon notions of biomimetics, and feature appropriately configured actuators, sensors, signal-processing capabilities and control algorithms which enable the materials to respond autonomously to external stimuli. Smart materials will have the capability to select and execute specific functions intelligently in response to changes in environmental stimuli. For example, these innovative multifunctional materials may exhibit homeostasis, the tendency of an organism to maintain normal internal stability by coordinated responses systems that autonomously compensate for environmental changes. This ability may be complemented by several other capabilities that are characteristic of intelligent systems, such as self-diagnosis, self-repair, self-multiplication, self-degradation, and self-learning. Furthermore, these features may be augmented by capabilities for anticipating future challenges and missions and the ability to recognize and discriminate. It is clearly evident, therefore, that all aspects of our lives will be significantly touched as the development of smart materials impacts industries as diverse as automotive, aerospace, defense, biomedical devices, advanced manufacturing, robotics, industrial machinery, sporting goods, high-precision instruments, highways, buildings and bridges.

These material functions of structure, actuator and sensor are currently incorporated into a smart structure in a discrete global sense. Thus, for example, a current generation smart structure might feature a load-bearing graphite-epoxy, fiberous polymeric structural material, in which are embedded piezoelectric discs for sensing purposes and embedded shape-memory-alloy wires for actuation purposes. Research is currently being pursued on embedding these material functions of sensor, actuator, and structure at a much more local level. For example, carbon fibers may be coated with piezo-electric materials in order to synthesize a smart composite material which has distributed actuator and structural properties at length scales comparable to the diameter of the fiber. Similarly, electro-rheological fluids may be embedded within hollow fibers which may be employed in the structure for reinforcement or sensing, for example.

In the future, the current methodology of large scale macroscopic and mesoscopic integration of structural, sensory, and actuator materials will be replaced by the integration of the microstructural properties at the atomic scale, in order to synthesize somewhat more homogeneous substances as shown in Figure 2.6. Typically these techniques will be employed at regions with dimensions that are too large to be considered to be at the inter-atomic level but too small to be considered at the solid-state level. This concept has been referred to by several scientific terms such as *micro-composite materials*, *mesoscopic materials*, *hybrid materials*, *structurally controlled materials*, and *engineered materials*. When this technology has been perfected, the materials scientist will be able to synthesize, design and create three-dimensional atomic arrangements which will render obsolete the

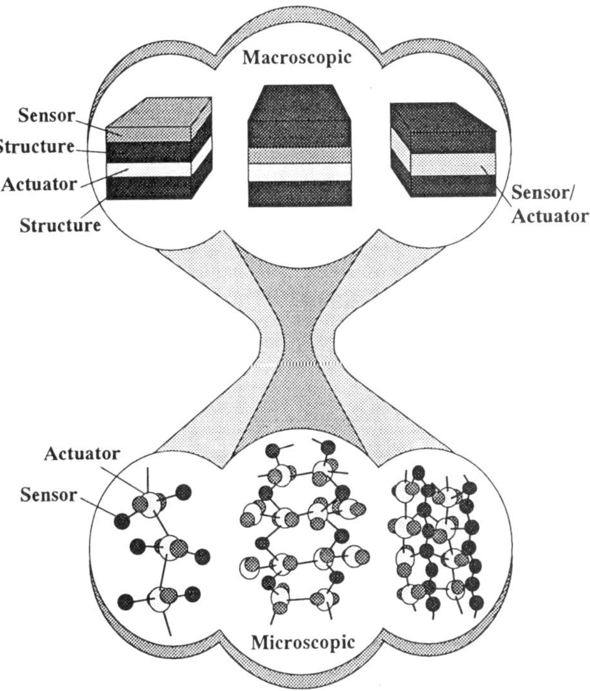

Fig. 2.6 Synthesis of macroscopic and microscopic substructures.

categorization of materials into such groups as insulators, metals, polymeric materials and biomaterials, for example. *The authors have the clarion conviction that as the structural complexity of materials increases, the coupling between design, analysis, and manufacturing processes becomes more and more inextricably intertwined. Therefore the integration of actuators, sensors, processors, and structures for smart materials applications will mandate the evolution of sophisticated manufacturing-process-driven design, analysis, and synthesis methodologies.* It is anticipated that along with quantum leaps in manufacturing technologies, manufacturing-process-driven analysis and design technologies, packaging and environmental considerations will need to be addressed for the successful evolution of this class of smart materials.

Furthermore, in the context of intelligent materials there is considerable focus on sensors and actuators which are of course discrete materials with discrete functional properties. With the development of techniques for designing and manufacturing materials at the atomic level these terms will again become somewhat obsolete. After all, the basic units of life in biomaterials, the cells, monolithically unite all of the structural, sensory and actuator functions in a truly integrated system.

The current generation of smart materials and structures incorporate one or more of the following features:

1. *Sensors* which are either embedded within a structural material or else bonded to the surface of that material. Alternatively the sensing function can be performed by a functional material which, for example, measures the intensity of the stimulus associated with a stress, strain, electrical, thermal, radiative, or chemical phenomenon. This functional material may, in some circumstances, also serve as a structural material.
2. *Actuators* which are embedded within a structural material or else bonded to the surface of the material. These actuators are typically excited by an external stimulus, such as electricity in order to either change their geometrical configuration or else change their stiffness and energy-dissipation properties in a controlled manner. Alternatively, the actuator function can be performed directly by a hybrid material which serves as both a structural material and also as a functional material.
3. *Control capabilities* which permit the behavior of the material to respond to an external stimulus according to a prescribed functional relationship or control algorithm. These capabilities typically involve one or more microprocessors and data transmission links which are based upon the utilization of an automatic control theory.

Materials with the above features are indeed worthy of being described by the adjective ‘smart’, as defined by *Webster’s Third International Dictionary of the English Language*, which states that the meaning of ‘smart’ is: *having or showing mental alertness and quickness of perception, shrewd informed calculation, or contrived resourcefulness, marked by or suggesting brisk vigor, speedy effective activity, or spirited-liveliness.* Clearly, materials featuring control capabilities possess 'mental alertness', and some will certainly be 'informed' and 'resourceful' within specified limitations. Materials featuring sensing characteristics have the opportunity to demonstrate an 'informed' response along with 'quickness of perception'. Finally, materials featuring actuator functions possess 'spirited-liveliness' characteristics. Thus 'smart' materials clearly exist today in the arsenal of weapons for deployment by the materials scientist but 'intelligent' materials are an order of magnitude more sophisticated than smart materials because 'intelligence' is associated with learning, abstract thought, and the ability to think and reason. These capabilities have not been demonstrated at this time and they shall be the focus of much research and development during the coming decades.

Again upon consulting *Webster's* dictionary the definition of 'intelligent' is: *the available ability to use one's existing knowledge to meet new situations and to solve new problems, to learn, to foresee problems, to use symbols or relationships, to create new relationships, to think abstractly*; *ability to perceive one's environment, to deal with it symbolically, to deal with it effectively, to adjust to it, to work toward a goal*; *the degree of one's alertness, awareness, or acuity*; *ability to use with awareness the mechanism of reasoning whether conceived as a unified intellectual factor or as the aggregate of many intellectual factors or abilities, as intuitive or as analytic, as organismic, biological, physiological, psychological, or social in origin and nature*; *mental acuteness*.

The noun, 'intelligence', provides the basis for the definition of the adjective, 'intelligent', which is defined as *possessing intelligence or intellect*; *having the power of reflection or reason*; *having or indicating a high or satisfactory degree of intelligence and mental capacity or powers of perception, consideration, and correct decision*; *revealing or reflecting good judgement or sound and comprehensive thought*; *marked by quick active perception and understanding*.

The synthesis of materials with these significant capabilities is an extremely challenging undertaking which involves the integration of diverse scientific and technological disciplines as illustrated in Figure 2.7. However, research in many diverse disciplines is currently being prosecuted from different premises in order to achieve this goal of creating materials and substances, with the innate ability to autonomously respond in an intelligent manner to dynamically-changing environmental conditions. Thus, for example, research programs have been undertaken to create materials with intelligent functions to replace traditional software algorithms and electronic circuitry in the fields of neuro-computers and artificial intelligence.

Furthermore, materials possessing intelligent functions and also bio-

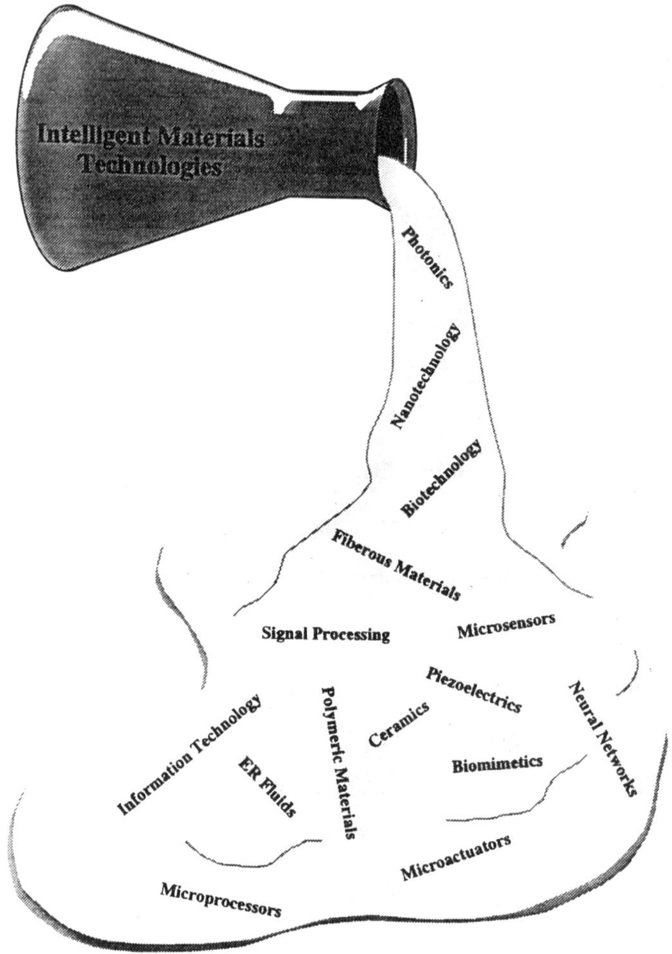

Fig. 2.7 The diverse areas of intelligent materials.

compatibility are the subject of intensive research efforts. The focus of this research is to develop a new generation of autonomous, lightweight, compact *in vivo* devices for diagnostic purposes and treatment, such as biosensors and drug delivery systems, and also a new generation of prosthetic materials for bones, blood vessels and artificial organs, for example, which will render obsolete the current generation of open-loop externally controlled devices.

Other research efforts are focused on the development of a different class of intelligent materials for deployment in the severe environments of atomic energy, space and aeronautics where the relevant intelligent functions include self-diagnosis, damage assessment and self-repair in addition to high reliability. Intelligent materials possessing harmony with the environment and humanity will also be created in the near future due to growing environmental concerns in Western Europe and the United States. Such materials will be associated with the mundane accessories of life, such as clothes and the other daily products, and they will be user-friendly and provide convenience and enhanced comfort whilst being compatible with the relevant ecosystems. While the skills and experience of the individual researcher may be responsible for breakthroughs in the different sub-disciplines of the broad field of smart materials and structures, it is anticipated that collaborative programs involving inter-disciplinary groups of researchers with diverse backgrounds, will be an essential ingredient for the creation of viable materials technologies for deployment in practice for the betterment of humankind.

Thus the fundamental scientific and technological expertise resident in fields such as physics, chemistry, and biology for example will be responsible for the creation of new classes of intelligent materials. These advances will be attributed to the synergistic interactions between researchers in these fields, which will result in enhanced methods of analysis based on advancements in the integration of material structures and functions, using superior design and control methodologies, permitting materials to be synthesized at both the atomic and molecular levels. Indeed such capabilities are already being exploited in the creation of semiconductor superlattices which feature precisely controlled structural configurations. These successes are part of a broader initiative involving hybrid materials, where different classes of organic and inorganic compounds are combined at the atomic and molecular levels with other materials exhibiting the desired structural properties.

The innovative functions that will characterize these new classes of intelligent materials will be similar to those associated with biological systems. This realization has motivated research efforts to scrutinize biomaterials and to develop a superior knowledge of their intrinsic behavior in order to apply these generic concepts to the design of intelligent materials. Thus, biological systems are generally regarded as a reference state, or datum for measurement, for evaluating the properties of an intelligent material because these naturally occurring systems have evolved during the previous millennia. Thus these biological materials have continuously adapted to a dynamically evolving environment, while developing innate properties of homeostasis to provide protection while ensuring both reproducibility and survivability. The new classes of intelligent materials should also display these characteristics while maintaining harmony with society, and the environmental ecosystem.

The concept of intelligent materials, their definitions, and also the interplay of the distinct scientific disciplines relevant to this embryonic field are the theme of this chapter. One of these themes is that intelligent materials can be classified by three categories of materials. The three categories are intelligence associated with the most primitive functions of materials; intelligence intrinsic to materials; and intelligence from a human perspective.

2.1 Primitive functions of intelligent materials

Intelligence at the most primitive levels in materials may be classified by three primitive functions associated with sensing, actuation and processing. Embedded in these primitive functions are energy conversion mechanisms, and information-transfer mechanisms involving fundamental concepts at the atomic and molecular levels in chemistry and physics. This theme is pictorially presented in Figure 2.8.

The sensor function associated with an intelligent material is typically employed to detect and monitor information characterizing the external environmental stimuli imposed upon a material or else to measure the behavior within the material itself. The processor function processes and evaluates detected information provided by the sensor function in concert with information already memorized in the material, which requires the multiple functions to be systematically coordinated. This processor function is also responsible for recording information and retrieving relevant data-sets concerning previous experiences and judgmental algorithms for example. The final function of the triumvirate of functions at this level is the actuator function or effector function. This function provides appropriate actions in response to the data furnished by the sensory function to the processor function.

At the next level in the structure of this class of materials, the systematic information network function systematically orchestrates the transmission of information between the sensors, processors, and actuators in order to achieve a desired response. The energy conservation function supplies the energy necessary to operate the processor, sensor and actuation functions by utilizing energy sources in the neighboring environment or else from energy sources within the material itself. It may also be necessary under these circumstances for the intelligent material to feature an energy-conversion function, in order to accommodate situations where the form of energy supplied by the environment is different from that required for actuation or processing. Finally, at the next level in this class of intelligent materials, a comprehensive understanding of the physical and chemical structures of the substances inherent in these materials, are required at the most fundamental of levels in order that techniques can be established for designing and controlling these materials. These techniques involve a variety of structural considerations such as the transfer of ions, and changes in crystalline structure, molecular structure, electronic structure and grain boundaries.

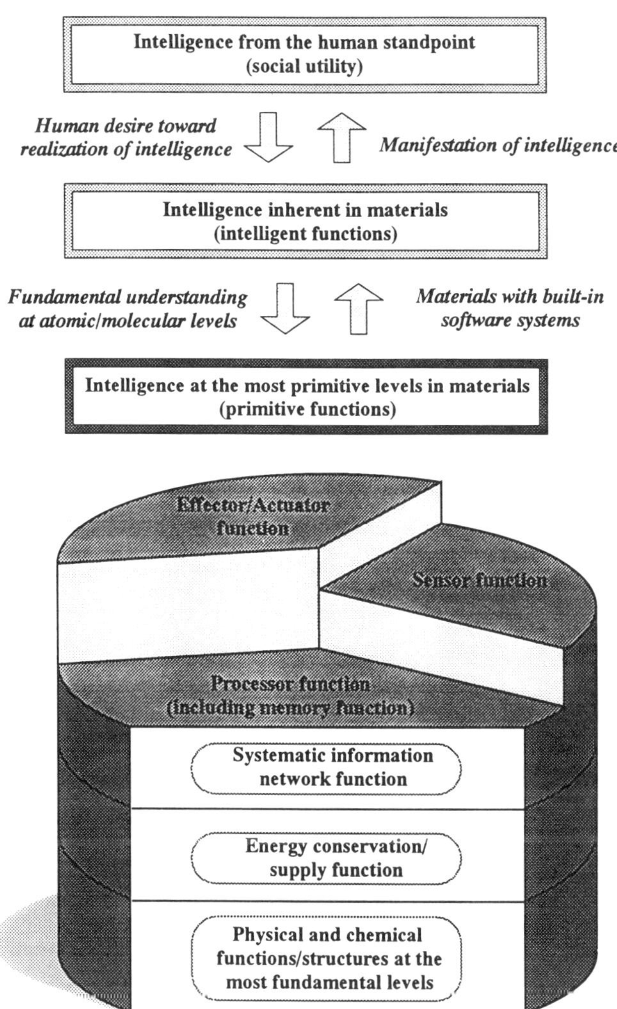

Fig. 2.8 The primitive functions of intelligent materials.

2.2 Intelligence inherent in materials

Intelligence inherent in a naturally-occurring material is typically a property unique to a material and does not require human involvement. As the field of smart materials and structures matures, greater emphasis will be placed upon the creation of new synthetic materials requiring the synthesis of compounds with inherent functional properties. The intelligence characteristics associated with this activity will typically include numerous autonomous functions such as self-degradation, self-learning, self-diagnostics, and regeneration. The rusting of iron in humid environments may be considered as a simple form of the self-degradation characteristics of the material. An example of a material with a regeneration function is the semiconductor *InP* which upon suffering a reduction in its properties when subjected to radiation, returns to its original state unlike standard semiconductors which suffer irreversible damage under these conditions. Other functions could include the ability to recognize and subsequently discriminate, redundancy at the most primitive level of computational, sensory, and actuation functions, hierarchical control functions and predicting the future based upon sensory data prior to developing an appropriate response. The framework of this sub-discipline is presented in Figure 2.9.

These diverse intelligence attributes are dependent upon a more basic set of intelligence properties which include homeostasis, the tendency of an organism to maintain normal internal stability by coordinated responses of systems that autonomously compensate for environmental changes, appropriate feedback mechanisms, and also the ability to respond to an external stimulus in an appropriate time-frame and in an appropriate manner. Examples of these latter traits include materials that can evaluate current trends and circumstances in order to predict future conditions. Another classification includes materials that possess time-dependent properties which may be autonomously tailored to respond to the external environment. Thus, for example, a material which has been damaged and is undergoing a self-repair process may reduce its level of performance when subjected to external excitations from a superior performance category, to an adequate performance category in order to survive.

Intelligence inherent in a material therefore comprises a number of categories which can be defined utilizing generic concepts pertaining to biological systems, and biomimetics. The following paragraphs highlight the principal intelligence features inherent in biological materials, and also the interpretation of their attributes in the context of materials science.

Autolysis

When the tissues and cells of biological systems are destroyed or cease to receive nutrition and die, they decompose. Thus intelligent materials with this attribute would immediately decompose upon completion of their useful life, and be assimilated within the environment.

Redundancy

Biological systems typically possess some degree of redundancy in their structures and functions in order to survive. Generally the efficiency of the system is inversely related to the degree of redundancy. Thus intelligent materials with this attribute would feature redundant functions that are not employed under normal working conditions. As the external

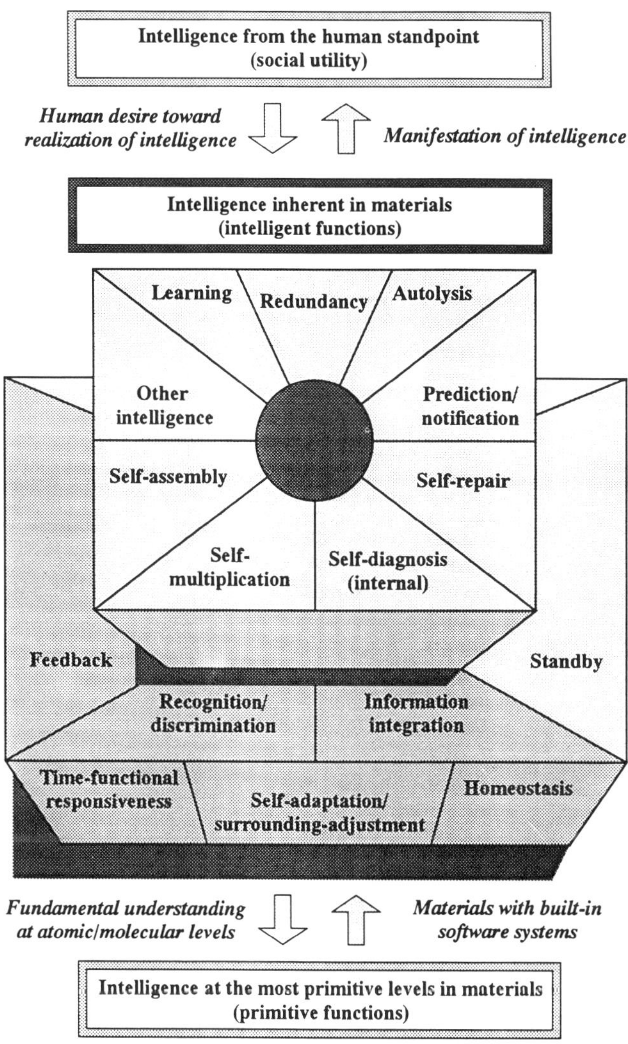

Fig. 2.9 Intelligence inherent in materials.

environmental conditions change, or as the material's structure changes, the material microstructure in other regions would change in order to replace the original functions and maintain equilibrium.

Mitosis

Self-breeding, self-multiplication, or growth in a biological system involves a cell creating similar cells by mitosis to replicate them. To mimic these phenomena, intelligent materials would be self-producing with the additional requirement of automatically halting this activity when prescribed conditions have been achieved.

Self-repair

The biological self-repair function is closely related to the growth or self-multiplication function because it involves the various phenomena involved with restoring damaged living material to its state of normality. Typically this involves processes associated with the DNA recovery, the recovery of tissue to the cell level and also to the molecular level. Intelligent biomimetic materials must be able to autonomously perform appropriate self-repair functions in order to restore regions of degradation caused by extreme environmental conditions, for example.

Prediction

Some biological systems can predict the future, by learning from past experiences, and take appropriate actions based upon sensory data and knowledge. Classes of intelligent materials would also possess the ability to respond appropriately to changing environmental stimuli by capitalizing on knowledge and experience distilled from previous experiences.

Learning

The predictive capabilities of biological systems are closely associated with the ability of these systems to learn. While some of these capabilities are innate, others are developed through experience with the local environment and may involve inductive logic wherein general conclusions are distilled from the observations of a natural event. Intelligent materials would also feature these capabilities in order to respond more appropriately to external stimuli through the recollection of previous actions involving sensory data, processing of information, and subsequent actions and response.

Autonomous diagnosis

The ability to self-diagnose problems, degradations, malfunctions and judgmental errors by comparing current performance or conditions with past performance or conditions, for example, is a health-monitoring feature of some biological systems. Classes of intelligent materials would be able to monitor their state of well-being, or health, in order to determine the effect of a defective region on the ability of the material to function properly and the state of degradation.

The previous listing of biological functions for intelligent materials presents a group of somewhat global, high-level functions for different classes of innovative materials. Embedded within each of these biological phenomena is another set of more basic intelligent functions concerned with sensing, processing, and actuation capabilities as presented in Figure 2.9. These intelligent functions are associated with the systematic cooperation of a multitude of other functions such as sensory feedback and information processing, in addition to homeostasis and the time and functional response to external stimuli. The relevant categories of this class of intelligent functions are synoptically listed below.

Feedback

The state of homeostasis that is fundamental to the survival of numerous biological systems, is dependent upon closed-loop feedback where sensory data at both the input and output state of the system are compared and utilized in the control scheme. This ability to utilize closed-loop feedback control algorithms should be replicated in intelligent materials in order to develop desired conditions and functions where sensors, processors, and actuators are orchestrated in unison.

Information integration

Biological systems generally evolve by storing the integrated experiences of time prior to transferring this information to subsequent generations of the species in the form of genes, for example. Intelligent materials should mimic this evolutionary strategy by featuring innovative functions which integrate and maintain memory-banks of information.

Standby

Biological systems frequently maintain a state of readiness for action. Intelligent materials should also maintain this condition of alertness.

Recognition

Creatures frequently embody functions that permit them to recognize similar members of the species in a discriminatory manner. Classes of intelligent materials should be established with the ability to recognize and discriminate between information contained in different sensory data-sets. These innovative actions should embody both analytical skills and also a more global, fuzzy level of interpretation of the information.

Responsiveness

Classes of intelligent materials should be developed which feature innovative time-dependent functions, whose responsiveness is governed by the dynamically changing external environmental conditions and the role mandated of the material.

Homeostasis

Biological systems are typically subjected to dynamical external conditions and they feature continuously changing internal structures. These systems survive and maintain stable physiological states by coordinated responses that autonomously compensate for these ever-changing environmental conditions. Intelligent materials could mimic these capabilities by incorporating innovative functions which permit them to autonomously change their behavior in response to dynamic environmental conditions.

Adaptation

The impact of gradually-changing environmental conditions on biological systems is manifest as the evolution of the physiological state to permit the system to adapt to the new environment. Thus intelligent materials will require innovative functions which will be able to change their behavior in response to environmental changes in an optimal manner.

2.3 Materials intelligently harmonizing with humanity

The intelligence of a material from a human perspective involves the circumstances in which a material is employed and its impact upon the environment. Thus new generations of intelligent materials must be synthesized in the context of harmony with the environment, the optimal utilization of resources for economy, reliability and compatibility with biological systems. This level of intelligence will need to be founded on basic concepts of both analytical and synthetic judgement, rationalism, logical thinking, and congruity. The scenario is presented in Figure 2.10.

Thus materials possessing intelligence in the context of interactions with *Homo sapiens sapiens* must possess a variety of diverse properties. They must be capable of functioning harmoniously with both the natural environment and also humanity, while maintaining human friendliness through their ability to adapt to the dynamically changing conditions under which they must operate. They must also be able to be manufactured and operate reliably in an optimal manner within the context of harmony with the local ecosystem and humanity. Thus these intelligent materials must be able to be manufactured by effectively utilizing natural resources without waste, and the specific innovative functions of these materials should be economically developed at minimal cost.

To this state of global harmony between these materials and their environment there must also be adjoined properties of analytical judgement, comprehension and rationalism. Thus these intelligent materials must be able to develop viable reactive strategies by processing data that characterizes the external environmental conditions. These reactive strategies should be based on algorithms featuring systematic recognition, analytical reasoning, rationality and also fuzzy logic in order that appropriate strategies are developed.

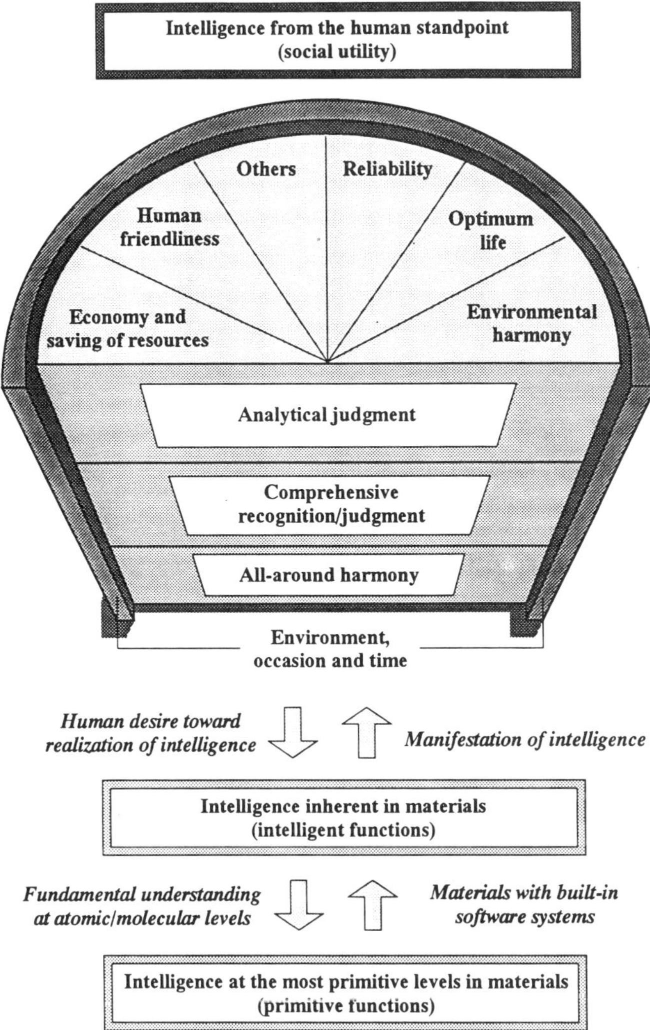

Fig. 2.10 The characteristics of materials intelligently harmonizing with humanity.

2.3.1 Examples of intelligent materials

The previous sections of the chapter have presented reasonably detailed discussions on the different ideas and concepts of intelligent materials within a generic framework. In an effort to stimulate further cogitations on these innovative classes of materials, a series of illustrative examples in diverse scientific fields will now be presented. It is anticipated that these diverse classes of materials and the requirements of continuous adaptation dictated by the changing environmental conditions, will require the implementation of new design-for-manufacture strategies to create these materials along with the careful orchestration, and possible coalescence of the sensing, processing and actuation functions in order that these materials fulfill their multiplicity of roles in these fields.

Structural Materials

1. Intelligent materials which may be employed over a very wide temperature range up to ultra-high temperatures should be developed for aerospace applications. The materials would feature self-adaptation and environmental adjustment properties, which would be manifest by their ability to undergo a transformation in composition which would be triggered by the external thermal environmental conditions.
2. Other classes of intelligent materials would be able to suppress the state of damage in aircraft and other transportation systems by employing self-diagnosis functions, prediction functions, self-repair functions and notification functions to indicate to the operator their state of residual life by changing color.
3. Intelligent materials employed in military vehicles and other systems with the potential for damage by projectiles, should be able to recognize the rate of change of stress in a body prior to rapidly generating an appropriate response. The material must possess inherent redundancy and be able to recognize the phenomena prior to discriminating between a potential shock stress and a static stress.
4. Materials subjected to repeated stress cycles, such as parts for transportation systems, typically fail in fatigue because of crack propagation phenomena. Innovative materials should be developed to restrain the propagation of cracks by developing compressive stresses at the crack tip, through a volume change from stress-induced phase transitions, or else reducing the stress concentration at the crack tip. Such materials would feature redundancy and discriminatory characteristics.

Electrical materials

1. Homeostasis can be demonstrated in intelligent electrical materials which possess the ability to regulate their resistance as a function of the local ambient temperature, and hence the magnitude of the electrical current conducted can be maintained at an optimum value.
2. Self-repair characteristics can be demonstrated in intelligent electrical materials, such as those employed in varistors, which possess the ability to detect the oxygen-deficiency in the material, prior to initiating the supply of oxygen to conduct the repair process.

Bio-compatible materials

1. Standby and recognition characteristics are necessary for materials being developed for micro-capsule, drug-delivery systems containing anti-cancer drugs for example. These materials must be able to first recognize cancer cells, prior to coalescing with them and subsequently releasing the appropriate anti-cancer drugs.
2. Autolysis and growth characteristics are necessary for intelligent bio-compatible materials which are implanted in human beings. Such characteristics are necessary for artificial bone materials, for example, in order that they mimic the behavior of these sophisticated natural biomaterials.

Miscellaneous materials

1. The food packaging industry, which generally harvests fruit and vegetables before they have fully ripened in an effort to present ripe produce in the retail stores, would significantly benefit from thin-film wrapping materials that embody adaptation characteristics and also the ability to respond to local environmental conditions. These wrapping materials would not only possess the ability to optimally control the maturing of vegetables, for example, by controlling the emission of ethylene gas, but they could also predict the window of opportunity for consuming the product in the most palatable condition.
2. Several segments of the economy would benefit from innovative adhesive materials which feature time-dependent functional responsiveness. These materials would feature adhesive coatings which would only maintain their adhesive properties for a prescribed time-period, prior to suddenly undergoing a severe attenuation in their adhesion in order that the adherends would easily separate. A material with these characteristics could be employed to develop a new generation of medical adhesive tapes which could be painlessly removed from hairy skins. Such products would also contain the ability to monitor and evaluate the health of the wound, so that upon achieving a satisfactory state of skin repair, the properties of the adhesive film would be attenuated, thereby permitting the adhesive tape to be removed at the optimal time.
3. The clothing industry and the housing industry would both benefit from materials whose thermal conductivity and porosity vary as a function of temperature and humidity. Thus, clothes fabricated in these materials would enable the individual to optimally maintain a state of comfort commensurate with the ambient conditions. They would be neither too hot nor too cold. Similarly, homes and office complexes would be able to provide a comfortable air-management environment by exploiting the capabilities of these materials with adaptation properties, ambient-adjustment characteristics, and feedback.

The introductory section of this chapter focused upon the evolution of materials science from structural materials through functional materials to intelligent materials. The subsequent sections utilize the same philosophy to discuss typical examples of anticipated classes of intelligent materials with structural properties and also functional properties with embedded intelligence.

2.3.2 Materials with autonomous dynamically-tunable properties

Materials should be developed with properties that are autonomously changeable in response to their environmental conditions. Two categories of materials may be considered in this category:

1. Materials whose composition or structural properties autonomously change in response to the operating environment or service conditions.
2. Materials whose non-structural properties autonomously change in response to the operating environment or service conditions.

Examples of the first category of variable-structure materials include compounds whose structural composition changes in response to external electromagnetic fields or radiation, or the structural degradation associated with oxidation. These compounds may be synthesized to trigger an appropriate re-ordering of the structure so as to improve the resistance to the potentially catastrophic external stimuli. These materials would belong to a category of stimulus/structurally-reactive materials. An example of a somewhat simplistic class of these materials are those materials employed in thermistors, which are temperature sensors featuring a material whose resistance is temperature-dependent. Thus as the temperature increases, the properties of the material change in direct response to the change in this environmental parameter. Alternatively, materials may be synthesized whose chemical constituents change in response to the operating conditions or environment in order that the material may undertake a self-repair process or else decompose.

Materials belonging to the second category may manifest their intrinsic intelligence in the context of self-learning, the ability to recognize, the ability to predict, self-diagnosis, the ability to discriminate, and selective redundancy. Examples of these classes of intelligent materials include materials whose electrical or physical properties are load-dependent.

Thus materials could feature electrical switching characteristics which would be load-dependent or time-dependent. Other examples include materials whose ultimate strength in highly-loaded regions of a structural member increase in response to the magnitude of the stress levels and external loading, while the ultimate strength in the lower stressed regions of the member is reduced accordingly.

These load-sensitive materials could be considered as a subset of materials whose properties are dependent upon external stimuli. Thus the environment in which the material is situated could be responsible for changing the Curie temperature of a piezoelectric material, the hysteresis profile of a shape-memory alloy or the number of fatigue cycles for failure of a structural member. Smart structural materials could be created whose surface color or other visual appearance characteristic, would change in response to the extent of damage attributed to mechanical loading or electromagnetic radiation, for example.

2.3.3 *Materials with autonomous dynamically-tunable functions*

In analogy to materials whose structural properties are changed in response to external stimuli, the functional properties of intelligent materials should also be dynamically-tunable in response to their environment. Typically this class of intelligent materials would possess the ability to recognize and discriminate, self-learn, self-diagnose, and the ability to react in a predetermined way to an external stimulus.

Examples of these materials include compounds whose optical characteristics vary according to the wavelength and intensity of the incident radiation. This could be employed in the fabrication of intelligent windows which admit only visible light on hot days while rejecting heat radiation, for example. Other heat transfer characteristics of the glass would enhance heat removal from hot buildings in cool summer evenings while preventing heat loss in the winter. The gas permeation characteristics of materials could be made to be responsive to the environment in order to optimize the rate of diffusion of a fluid through the porous medium. The electrical properties of a material could be dynamically-tuned in response to the mechanical loading imposed upon it or the applied electrical potential. Such actions may be employed to automatically activate an electrical circuit. Alternatively, under the same conditions, the type of electrical signal imposed upon a material which can discriminate between many such signals could be employed to activate a device.

Materials with embedded systems represent a higher-order class of intelligent materials than the aforementioned materials with dynamically-tunable functions. These notions of embedded systems within a material can be considered in the context of embedding information in the material in the form of a transfer function with multiple inputs. Thus materials of this kind would be able to simultaneously detect a variety of external stimuli prior to scrutinizing these stimuli and coordinating an appropriate response. These materials could feature the ability to recuperate and restore their degraded sensitivity when damaged, and also adjust their sensitivity in response to changing environmental conditions. Some materials in this subclass may feature the ability to serve as intelligent catalysts which regulate, control, and monitor chemical reactions. Other classes of materials may be employed in the development of intelligent textiles which, when employed by the garment industry are able to simultaneously measure the local atmospheric conditions and the condition of the human being, in order to optimally control the comfort of the person through the active control of the heat retention or cooling properties of the clothing.

2.4 Intelligent biological materials

The complexity of the biological environment mandates that intelligent materials for biological applications, should generally feature a comprehensive array of capabilities. These would include self-diagnosis, self-learning, self-repair, self-degradation, selective redundancy and self-reproduction. At a more primitive level these materials would feature feedback, recognition, and discrimination capabilities and information accumulation characteristics and homeostasis. Typical applications would include materials behaving as drug delivery systems, materials replicating the human pancreas, liver, kidney and skin, and materials promoting bone growth.

2.5 Biomimetics

The emergence of the embryonic field of smart materials has stimulated an increased focus on biological materials as researchers attempt to mimic the features, characteristics and growth of these naturally-occurring materials which have been in existence for some four billion years. The field of biomimetics is based on the assumption that through evolution, nature has solved an optimization problem. Whereas the solution may be known in its entirety, caution is clearly in order since the object functions may be fuzzy, and the optimization criteria are not clearly and completely known. Upon reviewing the substances from which the skeletal tissues of plants and animals are formed, it is clearly evident that biological systems feature a very limited range of substances but the microstructural characteristics of these substances are extremely diverse. It may therefore be concluded that nature employs a limited group of generic materials comprising polysaccharides and proteins, which are subjected to subtle biological processing methods at largely ambient temperature and pressure, in order to synthesize a diverse range of mechanical and structural properties.

In sharp contrast to this philosophy, *Homo sapiens sapiens* operates at the other extreme. Since technological disciplines have far fewer constraints to comply with than the biological materials, exceedingly diverse manufacturing techniques have been developed during the previous millennia. As a result, technological manufacturing processes frequently involve high temperatures and pressures, and a wide variety of materials to initiate the process. Consequently there are now thousands of different commercially-available synthetic materials in the marketplace for utilization by the design engineer.

However, to achieve optimal performance of composite materials featuring a fiberous reinforcement and a matrix system, for example, it would appear that attention should be devoted to the mimicking of nature's manufacturing process where the different phases of the same material, namely the reinforcing fiber and the matrix, are formed simultaneously, which contrasts again with the technological approach of fabricating each phase separately prior to unification. This topic has some considerable significance because unlike the monolithic materials frequently employed in engineering practice, all biological materials, which are the result of evolution throughout the previous half a billion years or so, are composite materials featuring two or more phases. Thus these multi-phase natural systems, have contrasted very sharply with the commercially-available monolithic engineering materials until the past thirty years or so, which has witnessed the emergence of fiberous composite materials technology.

Biological systems are a rich source of ideas for individuals concerned with the synthesis of new materials and processes. Thus, for example, biological systems have inspired the evolution of the field of neural networks where a large number of imperfect and functionally slow nerve cells together perform complex tasks. These systems perform tasks on information in a parallel fashion, which differs from the sequential processing that is characteristic of the conventional Von Neumann computers, and they are able to program themselves to a limited extent.

A study of the microstructural properties and characteristics of biological materials provides a variety of ideas for replication in synthetic technological materials. For example, it is evident that these materials feature great morphological complexity and they provide high mechanical performance. Thus the focus of attention is the control of the interaction between the material constituents, such as the amino acids and the polysaccharides, in the synthesis of biomaterials with diverse characteristics ranging from the soft and extensible properties of skin and tendon, to the rigid properties of grass and insect cuticles. Biological materials which are soft and extensible generally have a high resistance to fracture and are difficult to tear. This is attributed to the collagen and elastin fibers re-orientating themselves in the deformed state relative to the undeformed state. Furthermore, the crack-stopper mechanism inhibiting fracture is enhanced by the much more severe re-orientation of the fibers at the crack tip, because of the much higher stresses and deformations in that region. Also, the collagen and elastin fibers are stretched across the crack tip to develop a smoother, less sharp safer profile, along with a new ensemble of fibers to be fractured if the crack is to progress. This concept provides much food for thought for materials scientists developing tough synthetic materials.

Continuing with this theme of material toughness, there has been recent research on enhanced methods of shelling macadamia nuts which are extremely difficult to crack. The reason for this characteristic of these nuts is that the nut shell comprises bundles of fibers, which are randomly orientated and occupy almost 100% of the shell volume to form a nominally isotropic structural medium on the global scale. A consequence of this structure is that there are no major lines of weakness along which a crack can propagate. Research results suggest that in order to replicate this complex structure, the fibers must be grown *in situ* rather than adopting a more conventional synthetic procedure, whereby the fibers are first manufactured and then appropriately packed and bonded together.

Bone is a multi-functional two-phase substance, comprising flexible collagen and brittle hydroxyapite, which is approximately 50% stronger in compression than in tension. Bone performs many different functions which may be grouped under the broad headings of skeletal homeostasis and mineral homeostasis. The first heading governs the growth and maintenance of the skeleton which is characterized by an assemblage of relatively rigid bodies, with muscle attachments to selected bones which facilitates the articulation and mobility of the skeletal system. Furthermore, the skeleton provides protection for the vital body organs, such as the heart, lungs and brain. Mineral homeostasis is dependent upon the high content of mineral ions in the bones to provide a constant concentration of essential ions in tissue fluid and in plasma. In addition there are critical interactions between bone marrow and the surrounding bone structure.

Bone is an excellent example of an intelligent biomaterial which features a microstructure that adapts to the environment. For example, the tibiae and humeri are characterized by spiral alignments of the flexible collagen fibers, which provide the torsional strength mandated by the loading imposed upon these bones, and they transition from optimally-configured, structurally hollow, lightweight, dense shafts to broad porous heads at articulating joints for the development of an efficient load-transfer mechanism, and the directional porosity of these regions permits blood to penetrate the structure without severely weakening it. These skeletal materials can resist stresses by constant microfracture, repair and remodeling, as they ensure skeletal homeostasis. Indeed it has been estimated that 4 percent of the adult skeleton is being remodeled at any given instant in time.

During life, the amount of bone tissue and its distribution around the medullary cavity are dynamically adjusted to accommodate the regularly occurring stress fields. This biophysical event involves an electrical component concerning the piezoelectric properties of the collagen protein-polysaccharides, and also the streaming potentials or generation of electrical charges by the movement of physiological fluid through the various canals in the porous structure, as the bone is subjected to pressure or bending. The control of the bone remodelling process is probably due to the electrical command signals, generated by the transformation of mechanical energy to electrical energy by the bone matrix and the vasculature. This signal dictates the timing of the bone growth phase and the creation of the bone microstructure, in order to develop the appropriate properties to react to the imposed loading mechanisms.

Wolff's law (1892) has traditionally been cited as the classical theory governing the architecture of osseous elements. This law has been subsequently modified to incorporate subsequent developments and may now be stated as: the elements in bone locate or relocate themselves in the direction of the external forces, and increase or decrease their mass in response to these forces. Indeed, there is a diverse group of tissues, such as cartilage, arteries, ligaments and tendons comprising structural biopolymers, that also dynamically-tune their architecture in response to dynamical forces.

This suggests that there must be a sophisticated control algorithm in these connective tissues for detecting and responding accurately to environmental stimuli. Figure 2.11 presents a schematic diagram of the situation involving transducers and different forms of energy. For bone to respond to the external loads, osteoblasts must be sensitive to variations in compressive and tensile stresses. Furthermore, mechanical energy is transduced into electrical

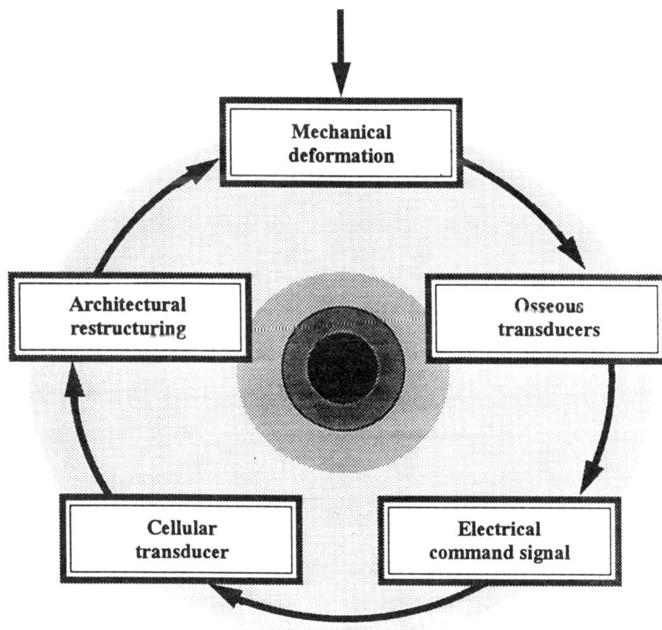

Fig. 2.11 Wolff's Law.

energy by the bone which changes the electrical environment in the cell and also the local behaviour. The bioelectric potentials associated with this process are considered to be the command signal for controlling bone cell activity and the orientation of their macromolecular structure. This architectural restructuring of the bone, in turn governs the mechanical deformation of the whole structure thereby closing the control-loop for this class of problems involving biophysics, biochemistry and bone physiology.

An example of this phenomenon of bone architecture occurs with malaligned fractures in young children, where over the course of time new bone is formed on the concave face of the deformity, while old bone is removed from the convex face in order to create a relatively straight structural member. Similar behaviour has been reported in the literature concerning the healing of fractures *in vitro*, and this process has been enhanced by the application of external electric fields to stimulate bone growth through streaming functions and the piezoelectric effect.

Bone is not only a self-repairing, dynamically adaptive material operating in an unstructured environment, but it is also a tough biomaterial. This toughness is partially attributed to the flexible collagen fibers featured in this class of composite materials, but also there are two other reasons for the crack-stopping capabilities of this material. Firstly, both the cementum rings and osteocyte lacunae surrounding osteons serve as crack stoppers. Secondly, the hydroxyapatite phase in bone occurs in very small crystallites, rather than in large domains, consequently cracks are halted at the junction between these distinct regions where the interfacial bonds are relatively weak.

Bones have also an important role to play in the mineral homeostasis of the human body in which calcium, phosphate and magnesium ions must be maintained in an appropriate and time-variant concentration for the viable functioning of the delicate cellular mechanisms throughout the body. Calcium is probably the most important element to control in plasma, in both extracellular fluids and also intercellular fluids. This is accomplished by a complex system of parameters acting upon excretion from the blood stream by the kidney, on the absorption from the gastro-intestinal tract, and on the skeleton. Indeed bone is the ultimate reservoir for calcium, and in states of calcium deficiency, which typically occur during the early morning hours, calcium ions are mobilized from the skeleton to maintain the calcium concentration of plasma. Furthermore, the regulation of blood calcium, magnesium and phosphates is accomplished by cells on bone surfaces such as the osteoclasts and the osteoblasts. Clearly, bone is a complex multi-functional material which, upon further study, could provide a rich source of ideas for materials scientists developing intelligent materials.

The skin of animals and human beings is a multi-functional material which could also provide materials scientists with a profuse fountain of ideas pertaining to the design of intelligent materials. In human anatomy, the skin is the largest of the body organs which protects the body in a variety of different ways in addition to serving several important functions. It is a tough, flexible, self-repairing organ of complex layers of tissue that is waterproof; contains sensors; prevents chemicals and bacteria from entering the body; facilitates excretion and secretion of substances; prevents the escape of essential body fluids; serves as part of the body's temperature regulation system; protects against injury, and features a self-protecting mechanism to prevent damage by ultra-violet radiation. A measure of the complexity of this class of multi-functional materials may be obtained by reflecting upon the fact that a quarter of a square inch of skin will typically contain approximately 100 sweat glands, 3 feet of blood vessels, and 12 feet of nerves.

The skin comprises three layers of tissue: the epidermis, which is the outer layer; the dermis; and the hypodermis. The epidermis, which is the principle protective barrier is only about 0.005 in thick and is supported along an interlocking corrugated boundary by the dermis which, is a fiberous layer some 20 to 40 times thicker than the epidermis depending upon the location in the body. The innermost layer of the skin is the hypodermis, which is thicker than the other two layers and contains fat cells arranged in lobules with collagen boundaries. These fat cells provide thermal insulation for the inner parts of the body against temperature extremes, they provide shape, they provide energy-dissipation devices to cushion the skin against impactive injuries, and they function as energy storage sites for the entire body.

The surface of the epidermis comprises a layer of flat dead cells containing keratin, which is a tough waterproof protein. These cells, which originated at the boundary of the dermis and the epidermis as living cells, die as they approach the surface of the skin, where they become separated from the body, to be replaced every 30 days or so by the subsequent layer at the surface which is in a state of dynamic equilibrium. The keratin not only ensures that the outer surface of the skin is tough, but it also prevents the diffusion of fluids through the epidermis.

While the epidermis is devoid of blood vessels and nerves, it contains granules of the pigment melanin, which is responsible for the freckles and suntans that characterize fair-skinned people exposed to sunshine for prolonged periods. This response of the skin to sunshine is a defense mechanism for protecting the surface of the body from potential damage, resulting from overexposure to the ultra-violet radiation from the sun.

The dermis structurally comprises a network of collagen protein fibers that provide the skin with flexibility, strength and toughness. These fibers are densely packed near the interface with the hypodermis but they are less densely structured in the upper dermis at the boundary with the epidermis. The spaces between the collagen fibers are filled with a complex assemblage of blood vessels, nerve endings, sweat glands, hair follicles, and an amorphous material which is partially responsible for healing skin injuries.

The dermis also contains millions of sensitive nerve endings for sensing pain, temperature, itch, pressure and touch. All of these sensors are mechano-receptors that trigger a signal to the central nervous system when activated. However, these sensory systems assume different forms. For example, touch receptors are bulbous in shape, while other nerve endings are networked near the follicles of hairs which are triggered when the hair is mechanically deformed. Nerve endings are particularly dense on the fingertips and in the palms of the hands.

In addition to this network of sensors, the dermis features a multitude of tiny capillaries which provide nourishment to the growing skin while at the same time regulating heat loss from the body. This latter function is accomplished by controlling the blood flow through these narrow tubular members. When the skin is exposed to cold temperatures, the body reduces the skin temperature, and hence heat transfer by constricting the diameter of these surface blood vessels in order to reduce the flow of warm blood through these extremities. Furthermore, under these conditions, the muscles surrounding the hair follicles contract in order to erect the hairs projecting from the skin until they are approximately orthogonal to the surface. In this configuration a boundary layer of trapped warm air is developed over the skin surface to provide further insulative thermal protection. Conversely, heat radiated from the skin is substantial when the capillaries are dilated and there is a high blood flow through these vessels. This complex behavior is controlled by the hypothalamus in the brain which operates in conjunction with the autonomic nervous system.

The dermis also features a second type of cooling mechanism. This mechanism is associated with the secretion of a dilute salt solution through the eccrine sweat glands which are narrow ducts with openings at the surface of the epidermis. The subsequent evaporation of this saline solution to the neighboring environment cools the body. The different classes of sweat glands are also responsible for the secretion of urea, sebum, lactic acid and body wastes associated with spices for example.

A hydrolipid layer comprising a thin film of water and fat-based substance protects the skin from dehydration. In addition, this layer prevents foreign substances, harmful bacteria, viruses and fungi from penetrating the skin. Thus it is clear that the skin is a complex, self-replenishing multi-functional material which serves diverse functions at both the local and global levels within the context of the body function. The diverse sensory and structural properties of this body organ should provide biomimetic philosophies for replication, by materials scientists developing macromechanical and micromechanical models for intelligent materials.

Living organisms are continuously bombarded by molecules in the environment which may be conducive or destructive to the sustenance and growth of the organism. At one level, intelligence pertains to the ability of the organism to accurately decipher and respond to the sensory input in a timely and organized manner. The outer surface of cells are outfitted with receptors, which are molecular antennas that detect incoming messages and activate signal pathways through transduction mechanisms that ultimately regulate the cellular processes. Sensory transduction processes are innate to living organisms and are not dependent on the existence of a nervous system.

Living organisms are complex chemical systems, and chemical stimulus response functions are evident at all levels of organization. For example, several organisms combat the cytotoxic effect of heavy metals by synthesizing metallothionine, which is a sulfur-rich protein that consists of a single peptide chain of 61 amino-acids which bond as high as six metal ions per molecule. Even micro-organisms have the capability to interact with their environment through sensory and motor adaptations, which are usually referred to as tactic responses. The chemostatic responses of bacteria allow cell movement towards or away from specific chemical signals from the environment. A simple level of intelligent behaviour is exhibited by bacteria in chemostasis, when bacteria are attracted towards higher concentration of nutrients such as sugars or amino acids and repelled away from high concentrations of toxic chemicals, for example. The movement of an organism towards, or an orientation towards a light source is well-known for a broad class of organisms ranging from bacteria to higher plants. The greening of plants and the tanning of the human skin in the presence of light depends on photochemical processes. Some organisms are also known to avoid light, whereas certain aquatic bacteria are known to be magnetotactic. These bacteria have a tendency to travel along magnetic field lines and they are therefore, influenced by changes in the earth's magnetic field. Magnetotactic bacteria synthesize a chain of magnetosomes, which are small particles consisting of cellular axes of mobility. The magnetosome particles are thermally-stable, single magnetic domains featuring crystal sizes ranging from 400 to 1200 Å. Brief exposure to an intense monophasic magnetic pulse of a few hundred Gauss opposite to the ambient field, results in the reversal of the magnetosome chain and hence the motion of the bacteria in the opposite direction. Such magnetic crystals have been found in several different organisms including pigeons.

Taste and smell are probably phylogenetically the oldest senses. Whereas receptor cells in the silk moth respond to single-molecule stimulations of pheromones, a typical animal requires the activation of about 200 cells in each antenna to start its motor response. Similarly, echolocation is an active information system in bats, which permits them to emit ultrasonic signals in the 20 to 140 kHz range into their environment, approximately every 100 m/s for detection, localization and identification of their prey and also for orientation on sensing a prey or an obstacle. The rate of signal emission is substantially increased, and the intervals between signals and its length are shortened continuously to 10 m/s. This system can detect echoes from nylon threads of 60μm thick, for example.

The vision system in living organisms has the ability to detect a single photon and also adapt to brightness levels varying by as much as a factor of 10^8. The range of the electromagnetic spectrum and the sensitivity of the visual system depend on the properties of the photo-receptors in the retina and also on the signal processing carried out by the higher-order neurons.

In vertebrates, rods and cones are the two kinds of photoreceptor cells that function in bright light and provide color vision, and high visual activity. Rods are many more in number, but they function only in dim light and are not involved in color vision. Rods are slower than cones and they saturate at light intensity levels which are sufficient to excite the cones. The rods feature a photosensitive molecule called rhodopsin, which consists of opsin, a colorless protein and polyene chromophase. Nature employs the same retinal molecule in the color sensitive cones to provide different absorption properties.

The previous discussions have focused on the role of bones as typical intelligent biomaterials; also the variety of intelligence manifested by living organisms to cope with external stimuli. In order to understand the nature of intelligence, attention is now focused on the ultimate organism, the human being.

In order to appreciate the complexity of the human organism we can start out by considering that there are more than 50 trillion cells that make up the body. These cells work in superb conjunction with one another to orchestrate the countless functions internal and external to the organism. Each cell is almost similar in structure featuring an outer membrane, the cell wall and a mixture of water and swirling chemicals. The nucleus of each cell protects the tightly twisted coils of DNA. Nature has adopted such an elegant and simple design for DNA that it can accomplish its very distinct functions in humans and gorillas by featuring a 2% observable difference in genetic structure. Furthermore, the simple structure of a liver cell, for example, belies the fact that it has at least 500 distinct functions. The mechanism for detecting and repairing damage in human beings is clearly resident somewhere in the complexity of over 50 trillion cells. What is the mechanism by which the body detects damage and knows how to repair the damage? What permits almost every human body to heal a cut or fight a flu, but allow only a few to fight cancer or AIDS?

The infliction of a wound immediately results in blood cells rushing to the wound site that needs to be healed. The blood cells do not reach there at random: they ‘know’ where to go and ‘what to do’ – which is to form clots. This whole process is undertaken absolutely spontaneously and completed with perfection. Although it is intuitively obvious that intelligence is at work in this healing process, the search for understanding behind this process will not uncover a minute hormone, messenger enzyme or protein strand which has this intelligence exclusively embedded in it.

Man-made drugs have tremendous difficulty in competing with 'healing' chemicals produced within the body. This is mainly due to the uniqueness of the receptors in each cell wall. These receptors are comprised of complex molecular chains with the last link open-ended which can be completed only with a specific molecule type. The hormones, enzymes and other biochemicals employed by the body have the appropriate molecular structure and the knowledge and ability, to distinguish between the multitude of available sites within the cells. The body has unparalleled capability to produce painkillers, antibiotics, diuretics, etc. in perfect quantities, with no side-effects and perfect strategies for employing them as part of its embedded intelligence.

The human body continually renews itself through the processes of respiration, digestion, elimination, etc. and through continual contact exchange with its environment. For example, typically 98% of the atoms in the human body were not there a year ago. The skeleton renews itself every three months, whereas the skin, liver and stomach lining renew themselves every month, every six weeks, and every four days, respectively.

Humans lose more than a billion neurons out of an average total of 15 billion during their lifetime, however this significant loss of neurons on a regular basis is adequately compensated for by the development of dendrites, which are branchlike filaments connecting the nerve cells to one another. Nerve cells typically feature a bulbous core from which emanate a multitude of thin arms called axons. Each axon ends in a bunch of filaments called dendrites. Dendrites serve as interfaces for the transmission of signals between adjacent neurons. A typical neuron may feature dendrites varying from as low as ten in number to as high as a thousand. However, depending on the needs of the nervous system a typical neuron can grow a number of new dendrites thereby significantly enhancing the number of ports of communication and the amount of data/signals transmitted. Interestingly enough, experimental evidence has clearly demonstrated that this formation of dendrites continues even in old age. In the case of newly developed dendrites the process of thinking, memorizing and other aspects of mental activity create the new tissue. Furthermore, with advances in age the multiplication of dendrites and the geometric increase in the number of connections between the neurons not only compensates for the loss of 18 million neurons per year, on an average, but possibly also leads to enhanced 'wisdom' with increasing chronological age.

Discussion of intelligence in the human body clearly leads to the question as to where 'intelligence' is resident in the human body. Every cell in the body has manifested from the double strand of DNA at the moment of conception. Each cell therefore has embedded in it intelligence, as a built-in feature and all the infinite capabilities of DNA. The DNA itself is comprised of the same structural blocks as the neurotransmitter that it creates and regulates. Thus the DNA creates a multitude of its own variant thereby functioning like the most sophisticated Von Neumann machine. It is therefore tempting to think of DNA featuring billions of genetic bits as an intelligent molecule. However, DNA is itself constructed from strings of sugars, amines and other basic building blocks. If these elementary building blocks were not intelligent then how could an organized integration of these building blocks lead to an intelligent entity.

Clearly the sugars and amines are comprised of carbon and hydrogen atoms, therefore for these sugars to be 'intelligent' their intelligence must be derived from the atomic building blocks. This line of reasoning leads one to conclude that there is possibly no single residence for intelligence, and that intelligence must be manifested everywhere within the organism.

The preceding discussion on some aspects of the human body clearly points out the importance of intelligence in the routine bio-chemical-mechanical functioning of the human body. These aspects of intelligence are very distinct from the common abstract 'higher' notions of intelligence involved in learning and decision-making, for example. Although these higher notions of intelligence are clearly critical at the global system level, the concepts of intelligence discussed herein will be clearly very critical in the development of man-made intelligent materials.

These discussions on the notions of intelligence clearly suggest that intelligence is resident everywhere in the human body. Furthermore, this innate intelligence is significantly superior to any substitute from the outside world. Intelligence clearly integrates the biological, chemical and mechanical elements of the human body and provides it with form, order, direction and function, for example. Without intelligence, these elements of the human being would be undirected and chaotic.

The brain circulates intelligence throughout the body. For example, in addition to the neurons which have fixed locations in space, the monocytes of the immune system travel throughout the bloodstream which grants them access to every cell in the body. The immune system has the unique ability to mirror the signals of the nervous system, permitting the immune system to transmit and receive signals that are as diverse as those of the central nervous system. This ability is carried out by monocytes which essentially serve as a function of circulating neurons.

In essence, these discussions clearly focus on the argument that intelligence in humans is embedded at the cell level. Localized intelligence in the form of DNA at the cell level has been well-known since the early 1950s. This localized intelligence is essentially fixed in the sense that the DNA chemical structure is one of the most stable in the human body which permits the genetic traits to be preserved and transmitted from generation to generation. However, the intelligence manifested by neurotransmitter and neuropeptides corresponds to the transient intelligence of the human mind. These intelligent chemicals are not only generated by the brain but also by the immune system. These notions of 'steady-state' and 'transient' intelligence provide a material basis for intelligence at the cell level in human beings. Clearly, the human mind is not neatly quarantined in the human brain but projected everywhere throughout the human body. The human body does not have a single source of intelligence, rather the whole body is a web of intelligence. Consider, for example, the intelligence which is employed by the body when it repairs a broken bone. The healing process is far too complex and time consuming involving a myriad of very well orchestrated processes, only a few of which are imperfectly understood by modern science. Furthermore, there is tremendous evidence in modern medical science which suggests that early internal manifestations of diseases, including cancer, are probably detected and combatted within the body on a regular basis, resulting in no perceived manifestation of the disease at a gross external level. Disease may indeed be a breakdown of intelligence at some level in the organism.

3

Smart materials and structural systems

A cursory review of the history of materials science will certainly reveal the profound influence of this scientific discipline upon the evolution of civilization during the millennia. Thus, it is therefore inevitable that the new generation of smart materials and structures technologies featuring at the most sophisticated level a network of sensors and actuators, real-time control capabilities, computational capabilities and a host structural material, as presented in Figure 3.1, will not only have a tremendous impact upon the design, development, and manufacture of the next generation of

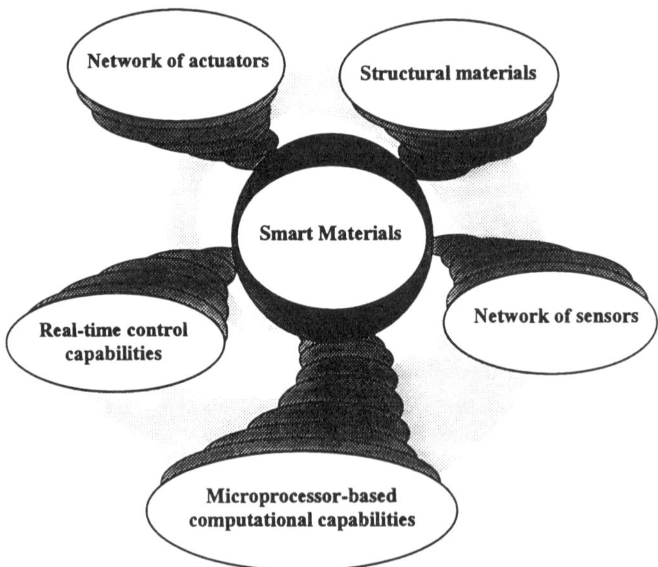

Fig. 3.1 The principal ingredients of a premier class of smart materials.

products in diverse industries but also the economic climate in the international marketplace. Those countries with the ability to harness and coordinate the diverse technologies associated with smart structural systems, will reap a bountiful harvest while those countries which adopt a less aggressive posture will surely be afflicted by a draught.

The applications for these new generations of smart materials and structures will be diverse, but a common denominator for the deployment of the most sophisticated class of systems featuring sensors, actuators and microprocessors will probably be the unstructured environment in which a system must operate. Thus the uncertainty associated with the behavior of the relevant external stimuli which govern the system response relative to prescribed design criteria will largely dictate the deployment of smart materials and structures. The decision-making algorithm is presented in Figure 3.2 from which it is evident that the necessity for synthesizing materials and structures with autonomous self-adapting, self-correcting characteristics is governed by the desire to achieve optimal performance at all times under variable service conditions and while operating in unstructured environments. There are several characteristics of these smart materials and structures which have been the foci of research activities, and these include changing the mass-distribution, the stiffnesses, and the energy-dissipation characteristics for vibration-control purposes. This work has permitted engineers to synthesize systems with controllable vibration amplitudes, natural frequencies, resonances and transient response settling times. Other work has focused on actively changing the geometries of structures.

This generic research on smart materials and structures will profoundly impact the next generation of products that must operate under variable service conditions in unstructured environments. Consider, for example, the uncertainties imposed on the safe operation of commercial aircraft in winter climates where weather conditions can dictate the operation of the airplane. The formation of ice on the wings, fuselage and control surfaces of the machine between departure from the terminal jetway and the commencement of take-off can adversely affect performance dramatically, as evidenced by the 1982 tragedy involving Air Florida Flight 90 which crashed into the Potomac River upon take off from Washington DC's National Airport. In winter conditions, delays in taxiing and take-off, rapidly deteriorating weather conditions, and commercial pressures to adhere to published flight schedules all compound to create an unstructured environment for the safe operation of commercial aircraft. Under these conditions, the flight crew are currently required to evaluate the flight-worthiness situation on a largely empirical basis prior to determining whether they should commence take-off, or return to the terminal for an additional de-icing treatment in order to limit ice build-up on the aircraft. The next generation of aircraft operating under these conditions would benefit from smart structures which would determine the extent of ice build-up on critical

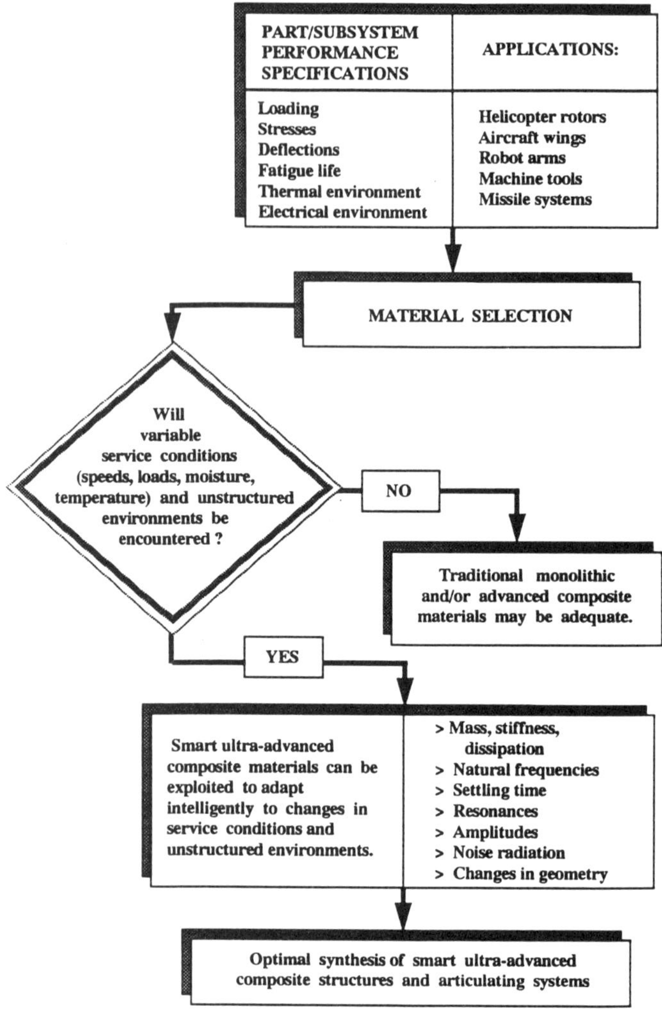

Fig. 3.2 A decision-making algorithm for material selection.

regions of the aircraft. An array of sensors on the surface of the aircraft could measure the weight of ice, for example, and provide information to the flight deck for incorporation in flight-worthiness checks. This type of smart structure is only one of several classes of smart aerospace skins initiatives being undertaken at this time.

The field of smart materials and structures is very broad and will ultimately include a variety of diverse disciplines including biotechnology, neural networks, photonics, nanotechnology and artificial intelligence.

However, at this time, significant developments have occurred in a somewhat limited area of this diverse field because teams of engineers and scientists have focused on employing a network of embedded sensors, microprocessors, and an array of dynamically-tunable actuator materials interfaced at a global level with traditional structural materials. In this novel class of structural systems, the structural materials provide the skeleton of the system; the network of actuators provide the muscles to make things happen; the network of sensors and data transmission systems provides the nervous system, to monitor and communicate the characteristics of the external stimuli to the microprocessor-based computational facilities; and the microprocessors provide the brains which ensure the optimal performance of the overall system in the presence of variable and unstructured stimuli.

These smart systems are typically fabricated from a range of off-the-shelf materials and discrete devices whose properties are well known, rather than fabricate systems with new materials synthesized at the microscopic level. Thus each discrete material in the current generation of smart materials and structures satisfies a discrete functional requirement of the overall system in order to provide sensing, actuation or structural properties. The current generation of smart materials and structures generally exploit the actuation characteristics offered by electro-rheological (ER) fluids, piezoelectric

Actuator / Characteristics	**Electro-strictive Materials**	**Electro-Rheological Fluids**	**Magneto-strictive Materials**	**Nitinol Shape Memory Alloy**	**Piezoelectric Ceramic**
Cost	Moderate	Moderate	Moderate	Low	Moderate
Technical Maturity	Fair	Fair	Fair	Good	Good
Networkable	Yes	Yes	Yes	Yes	Yes
Embedability	Good	Fair	Good	Excellent	Excellent
Linearity	Fair	Fair	Good	Good	Good
Response (Hz)	1-20 000	0-12 000	1-20 000	0-5	1-20 000
Maximum Microstrain	200	——	200	5000	200
Maximum Temperature (°C)	300	300	400	300	300

Fig. 3.3 Actuator candidates for smart materials.

materials, shape-memory-alloys (SMAs), magnetostrictive and ferro-electric materials, for example, but there are a variety of actuator candidates, and the characteristics of some of these actuator materials are documented in Figure 3.3.

3.1 Actuator materials

Actuator materials are one of the principal ingredients of several important classes of smart materials and structures. These actuator materials are typically employed to dynamically tune the global mechanical properties of the structure, or else to dynamically tailor the shape of the structure in an orchestrated manner. The subsequent pages document synoptic qualitative descriptions of the current generation of prominent candidate actuator materials.

3.1.1 Electro-rheological fluids

Electro-rheological (ER) fluids are typically suspensions of micron-sized hydrophilic particles suspended in suitable hydrophobic carrier liquids, that undergo significant instantaneous reversible changes in their mechanical properties, such as their mass distribution, and energy-dissipation characteristics, when subjected to electric fields. When these fluids are embedded in voids in structural materials, the imposition of an appropriate electrical field on each voidal domain permits the mechanical properties of the embedded fluid to be actively controlled, which enables the global properties of the structure containing the ER fluid domains to be controlled. The voltages required to activate the phase-change in ER fluids are typically in the order of 1–4 kV/mm of fluid thickness, but since current densities are in the order of 10 mA/cm^2, the total power required to change the fluid properties is quite low. Furthermore, the response-time of ER fluids to an electrical stimulus is typically less than one millisecond.

3.1.2 Shape-memory-materials

Shape-memory-alloys are metals which, when plastically deformed at one temperature, will completely recover their original undeformed state upon raising their temperature above an alloy-specific transformation temperature. A characteristic of these alloys is that the crystal structure undergoes a phase-transformation into, and out of, a martensitic-phase when subjected to either prescribed mechanical loads or temperature. During the shape-recovery process, these alloys can be engineered to develop prescribed forces or displacements. The transformation temperature of these materials can be selectively-tuned over a broad temperature range to suit each application, and plastic strains as high as 6% can be completely recovered by subjecting them to heat. SMAs have been employed as both actuators and sensors in smart materials and structures applications. Shape-memory phenomena are not limited to the metals. Shape-memory-plastics have also been developed. The commercial marketplace includes the shape-memory-plastic 'Norsorex' which features the polymer polynorbornene, and 'Zeon Shable' which is a polymer blend featuring polyester.

3.1.3 Piezoelectric materials

Piezoelectric materials are solids which generate a charge in response to a mechanical deformation, or alternatively they develop mechanical deformation when subjected to an electrical field; consequently they are employed as either actuators or sensors in the development of smart structures. Piezoelectric ceramic materials are typically employed as actuators while polymeric piezoelectric materials are typically employed as sensors for tactile sensing, temperature sensing, and strain sensing.

3.1.4 Magnetostrictive materials

Magnetostrictive materials are solids which typically develop large mechanical deformations when subjected to an external magnetic field. This magnetostriction phenomenon is attributed to the rotations of small magnetic domains in the material, which are randomly-orientated when the material is not exposed to a magnetic field. The orienting of these small domains by the imposition of the magnetic field results in the development of a strain field. As the intensity of the magnetic field is increased, more and more magnetic domains orientate themselves so that their principal axes of anisotropy are co-linear with the magnetic field in each region and finally saturation is achieved. Terbium-iron alloys are typical magnetostrictive materials.

'Terfenol-D' is a commercially available magnetostrictive material marketed by Edge Technologies Inc. which incorporates the rare earth element dysprosium. This material offers strains up to 0.002, which is an order of magnitude superior to the current generation of piezoceramic materials. However, this class of mechano-magnetic materials does suffer from some disadvantages which must be carefully considered when synthesizing smart structural materials. These include the technological challenges of delivering a controlled magnetic field to a magnetostrictive actuator embedded within a host structural material; furthermore, the material generates a much greater response when it is subjected to compressive loads; and the power requirement for this class of actuators is greater than those for piezoelectric materials.

3.1.5 Electrostrictive materials

Electrostrictive materials are analogous to magnetostrictive materials since these materials develop mechanical deformations when they are subjected to an external electrical field. The electrostrictive phenomenon is attributed to the rotation of small electrical domains in the material when an external electrical field is imposed upon them. In the absence of this field, the domains are randomly orientated. The alignment of these electrical domains parallel to the electrical field results in the development of a deformation field in the electrostrictive material.

The ceramic compound lead-magnesium-niobate, PMN, exhibits a thermo-electrostrictive effect and the quadratic electrostrictive phenomenon is dependent upon the ambient temperature. This type of material, while exhibiting less hysteresis loss than piezoceramics, features a nonlinear constitutive relationship requiring some biasing in practice to achieve a nominally linear relationship over a limited range of excitation.

3.1.6 Thermal materials

Distributed thermal actuators exploiting the Peltier effect, which characterizes certain classes of semiconductor materials, have been employed to generate thermal gradients for controlling the dynamic response of flexible structures. These semiconductors act as heat pumps when a voltage is applied to them and this excitation is responsible for the development of a temperature gradient and hence deformation of the material. There are a number of environmentally-responsive polymers that are commercially-available, and some of the materials generate a mechanical deformation in response to a thermal excitation.

3.2 Sensing technologies

Several sensing technologies can be incorporated into the current generation of smart materials and structures, and some leading candidates for undertaking strain measurement are presented in Figure 3.4. These sensors have diverse characteristics. For example, it should be noted that while piezoelectric or strain gauge sensors provide local domain information, the fiber optic or SMA sensors generally provide an integrated or average global measure of the deformation field between two discrete domains in a structure. Thus the characteristics of each candidate must be carefully evaluated when synthesizing a smart structure.

Figure 3.4 presents information on fiber-optic sensors, SMA sensors, piezoelectric sensors and electro-mechanical sensors. The phenomena associated with sensors featuring piezoelectric materials and SMAs were

Sensor / Char-acteristics	Fiber-Optic Interferometer	Nitinol Shape Memory Alloy	Piezoelectric Ceramic	Strain Gauge
Cost	Moderate	Low	Moderate	Low
Technical Maturity	Good	Good	Good	Good
Networkable	Yes	Yes	Yes	Yes
Embedability	Excellent	Excellent	Excellent	Good
Linearity	Good	Good	Good	Good
Response (Hz)	1-10 000	0-10 000	1-20 000	0-500 000
Sensivity (Microstrain)	0.11 per fiber	0.1-1.0	0.001-0.01	2
Maximum Microstrain	3000	5000	550	10 000
Maximum Temperature (°C)	300	300	200	300

Fig. 3.4 Sensor candidates for strain measurement.

discussed previously. The measurement of strain by electrical strain gauges have employed several diverse principles including resistive, piezo-resistive, capacitive, inductive, piezoelectric and photoelectric phenomena. However, the resistive gauges enjoy overwhelming popularity primarily because of their small size and low mass. Foil strain gauges, or resistive strain gauges, operate on the principle of the change in electrical resistance of the gauge when it is subjected to mechanical deformation, and this change in resistance is typically measured using a wheatstone bridge arrangement.

Fiber-optic waveguides embedded in a structure provide the basis for several classes of optical sensor such as polarimetric sensors and interferometric sensors. The Mach-Zehnder interferometer belongs to this latter category of sensors and it features a device for splitting a coherent beam of light which is then transmitted through one or both fibers prior to being recoupled. If the fibers are deflected because of the deformation of the structural material in which they are embedded for example, then the length of the optical path changes which results in an interference pattern at the recoupler. This interference pattern provides a measure of the physical phenomenon being investigated such as mechanical strain.

The current practice in the smart structures community is typically to embed piezoelectric sensors or fiber optic sensors into prototype platforms, because this category of sensor can be manufactured with appropriate strength and size characteristics to avoid degrading the structural integrity of the material. Alternatively, intelligent sensors can be employed which integrate the sensing and actuation functions, and also the associated electronic-data-processing function, on a single integrated circuit chip. There are other classes of sensors, which are potential candidates for incorporation in smart structures, such as those involving capacitance, X-rays, and acoustical waveguides but these devices are associated with embryonic technologies that require further development in order that they can be embedded in the current generation of smart materials and structures.

3.3 Microsensors

Sensors are one of the essential ingredients of a smart structure since they provide the ability to monitor and measure external stimuli in addition to the subsequent behavior throughout the structural medium. Conventional sensors, such as thermocouples, accelerometers, and strain gauges are generally unable to serve in this capacity because of their significant size and weight characteristics, which are typically measured in inches and ounces. Consider for example a conventional accelerometer which typically features a mass in contact with a ceramic piezoelectric element. These micro-components are generally housed in a stainless steel enclosure weighing approximately one ounce or so, and measuring 0.5 in in diameter with a length of 0.75 in. A coaxial cable protrudes from the enclosure for connection to the associated charge amplifier. Clearly this is quite an assemblage.

These relatively large conventional sensors can also adversely affect the behavior of the structure being monitored. For example it is inappropriate to utilize a heavy accelerometer in measuring the dynamic behavior of a light flexible structure subjected to a time-dependent fluid loading. The protrusion of the accelerometer from the surface of the structure will modify the flow field, and hence the dynamical loading, and the additional weight of the accelerometer will change the dynamical properties of the structure being excited. It is evident, therefore, that the large conventional sensors are inappropriate under these conditions and these constraints of large mechanical, thermal and electrical inertias have been responsible for the development of a new generation of smaller sensors exploiting semiconductor technology.

Microsensor technology exploits the maturity of the semiconductor industry to manufacture precise three-dimensional structures on the surface of a silicon chip, measuring only a few microns. These sensors involve the fabrication of chips and integrated circuits which enable a sensor of dimensions one mm^2 to typically respond to changes in pressure, radiation, temperature, and acceleration.

There are numerous advantages for engineers to exploit this emerging technology. First a micron-sized sensor will not adversely affect the behavior of the environment in which it is located. Thus microsensors can be employed in very small regions where conventional sensors cannot be located due to their large size. Second, by integrating the sensing function and the processing function on a single silicon wafer then a 'smart' sensor can be developed which offers the designer great versatility. Third, the current generation of microsensors is extremely robust and durable relative to conventional sensors because of the inate mode of manufacture of these small semiconductor products. Finally, microsensors are cheaper than conventional sensors. This cost-advantage is attributed to their manufacture in an inexpensive material, the volume of material is very small, and they are mass produced using the same automated production technologies routinely employed in the semi-conductor industry. Having enunciated the advantages of microsensors, these micron-sized devices clearly have a potentially very significant role to play in the synthesis of smart engineering materials and structures.

3.4 Intelligent systems

The previous sections have been devoted to the discussion of two of the four essential ingredients of the most sophisticated class of smart materials; namely, actuators and sensors. The third essential ingredient is a control system featuring computational capabilities, in order to orchestrate the behavior of the actuators in a smart structure in response to the excitation data furnished by the sensors embedded in the smart structure. The field of automatic control is a mature discipline, consequently, comprehensive literature on lincar, nonlinear, adaptative, and optimal control systems is relevant to the synthesis of viable smart structures featuring intelligent control systems. To the knowledge-domain of automatic control, the designer of smart materials must consider adjoining theory and knowledge from the embryonic fields of artificial intelligence and neural networks. These fields embody notions of biomimetics, as with many other subfields of the smart materials discipline. Neural networks mimic the structure and capabilities of the human brain where approximately 12×10^9 nerve cells, or neurons, each have between 6000 and 60 000 dendritic connections with a signal carrying capacity. Furthermore, each processing element of a neural network is relatively simple, and because of the parallel structure of the network, the computational time for complex problems is very fast. This is especially true where approximate solutions are acceptable wherein rules of fuzzy logic can be exploited.

Networks feature unstructured decision making logic while being devoid of the precise rule-based computer program characteristics of expert systems. Furthermore, every decision is the result of complex interactions between the interconnected network of processors, and the neural network features self-learning characteristics because it incorporates previous experiences into the interconnection patterns of the neurons.

Artificial intelligence (AI) is a field of computer science focused upon the development of systems that can be programmed to act intelligently. Thus research on AI is typically directed towards determining and subsequently describing certain aspects of human intelligence which can then be simulated by some kind of machine. This is a broad undertaking since human intelligence involves evidence from a multitude of disciplines such as anthropology, biochemistry, biology, history, mathematical logic, neurobiology, political science, psychology, sociolology, and systems theory. A review of these disciplines has resulted in the distillation of three classes of systems that have been the focus of research in the AI community. These systems are the continuous, logical, and symbolic systems, and these research endeavors should be exploited by the designer of smart materials and structures.

3.5 Hybrid smart materials

The above muscular systems and sensory systems feature diverse characteristics, such as thermal operating range and dynamic response, consequently each class of actuator and each class of sensor have distinct advantages and disadvantages. Nevertheless, by judicious selection, the smart-materials designer can synthesize numerous classes of hybrid actuator systems and sensor systems, featuring several different classes of actuators and sensors to satisfy a broad range of performance specifications as shown in Figure 3.5. Typical experiment results for this class of hybrid smart systems are shown in Figure 3.6.

Consider the smart-structural system whose desired actuation and sensing characteristics are schematically shown in Figure 3.7. This illustrative example features a structure which must be controlled over a broad bandwidth from low frequencies to high frequencies. Typical requirements include vibration control at the high frequencies through the tailoring of damping, natural frequencies, and the amplitude of the dynamic behaviour of the structure, while at the low frequencies a shape or geometrical control requirement is imposed upon the structure. Since these requirements are extremely diverse, a hybrid smart material is proposed featuring two types of actuators and two types of sensors. Thus, for example, while the high

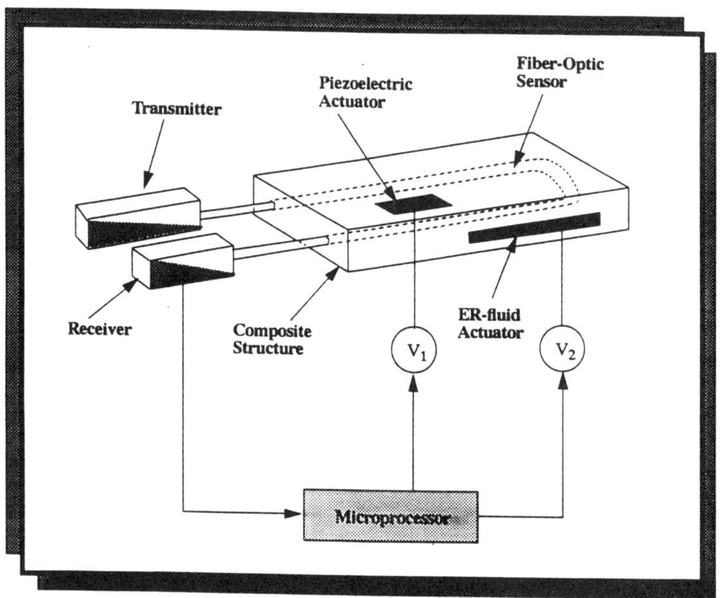

Fig. 3.5 Smart beam featuring a hybrid actuation and sensing system.

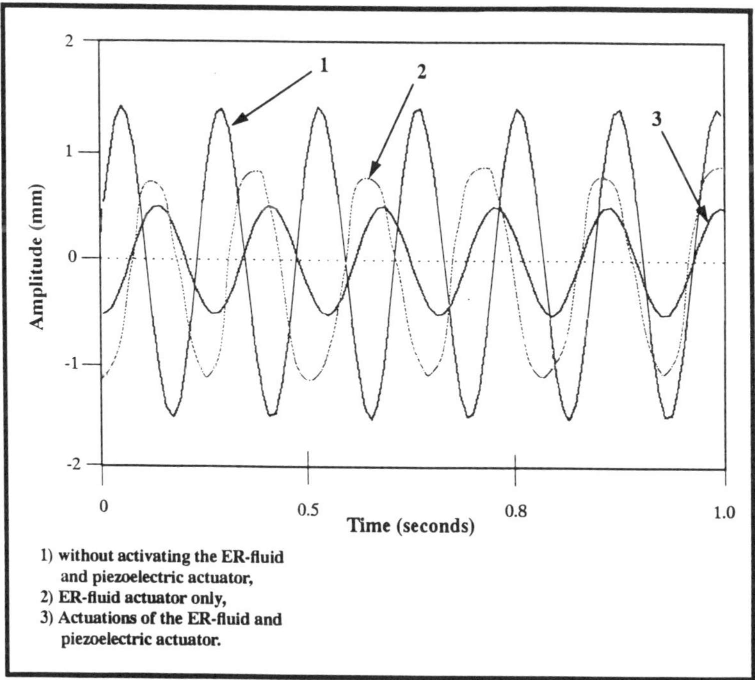

Fig. 3.6 The elastodynamic response characteristics of a hybrid beam featuring an ER fluid and piezoelectric actuators.

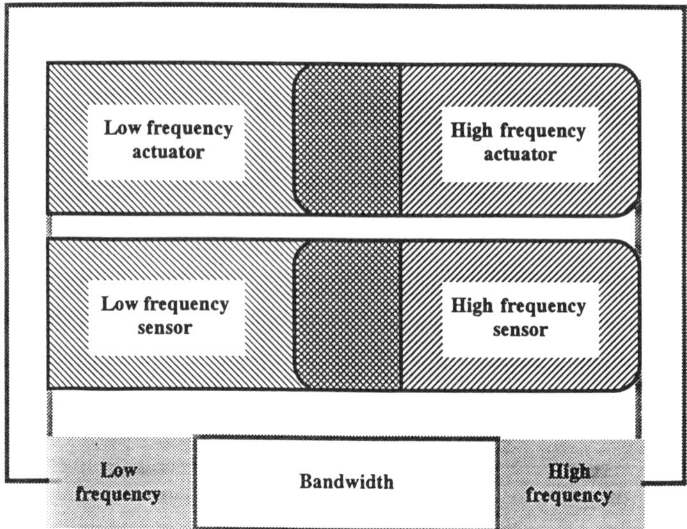

Fig. 3.7 Smart structures featuring hybrid actuation and sensing capabilities.

frequency actuator is able to provide actuation capabilities at high frequencies, and the low-frequency actuator is able to provide appropriate actuation capabilities at low frequencies, the domain of applicability for these actuation systems overlap in the mid-frequency range whereupon either actuator, or alternatively both actuators, may be activated.

In order to satisfy the low-frequency requirements, SMA actuators may be employed in conjunction with embedded fiber optic sensors in order to satisfy the shape control specification, while the high-frequency requirements may be satisfied by employing an array of piezoceramic actuators which are interfaced with an array of piezopolymeric sensors. A review of the actuator and sensor candidates suggests that some functional materials may be employed as either an actuator or as a sensor, while others are only able to perform one of these functions, as indicated in Figure 3.8. The dual versatility of these materials could be exploited in some smart materials and structures applications by employing the same material as a sensor and/or as an actuator during service. Thus, for example, a material serving as a sensor could continuously monitor the vibrational response of a structure, prior to becoming an actuator when the amplitude of vibration becomes excessive in order to alleviate this dynamical behavior.

Smart materials and structures in the foreseeable future will in the most general sense comprise an integrated set of discrete subsystems featuring structural materials, sensors, actuators, and microprocessing capabilities as indicated in Figure 3.9. The design philosophy presented in Figure 3.9 contains a very large multi-parameter search domain because of the very

	Actuators	Sensors
Acoustical Devices		X
Capacitive Devices		X
Electro-rheological Fluids	X	
Electrostrictive Materials	X	
Fiber-Optics Devices		X
Magnetostrictive Materials	X	X
Piezoelectric Materials	X	X
Shape-Memory Materials	X	X
Strain Gauges		X
X-ray Devices		X

Fig. 3.8 Actuator and sensor capabilities for smart materials and structures applications.

large number of decisions and design parameters associated with the properties and characteristics of sensors, structures, and actuators. Furthermore, the designer seeking a global optimal solution in the synthesis of an appropriate smart material for a particular application, must also address other crucial decisions concerning computational capabilities, networking issues, and appropriate control strategies.

The establishment of a set of viable analytical tools for predicting the behavior of smart materials and structures featuring embedded actuator materials, the characteristics of the structural materials, and also the characteristics of the interface regions and the characteristics of the piezoelectric actuators, is a prerequisite for hybrid optimal control strategies for smart mechanical and structural systems, featuring multi-functional smart actuation and sensing capabilities. Since these materials will typically be

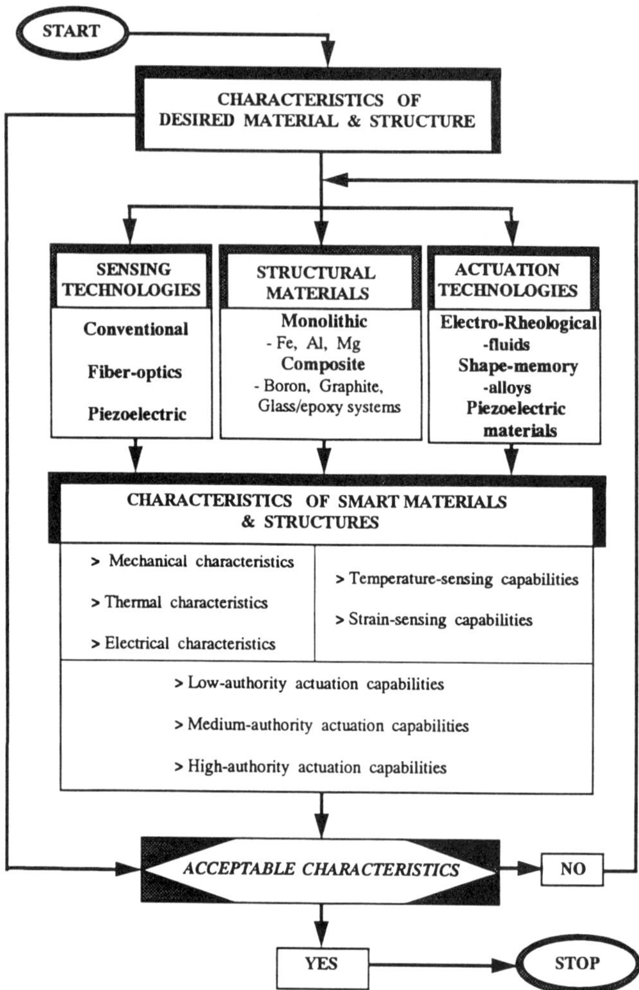

Fig. 3.9 An algorithm for synthesizing a smart material featuring structural materials, actuator technologies, and sensing technologies.

employed in engineering practice to satisfy various performance specifications, the tailoring of the structural characteristics by employing hybrid optimal control strategies is a crucial step in the development of the knowledge-base in this area.

Consider, for example, the task of controlling the vibrational response of a smart structure featuring embedded hybrid electro-rheological and piezoelectric actuators shown in Figure 3.5. By incorporating suitable simplifying assumptions, the governing equations can be formulated as

$$\mathbf{x} = \mathrm{A}\,\mathbf{x} + \mathrm{B}\,\mathbf{U} + \mathrm{B}_1\,\mathbf{V}_1 + \mathrm{B}_2\,\mathbf{V}_2$$
$$\mathbf{y} = \mathrm{C}\,\mathbf{x}$$

where **y** is the output vector and **x** is the state variable vector, where **A** is the plant matrix, **B** is the input matrix associated with the electro-rheological actuator, while $\mathbf{B}_2$ is the input matrix associated with the piezo-electric actuator. Vectors **U**, $\mathbf{V}_1$, and $\mathbf{V}_2$ provide the system inputs associated with the conventional, electro-rheological and piezo-electric actuators, respectively.

In the context of this mathematical framework attention may by focused on optimal hybrid control strategies for the following problem of finding the optimal location of distributed actuators incorporating electro-rheological fluids and piezo-electric materials which maximize the control influence matrices $\mathbf{B}_1$ and $\mathbf{B}_2$. For the matrix $\mathbf{B}_2$ associated with the piezo-electric actuator find the optimal location which maximizes the difference of the modal slope coefficients of the actuator segment, for example. This problem may be summarized as follows:

$$Maximize\left(\frac{d\emptyset_i}{dx}(x_i^{\,1}) - \frac{d\emptyset_i}{dx}\,(x_i^{\,2})\right)$$

subjected to geometrical and material constraints.

Similarly, for minimizing rise time, the problem reduces to the design of the controller inputs **U** and $\mathbf{U}_1$ which minimize the performance index

$$J = \int (\mathbf{x}^T\,\mathbf{Q}\,\mathbf{x} + \mathbf{U}^T\,\mathbf{R}_1\,\mathbf{U} + \mathbf{U}_1^{\,T}\,\mathbf{R}_2\,\mathbf{U}_1)\,dt$$

with state weighting matrix,

Q which is partitioned into a matrix $\mathbf{Q}_1$ and a matrix $\in\mathbf{I}$,
I is the identity matrix,
$\mathbf{Q}_1$ is associated with velocity state,
$\in\mathbf{I}$ is associated with displacement state,
$\mathbf{R}_1$ and $\mathbf{R}_2$ are the control weighting matrices.

Clearly, the optimal hybrid control strategies for minimizing settling time, rise time, overshoot time and tailoring the aeroelastic characteristics in the presence of external disturbances and bounded uncertainties, can be formulated in the analogous manner for a variety of actuator materials including magnetostrictive materials and shape-memory-alloys, for example.

The previous pages have focused upon the characteristics of actuators and sensors in isolation from the global integration issues associated with the design and manufacture of smart structures. Current philosophies involve the embedment of these actuation and sensing subsystems in advanced composite materials, but the designer of smart structures must carefully evaluate the processing parameters relative to potential actuator and sensor capabilities. The critical parameter in the manufacture of the common thermosets, thermoplastics and metal matrix composites is generally the processing temperature. Thus the designer must carefully consider the impact of the processing temperature on the sensors and actuators embedded within the structure. Figure 3.10 tabulates typical processing temperatures for a selected class of advanced composite systems and indicates the thermal environment that shape-memory-alloys, piezoelectric ceramics, and optical fibers can withstand without suffering permanent damage. The thermoset composites provide the smart structures designer with the greatest freedom of the current generation of structural materials for smart structure applications, because of the low cure temperatures and low processing pressures which are typically between 250° F and 650° F, and between 50 and 200 psi, respectively. Most of the commonly available actuators and sensors can be readily embedded in this class of structures, except for piezopolymeric films which are thermally limited to approximately 200° F. There are also a limited class of epoxy matrices which cure at room-temperature.

The processing of thermoplastic composite materials such as graphite/PEEK systems involves temperatures in the range of 600° F to 800° F with processing pressures similar to the thermosets. These elevated temperatures are somewhat high for the survival of fiber-optic systems because they exceed the upper bound on temperature for the fiber coatings. Nevertheless piezoelectric ceramic or shape-memory-alloy actuators and sensors can be employed in conjunction with thermoplastic structural materials.

Candidate Materials	**Processing Temperature (°F)**	**Optical Fibers**	**Piezoelectric Ceramics**	**Shape Memory Alloys**
Graphite epoxy	250 - 350	●	●	●
Graphite-polyimide	550 - 650	●	●	●
Graphite/PEEK	600 - 800		●	●
Metal-matrix composite	1000 - 2000			●

Fig. 3.10 Processing environment for various advanced composite materials, actuators, and sensors.

Classification	Sensors	Actuators	Sensors & Actuators
Type I	●		
Type II		●	
Type III			●

Fig. 3.11 Classification of smart materials and structures.

The highest processing temperatures for the different classes of advanced composite materials are experienced by the metal-matrix composites. These materials, such as the graphite fiber-reinforced aluminum composites, experience temperatures between 1000° F and 2000° F during manufacture. Under these extreme conditions shape-memory-alloys are the only potential candidates to satisfactorily perform actuation and sensing functions from the current generation of actuator materials.

The truly intelligent structural system learns and adapts its behavior in response to the external stimuli provided by the environment in which it operates. However, there are of course a wide variety of less sophisticated smart materials and structures which exploit the basic sub-disciplines as illustrated by Figure 3.11 which defines three classes of smart materials. These include materials with only sensing capabilities, materials with only actuation capabilities, and materials with both sensing and actuation capabilities at a primitive level relative to notions of true intelligence. One of the objectives of this chapter is to provide succinct descriptions of potential applications of smart materials and structures in the diverse industrial, commercial, and military segments of the economy in order to illustrate the benefits to be derived from deploying smart materials technologies in practice.

3.6 Passive sensory smart structures

A structurally integrated optical microsensor system for assessing the state of a structure has been termed by Professor R.M. Measures of the University of Toronto Institute for Aerospace Studies as a Type I passive smart structure. Typically the smart structure comprises a fiberous polymeric composite material in which is embedded an optical microsensor system. This embryonic field requires the integration of diverse technological disciplines ranging from materials science and composites manufacturing, through networks and artificial intelligence to microelectronics and fiber-

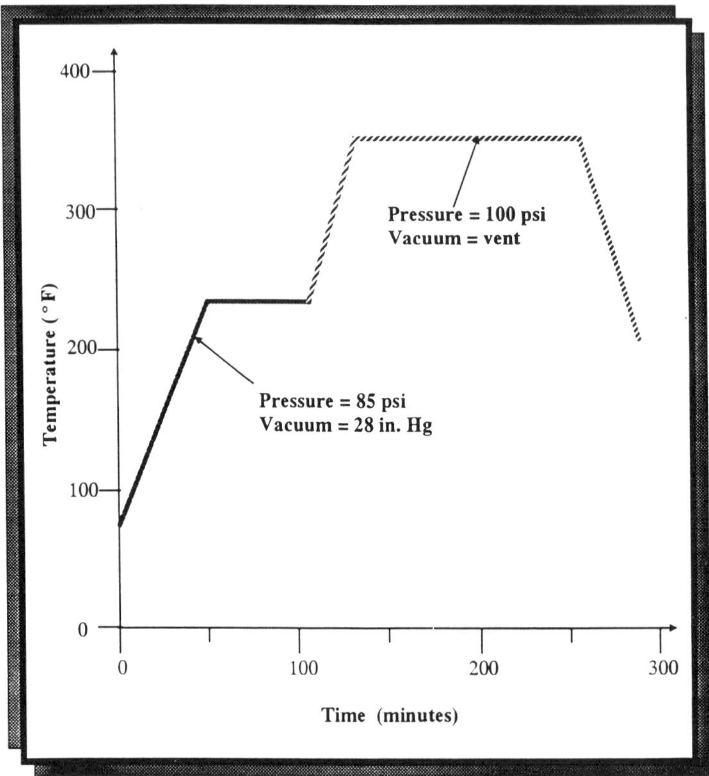

Fig. 3.12 Cure cycle for the Hercules 3501–6 epoxy resin system.

optics. With these basic raw ingredients, several different recipes exist for producing different classes of baked goods.

The first class of products involves the manufacture of fiberous polymeric composite parts which are typically cooked in an autoclave where the part is simultaneously subjected to both pressure and temperature in order to cure the matrix system. Figure 3.12 presents the cure cycle for a 3501-6 epoxy material that is typically featured in aerospace composite materials. Currently, this manufacturing process is undertaken using an open-loop control philosophy, where the heat and pressure profiles of the autoclave are prescribed by the manufacturer of the resin or polymeric material. Consequently, if the curing process is incomplete during manufacture, then the resulting mechanical properties of the part, such as the strength and the stiffness characteristics, will not comply with the design specifications for that part.

The curing process is by no means a simple straight-forward undertaking. For example, the degree of cure can vary dramatically in a large fiberous polymeric part because of the thermal characteristics of the autoclave; the exothermic phenomena associated with the curing of the matrix; the thermal diffusivity of the part material; which typically depends upon the layup and the fiber volume fraction, and also the geometry of the part. Variations in the batch properties of prepreg materials and also the period of time that the prepreg remains outside the cold-storage facility, may also be contributing factors. Since an incomplete state of cure in a region of a part

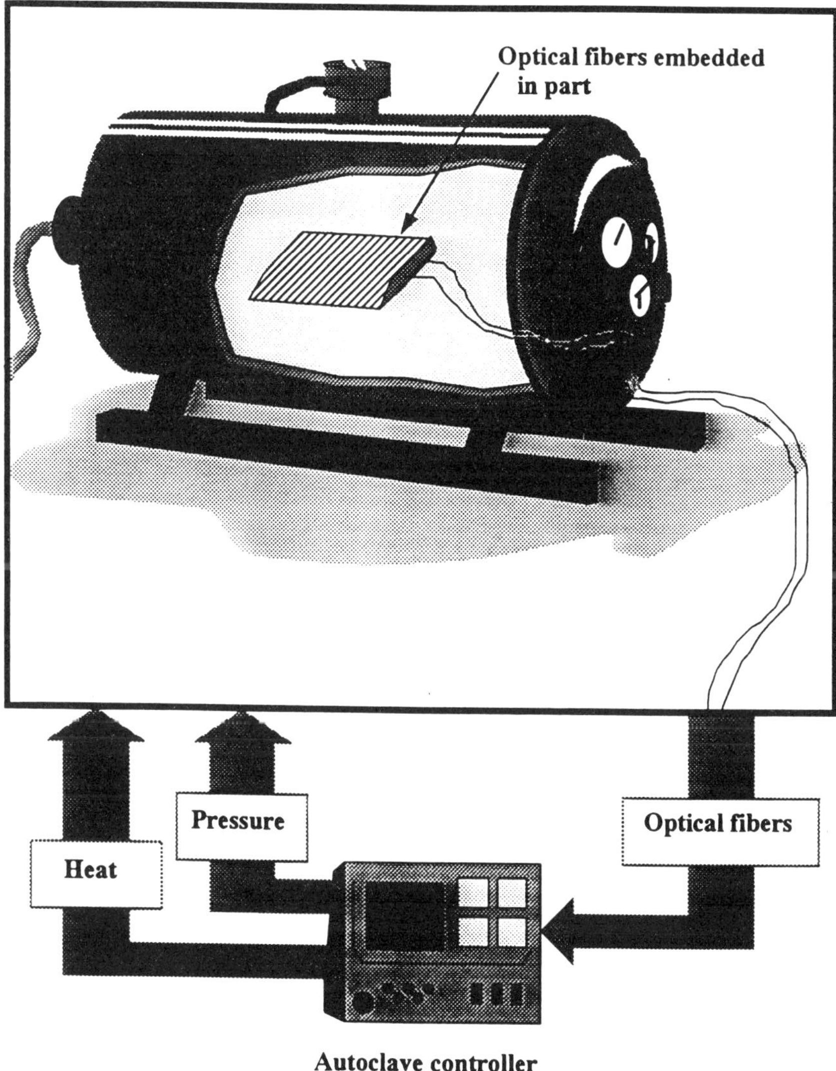

Fig. 3.13 Optical fibers embedded for autoclave cure monitoring.

will typically result in the part failing to comply with the design specifications, critical parts are typically subjected to cure cycles that are longer than the optimal duration in order to develop a state of overcure. This philosophy is clearly expensive.

A closed-loop control philosophy may be invoked for the manufacture of autoclave parts by embedding fiber optic sensors involving a network of thin optical fibers in the composite part. This processing philosophy enables the state of cure throughout the part to be continually assessed by a microcomputer-based control system, connected to the autoclave controller prior to tailoring the characteristics of the external stimuli imposed on the part by the autoclave, in a localized region of the part in order to achieve the desired state of cure. Figure 3.13 presents a schematic diagram of the system. This embryonic technology has broad ramifications for the fabrication of parts with complex shapes, and also high-performance composite parts where quality is of paramount importance. The technology would naturally impact all of the diverse industrial and commercial products that are currently exploiting the superior properties of polymeric fiberous composite materials.

In some situations the objective of the smart structure may be to simply ensure that a critical part for an aerospace system is manufactured to specification in an optimal fashion. The embedded sensors could monitor the state of cure, the magnitudes of the residual stresses, and also provide information on the size and location of defects. Thus, the part is only smart during the manufacturing cycle since it becomes dumb once the optical fibers are severed where they emerge from the composite structure.

In sharp contrast to this class of products featuring smart capabilities only during the manufacturing process, a second class of products would continuously employ this class of *in situ* photonic sensors throughout the processes employed to manufacture the part and also throughout the service life of the part. Thus, as depicted in Figure 3.14, the embedded sensing system would initially be employed to monitor the state of cure during the fabrication of a smart component featuring fiberous polymeric materials, in order to ensure that the part is processed correctly. Subsequently, the embedded sensing system would be employed to continuously monitor a number of critical parameters within the part during service, as illustrated by the dynamic stress characteristics of the robot arm shown in Figure 3.14 due to the loading generated by the end-effector tooling. If these stresses exceed prescribed criteria, for example, then the associated signals would be fed to the robot controller and the resultant mode of operation would be modified in order to restore the system characteristics to the prescribed work envelope. Finally, the embedded sensing system would be employed to monitor the structural integrity of the component. As shown in Figure 3.14, the embedded sensing system could be employed to monitor the structural integrity of structural members, such as aircraft wings, for example, by

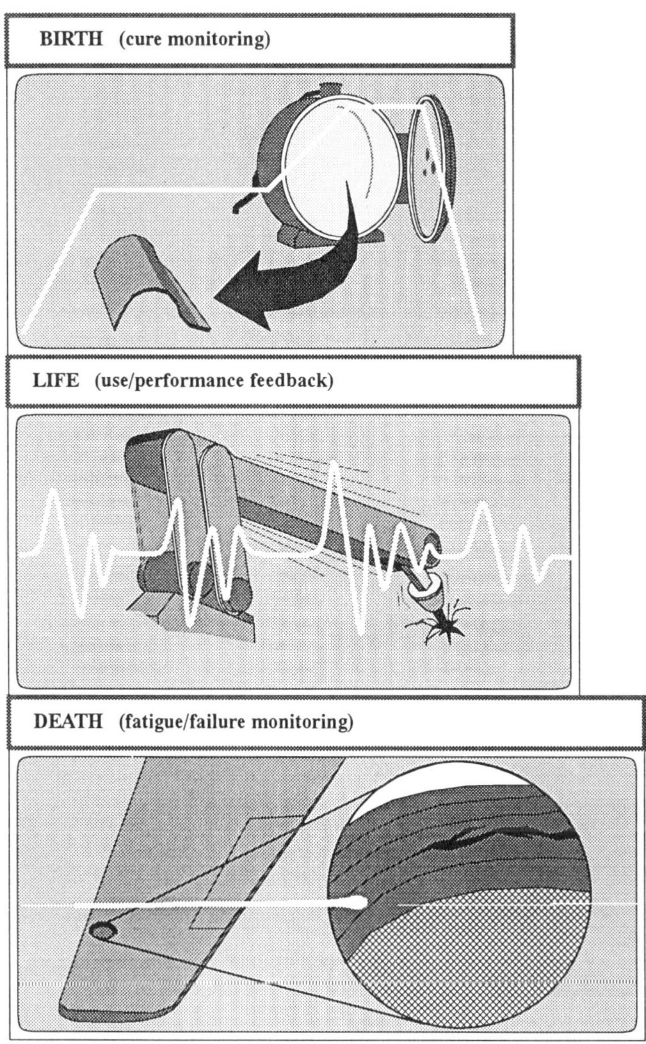

Fig. 3.14 The health-monitoring of parts utilizing embedded sensors.

detecting cracks and monitoring the propagation of these defects. Such a capability will be an invaluable tool for inspection and maintenance personnel.

This notion of employing sensors embedded within a structural member in order to monitor the properties of the part has resulted in the coining of the term 'health-monitoring,' and the relevant scenarios are depicted in Figure 3.15. Other terms include the cradle-through-grave philosophy. These terms are derived from the continual health-monitoring activities of

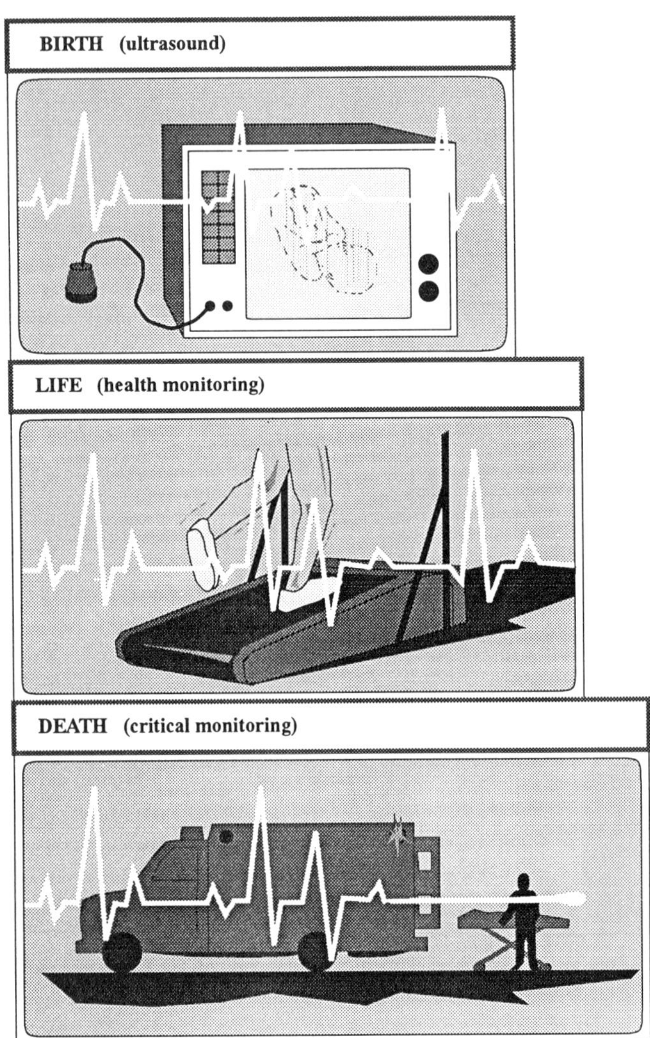

Fig. 3.15 The health-monitoring of *Homo sapiens sapiens*.

homo sapiens sapiens and they build upon notions of biomimetics. This field is a principal ingredient of many smart materials technologies. Prior to birth, human beings are often subjected to non-invasive techniques to monitor the status of the fetus prior to embarking upon corrective courses of treatment. These medical practices often continue through the birth process. Subsequently, throughout life the health of human beings is monitored in response to external stimuli of various forms prior to being subjected to corrective treatments. Finally, as depicted in Figure 3.15, in the scene involving the emergency care vehicle, the sensing systems can be employed to monitor the vital signs and record the failure of the principal organs such as the heart, liver, and kidneys.

It is clear, therefore, that the most sophisticated classes of smart structures will embody health monitoring capabilities as engineers endeavor to replicate biological systems. Products featuring health-monitoring capabilities would typically include pressure vessels and piping in industrial, military, and aerospace systems where the focus of attention would be the structural integrity of these shell-like structures in high pressure systems, or leakage in low pressure systems containing hazardous chemicals; the fuselages and wings of both commercial and military aircraft where structural integrity and vibration phenomena would also feature health-monitoring technologies. There are, of course, numerous other engineering situations involving components being subjected to fatigue environments, other classes of hostile environments, and other scenarios involving unstructured external stimuli, where this technology is also relevant.

The cost of installing a passive smart structure featuring a structurally integrated microsensing system must be carefully weighed, relative to the cost of part failure and subsequent loss of performance of the overall system containing the smart structural part. For example, consider a scenario in which a critical articulating member of a piece of production machinery is being subjected to a variety of unstructured external dynamic stimuli that are prematurely precipitating a fatigue failure of the part with the consequential loss of production. The deployment of this embryonic technology would provide a diagnostic tool whereby the status of fatigue cracks in the part could be monitored, and upon attaining a specific threshold, planned-maintenance procedures could then be initiated in order to replace the part at a convenient time, in order to avoid the subsequent loss of production associated with the previous generation of production machines.

There are numerous applications for passive smart structures in the field of civil engineering. Dams, buildings, bridges and other large concrete reinforced structures in earth-quake zones are typical examples. Of the 576 000 highway bridges in the US, at least 135 000 are structurally defective in some way. Thus, for example, the sudden unexpected collapse of the Mianus River Bridge on the Connecticut Turnpike in June 1983 with the associated tragic loss of life is no surprise to the US Department of Transportation. Research is currently focused on the development of coatings for steel bars reinforcing concrete bridge decks, substitutes for metallic reinforcement and the deployment of sacrificial cathodic protection. A passive sensing system to monitor the deterioration of critical regions of these structures may be another alternative.

The task of fitting an artificial limb to a patient is an activity currently undertaken using experience and art rather than science in order to achieve the desired qualities of function, comfort and cosmetics. The classical problem for the prosthetist is to develop a viable interface between the hollow socket of the prosthesis and the tissue of the residual limb, or stump. The situation is exacerbated by the changing volume of the stump which is partially attributed to variations in the fluid retention properties of the human body associated with diet. These dietary conditions can cause a loose fit in the socket or else discomfort to the amputee because of increased pressure at the interface between the residual limb and the socket of the prosthesis. What is ultimately required for this biomechanical application, is an adaptable smart material that autonomously responds to these variable service conditions in order to provide the optimal patient comfort in the interface region. These variable service conditions are typically characterized by changes in geometry and interface pressures when the muscles relax, for example, when the amputee is at rest, and furthermore, the same muscles periodically contract, but not simultaneously, during walking. The situation is further exacerbated by amputees of different ages with stumps with

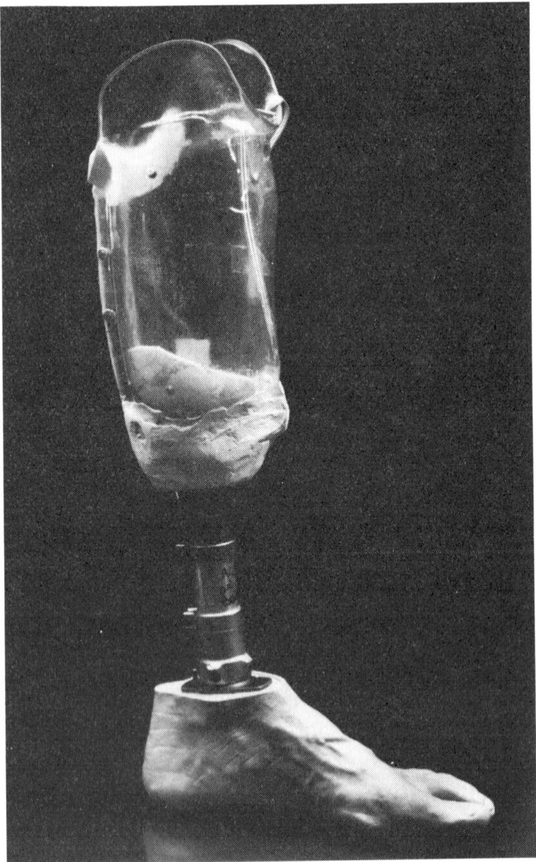

Fig. 3.16 Prototype smart lower limb prosthesis.

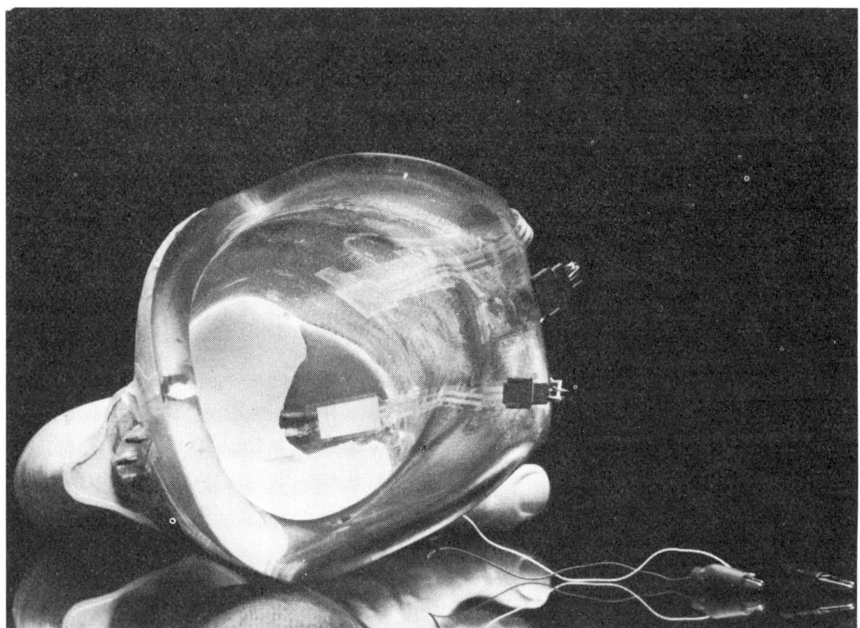

Fig. 3.17 Prototye smart lower limb prosthesis showing the piezoelectric sensors.

radically different characteristics. For example, those associated with a young person with a muscular stump, are different from those of a retiree who would typically have a flabby stump.

A development program has been initiated between the Institute for the Advancement of Prosthetics and the Intelligent Materials and Structures Laboratory at Michigan State University to address this problem. The first goal in this initiative was to measure the pressure distribution at the socket wall of a lower limb prosthesis as shown in Figures 3.16 and 3.17 which are photographs of an early prototype prosthesis retrofitted with sensors. Figure 3.17 shows two thin piezopolymeric sensors attached to the wall of the socket. These thin flexible transducers were selected in order to address the classical problem confronting prosthetists of measuring the normal pressure distribution at the stump-prosthesis interface. The selection of an appropriate sensor is constrained by the fact that the interface region between the socket and the residual limb is usually curved in several directions and the positioning of a sensor at the interface will modify the existing pressure distribution. The bonding of 0.028 mm thick piezopolymeric sensors marketed by Pennwalt Corporation addressed these concerns, and permitted dynamic measurements to be made in order to provide an enhanced scientific basis for the custom-fitting of prostheses for below-knee amputees.

3.7 Reactive actuator-based smart structures

The previous section focused attention on engineering products and systems featuring passive smart materials with embedded sensing capabilities but the structure was devoid of actuation capabilities. Conversely, there is a class of smart structures featuring smart materials which incorporate actuators but the system is devoid of sensory capabilities. This class of smart structures operates in a bang-bang mode, whereby the actuators are activated in an open-loop fashion because the structure does not feature a microprocessor and a control algorithm, and furthermore, sensory data is not employed to control the structural behavior of the smart material.

An example of such a system would be a large space structure deployed from the NASA space shuttle for an orbiting space station. The payload spatial envelope constraints imposed upon the designer of a large space structure by the shuttle cargo bay mandates that a high package density be achieved. Upon removing the payload from the cargo bay, the large space structure would be required to unfurl, to somewhat mimic the action of the petals of a flower exposed to sunlight, or change its configuration from a launch configuration to an orbiting configuration whose extreme dimensions may conceivably be of the order of tens of feet across. This transformation may be accomplished using various actuation technologies appropriately configured in the structure. Upon activating these actuators, the structure deforms and articulates until it is mechanically locked in the deployed configuration. Typical large space structures are space crane structural members, space platforms and large space antennas. These structures can feature trusses or membrane-like members.

A wide range of clamps, connectors, fasteners and pipe couplings are manufactured in SMAs. This class of TiNiFe or AiZnAl pipe couplings is employed on the hydraulic systems in US Air Force F-14 fighter aircraft, warships and submarines, and also underwater oil pipelines laid on the seabed. This application involves fabricating the coupling in a shape memory alloy with a transformation temperature lower than the operating temperature of the hydraulic system. The coupling is maintained at low temperature, by typically immersing the part in liquid nitrogen, prior to inserting the ends of the pipes to be joined together as shown in Figure 3.18. Upon warming the coupling, the coupling reverts to its former shape and grips the two pipes thereby generating a fluid seal.

Two plates can be readily bolted together, or rivetted together, when both sides of the plates are accessible to the appropriate tooling. When accessibility to the fastener is limited to only one side of the plates then the task of designing and installing an appropriate fastener becomes more challenging. Under these circumstances, SMA fasteners which exploit the unique characteristics of these metals provide one class of solutions.

These shape-memory fasteners are designed to have transition temperat-

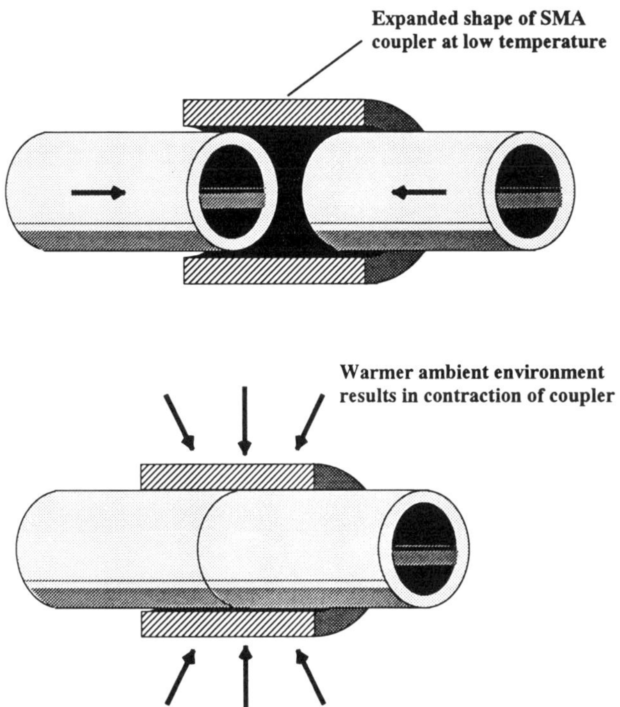

Fig. 3.18 Shape-memory-alloy pipe coupling.

ures below the operating temperature of the assemblage. Thus the fastener is manufactured with the split ends of the shank memorized at the ambient temperature state as shown in Figure 3.19. Upon drilling the appropriate clearance hole through the assembled plates to accommodate the fastener, the fastener is then cooled in liquid nitrogen and the split ends of the shank deform to yield the classical rivet-like configuration. The cooled fastener is then inserted into the holes in the aligned plates and upon being heated by the ambient air, the split ends of the shank again deform to the configuration shown in the lower diagram of Figure 3.19. thereby gripping the two plates to prevent relative motion.

Shape-memory-alloy components have also been developed for the cooling systems of automobiles. Figure 3.20 shows a sectional view of a thermally activated switching device. These applications, which focus on reducing energy consumption and on reducing the time for an automobile engine to attain thermal equilibrium, involve components in fan clutches and thermostats.

Shape-memory-alloy components have also been employed in diverse

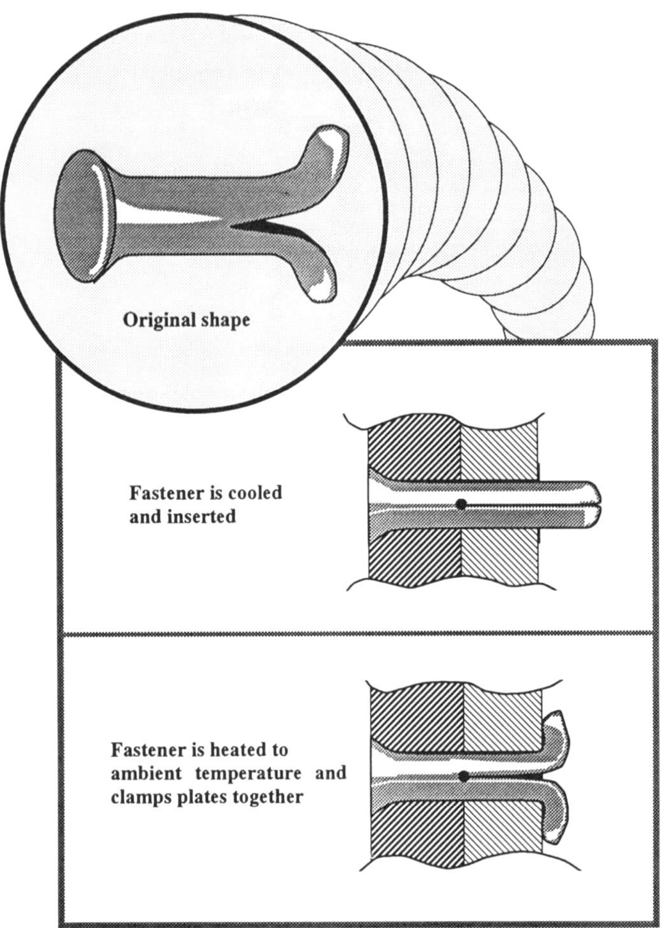

Fig. 3.19 A shape-memory-alloy fastener.

medical and orthodontic applications by appropriately designing the phase transition temperature relative to the temperature of the human body. Naturally with these applications, the material selection process is somewhat more demanding, because in addition to mechanical reliability considerations, the designer must also consider chemical reliability and biological reliability. Upon considering candidate shape-memory-materials, in the context of such parameters as biocompatibility, toxicity, *in vivo* degradation and decomposition for example, the vast majority of reported applications in this embryonic technology to date have employed nickel titanium (TiNi) alloys.

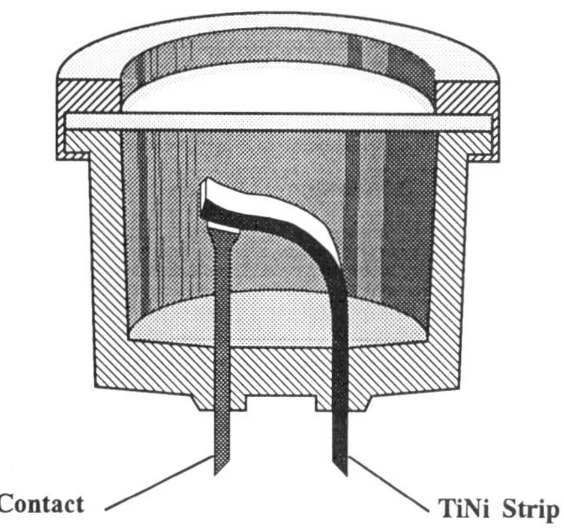

Fig. 3.20 A temperature fuse employing a shape-memory-alloy.

Exploratory research has been undertaken on the development of blood clot filters which are manufactured from TiNi wire in an effort to provide treatment of pulmonary thromboembolism. This medical condition is initialized when a thrombus develops in the vascular system, prior to becoming detached from the wall and being transported round the system of interconnected vessels and tubes. The standard treatment for this ailment is the administration of anticoagulant drugs, which of course hinder the cessation of internal hemorrhaging if this condition exists, or the removal of the thrombus by appropriate surgical procedures.

The proposed exploratory procedure involving TiNi wire requires the wire to be first trained to the blood-clot trapping coiled configuration prior to cooling and straightening the wire. Subsequently, the wire is then inserted into the *vena cava*, with the aid of a catheter, where it is heated by the blood flow and then reverts to the original blood-clot filtering configuration as shown in Figure 3.21. The procedure has been performed with animals using only local anesthesia and the results have been very satisfactory.

Orthopedic surgeons are developing a variety of procedures involving SMA components which are implanted in the body for extended periods of time. These applications include nails, pins, bone plates, stems for prosthetic devices and Harrington rods for the treatment of scoliosis.

Marrow pins, which are currently fabricated in stainless steels, are employed to assist the healing of fractured bones such as the tibia and femur for example. The surgical procedure involves inserting the pins into the bone marrow and exploiting the strength of the metallic component to complete the unification of the broken bone. This medical procedure, however, suffers from the danger of damaging the medullar cavity, since the

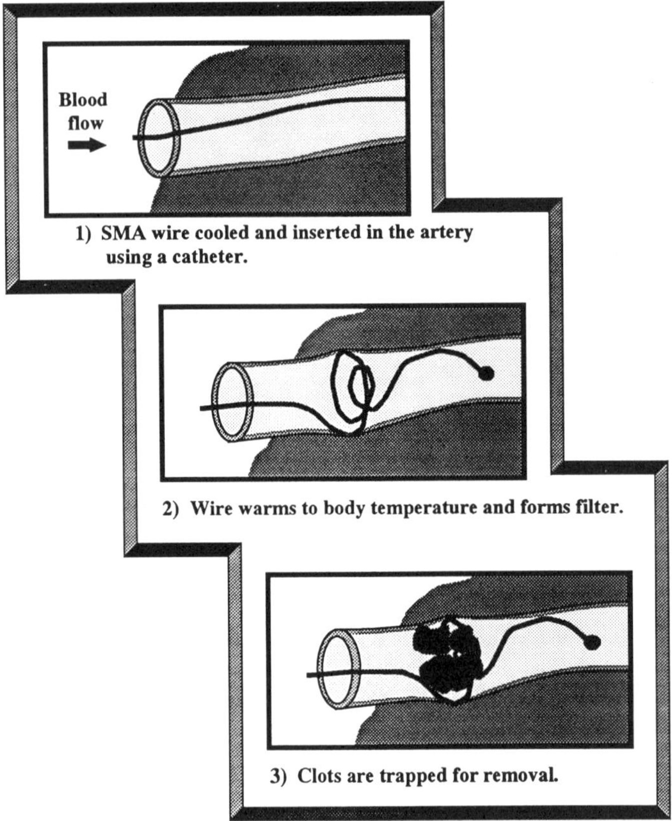

Fig. 3.21 Shape-memory-alloy blood clot filter.

cavity must be first reamed prior to force-fitting the marrow pins. A technique which exploits shape-memory-materials has been proposed to simplify the surgical procedure and also increase the robustness and reliability of the joint. Experiments have been performed with rabbits and sheep using pins which returned to their memorized three-dimensional shape upon being heated *in vivo*. The results have been very satisfactory.

Similar procedures have been employed to develop superior stems for artificial joints. The current practice is to employ an adhesive to bond the stems of artificial joints to the adjacent bones. This procedure, involving the bonding of dissimilar materials, frequently fails at the interphase region between the adherends. Looseness in this region delays the onset of bone growth, and the development of infections are a continual concern.

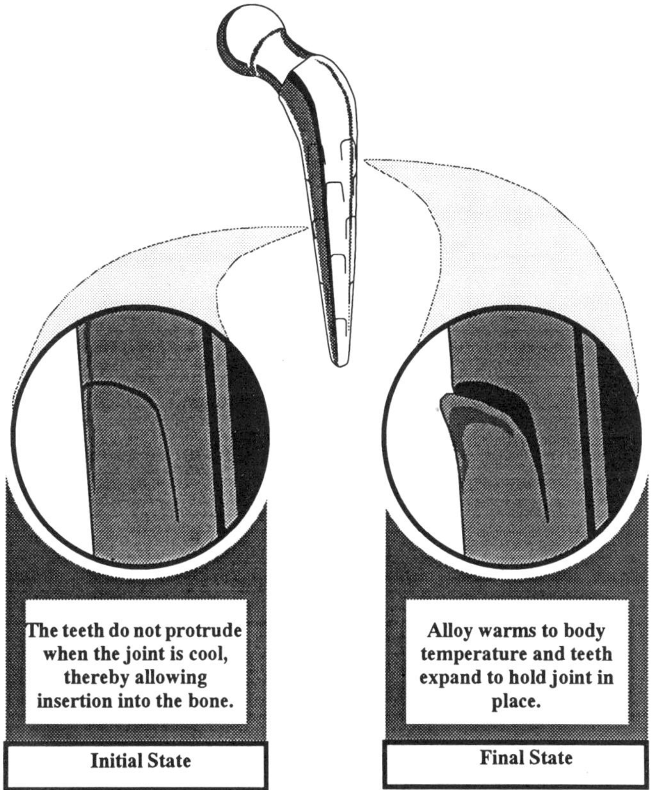

Fig. 3.22 A shape-memory-alloy artificial hip joint.

Prototype stems which feature a scale-like surface, somewhat similar to a human hair or the skin of a fish have been developed by Japanese researchers. Figure 3.22 presents an application where this approach is employed in the design of an artificial hip joint. The stems are cooled prior to insertion in the medullary cavity and the scales expand to exert a uniform force on the bone as they are heated by the body and return to their memorized shape. Trials involving sheep have been successfully completed and they clearly demonstrate that the scales function correctly within the adjacent bones, bone growth was good, and the stems were firmly fastened to the bones.

Scoliosis typically induces abnormal curvatures in the spinal column which not only subject the patient to severe pain but as these large motions develop, they can adversely affect the internal organs and mandate corrective surgery. This surgery typically involves the deployment of stainless steel Harrington rods which apply corrective forces to straighten the spine. These forces are largely uncontrollable and sometimes further surgery must be performed. Researchers have developed Harrington rods fabricated in TiNi which have successfully alleviated the necessity for follow-up operations and they also control the force exerted by the Harrington rod on the spinal column. This latter feature is accomplished by heating the structural member a few degrees above the body temperature using an electrical circuit. Preliminary results are promising for this new approach, but further research data must be collected prior to the deployment of this new generation of TiNi Harrington rods in practice.

There are a number of non-orthopedic procedures which can be enhanced by exploiting smart materials technologies. These procedures typically involve the insertion of long slender flexible cable-like devices into the body either through the natural openings, or else through surgical incisions permitting access to the vascular system. The current approach is to insert flexible members into the human body, whose mechanical stiffness properties cannot be actively controlled during this invasive procedure, and push the member through tubular vessels with their inherent natural variations in diameter, surface roughness, irregularities, and curvatures. This is a challenging undertaking since the stiffness characteristics of the medical instrument, or device, must be continuously tailored to the environment in order to ensure continual progress through the natural tubular system until the tip reaches the desired location within the body. This environment is continually changing because of the non-uniform curvature of the vessel through which the tip of the instrument is passing, in addition to the changing boundary conditions at the tip of the instrument. Failure to accommodate these variable service conditions results in the cessation of progress of the tip of the instrument and the invasive procedure can become an increasingly challenging task for the surgeon.

Consider, for example, the medical procedures associated with colonoscopes. These devices employ optical fiber technology to enable the physician to inspect the inside of the colon. The current generation of colonoscopes are inserted through the anus and employ a simple steering mechanism to facilitate the negotiation of the changing curvatures of the tube-like bowel. Unfortunately, the stiffness of the bowel wall is frequently very low when the organ is configured with a relatively large radius of curvature. Consequently, when the colonoscope is subjected to even greater axial loading by the physician in an attempt to ensure that the tip of the instrument progresses further along the bowel, the current generation of colonoscopes simply develop large lateral deflections due to buckling, and this increased pressure on the bowel wall imposes severe discomfort and pain to the patient.

These situations could be alleviated by the development of colonoscopes capitalizing upon smart materials technologies, which would feature stiffness characteristics that would be directly controlled and varied on command by the physician, and furthermore, they would possess the ability to change the boundary conditions at the tip of the instrument. The flexural stiffness of the long, slender colonoscope would need to be controlled and varied in discrete segments in order to enhance this procedure. Alternatively, the medical procedure could be modified by using first a sheath and then a colonoscope in order to inspect the colon wall. The sheath would be inserted first and would feature controllable variable stiffness characteristics which would facilitate the insertion process. The colonoscope would then be quickly inserted through the sheath. A distinct advantage of this procedure would be the ability to remove the colonoscope for cleaning upon partially completing an inspection procedure prior to quickly re-inserting to the same position in the bowel.

The fundamental notion of developing smart devices for invasive procedures for the medical profession has considerable utility. These procedures typically utilize devices which are long, slender, small diameter, flexible members of constant stiffness. The development of a new generation of devices which feature segmented variable stiffness characteristics whose behavior could be directly controlled by the physician during the medical procedure, would be of great benefit to both heart surgeons and also physicians practicing obstetrics and gynecology.

Percutaneous transluminal angioplasty is utilized in a myriad of vascular situations involving obstructive disease in the coronary, renal, mesenteric and peripheral arteries. The rationale behind the technique is to increase the luminal diameter of the diseased region in order to increase the blood flow. This goal is achieved by the high-pressure inflation of a distensible, or balloon tipped, catheter in the diseased region of the vessel in order to impose lateral pressure on the wall thereby increasing the diameter of the vessel prior to removal of the instrument. Figure 3.23 shows a photograph of one of these distensible catheters.

In the context of dilating coronary artery lesions, the basic equipment comprises a hollow guiding catheter approximately 100 cm long which is inserted into the brachial or femoral artery, prior to advancing the tip of the instrument until it is positioned in the region of the coronary artery to be dilated. This guiding catheter is fabricated as a composite material comprising three layers. The inner layer comprises the functional material Teflon, which has a low coefficient of friction in order to facilitate the subsequent insertion of the dilation catheter once the larger guiding catheter is correctly located in the vascular system. The middle layer comprises a fine wire mesh which is employed as a structural material to provide stiffness and a torque control capability. This inner layer ensures that the catheter has the relatively stiff shaft mandated by the necessity to ensure that the catheter tip

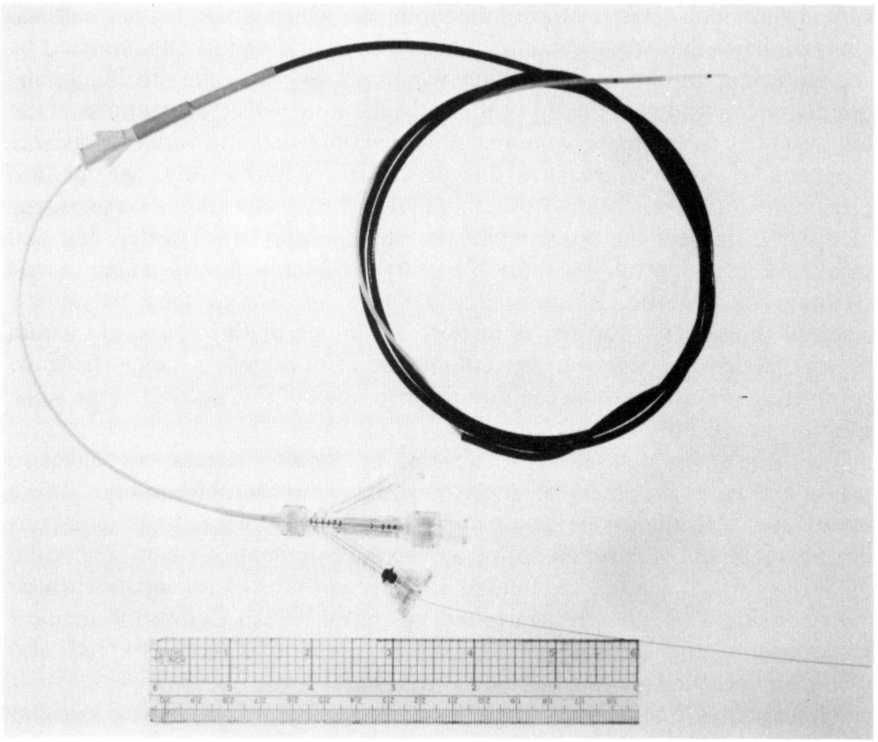

Fig. 3.23 Current generation distensible catheter for angioplasty operations.

is stable in the coronary ostium. The outer layer is polyurethane which is employed as a functional material to provide some memory capabilities. This task of positioning the tip of the catheter in the orifice of the coronary artery is a challenging undertaking because of the uniform constant uncontrollable stiffness of the device and also the significant variations in the curvatures of the vascular tubes and also their surface roughnesses. A smart catheter with controllable properties would significantly enhance this procedure.

Upon correctly positioning the hollow guiding catheter in the patient, a dilating catheter of length 140 cm and generally featuring two lumina is then inserted through the guiding catheter until the tip is positioned in the stenotic arterial branch. The balloon at the distal tip of the catheter is generally fabricated from polyvinylchloride or polyethlyene and is typically designed to withstand five atmospheres of pressure. A consequence of the inflation and subsequent deflation of the balloon is intimal splitting, plaque compression, the desaturation of liquid plaque elements and increased outer arterial diameter. This situation results in increased coronary blood flow and improved vessel caliber.

While there has been considerable success with percutaneous transluminal angioplasty and it has been employed in a wide variety of vascular beds, there are nevertheless a number of patients who cannot be treated by the technique, and furthermore, many patients develop re-occlusions later and require further treatment. These situations can generally frequently be treated by laser re-canalization in which the obstructive lesion is ablated. The clinical procedure is similar to balloon angioplasty and the CO_2 or Argon-ion laser beam is transmitted to the lesion along a catheter featuring optical fibers. This treatment, which utilizes equipment exploiting smart materials technologies, leaves a smooth surface on the vessel wall, unlike percutaneous angioplasty, which is characterized by restenosis.

Blocked fallopian tubes which are typically the result of endometriosis or inflammatory diseases of the fallopian tubes, uterus, or ovaries, cause an estimated 30% of female infertility by severing the connection between the uterus and the ovaries. Traditionally, this situation is treated by abdominal surgery and in-vitro fertilization which are expensive procedures. A new cheaper, safer outpatient procedure called transcervical balloon tuboplasty, has recently been developed which would only cost 40% of the conventional treatment. In this new procedure, a series of balloon-tip catheters are inserted through the cervix into the fallopian tubes. The balloons are subsequently inflated to increase the diameter of the tube in the blockage region thereby reopening the natural tube-like structure connecting the uterus and the ovaries. The development of smart catheters featuring members with segmented variable-stiffness characteristics which are directly controlled by the physician would significantly enhance this new catherization procedure.

3.8 Active sensing and reactive smart structures

This class of smart structures represents the pinnacle of sophistication for the short-term projections of smart structural systems. These systems typically feature a host structural material, a network of microsensors, data links, data-processing capabilities, and algorithms programmed into micro processors which appropriately activate a network of actuators for achieving a desired effect in an autonomous fashion. This desired effect typically focuses upon either changing the shape of the structure in a controlled manner, or alternatively, changing the mass, stiffness, or energy-dissipation characteristics of the structure in a controlled manner in order to tailor the static response of the structure, or alternatively depending upon the application, the dynamic response characteristics of the structure such as natural frequencies or mode shapes.

The defense industry is a logical sector of the economy that will naturally exploit the field of smart materials and structures because of the potential payoffs offered by smart materials technologies. There are a number of situations where smart materials technologies could play a crucial role in military systems and these include shock and vibration control; countermeasures to defeat armor-piercing weapons; the enhancement of the resistance to battle-field damage; the ability to monitor a wide range of environmental conditions and identify potential threats to both man and machine; and the enhanced maintenance and life prediction capabilities of equipment. Some of these programs have already been initiated.

For example, the United States Navy is developing a quiet self-contained actuator technology for broad deployment which utilizes SMA materials. These applications employ a nitinol alloy which is available in the form of wires in the range 0.08–0.25 mm in diameter. Current results suggest that future applications of this innovative actuator technology may be limited by the load capacity of the actuators, the life cycle duration and the thermal cycling frequency of the wires. In order to address these concerns, the next generation of SMA materials are anticipated to feature the following characteristics: transition from the austenitic phase to the martensitic phase and back again over a narrow temperature range, which may be as low as 1° C; transition temperatures may be as high as 150° C; life cycles of 10-100 million cycles without degradation of performance and strength; working strengths will exceed 40 000 psi; wire stock with diameters up to 2.5 mm; and nitinol stock will be developed in other geometrical configurations featuring large surface to volume ratios which may include foil, flat plate stock, or wire rope.

Another program focuses upon the development of an active damper utilizing electro-rheological fluids for a missile flight-control fin actuation system. The program features air stable polymer fins that can be electrically switched between electrically conductive and nonconductive states for microwave shielding applications. Test results currently indicate that the system is switchable over 100 000 times with little degradation in properties in either state. Development programs have focused on optimizing the absorption and transmission characteristics, the switching speeds, the air stability and the reversibility. In addition, measurements of microwave absorption in the conductive and nonconductive states at the X-band range have been undertaken.

Structures which must operate autonomously in space are also ideal candidates for the incorporation of several types of smart structural systems, because of the variable service conditions and the nature of the unstructured environment. The deployment of large space structures such as large robotic devices, platforms, telescopes, or solar arrays from the confinements of the payload envelope of the launch vehicle have already been discussed briefly. Constraints imposed by the launch vehicle upon payload weight mandate that these large space structures be lightweight, and a consequence of this is that they are somewhat flexible. Thus once the spacecraft is on station in orbit, engineers are then confronted with the tasks of shape control of the structure and also vibration control.

Vibration control situations would typically occur as a result of meteor impact, and in the case of an orbiting laboratory as a consequence of imperfect docking between a space shuttle which may be returning with supplies, or an astronaut may be exercising on a tread mill. These classes of transient and dynamic responses could be controlled by hybrid schemes of actuators controlled by a network of sensors throughout the structure.

Other large space structures also involve geometrical control problems. Consider, for example, the proposed NASA large deployable reflector program which focuses on the development of an orbiting far-infrared astronomical observatory with a diffraction-limited wavelength range of 30–50 microns. The principal structure on the observatory is a 60 foot diameter paraboloidal reflector whose surface geometry must be accurately maintained to within a few microns when in the observation mode. This is a challenging task when considered in the context of manufacturing errors, creep of the structure, thermal gradients and furthermore the telescope is typically in the observation mode for 20 minute time-intervals. Structural members incorporating embedded piezoelectric materials have been proposed to control the critical geometry of the reflector.

Aircraft continually operate in unstructured environments because of uncertainties in the weather conditions, turbulence, temperatures, payloads, and the duration of the flights. Several smart structures programs have been initiated for both commercial aircraft and also military airplanes. The focus of these 'smart skins' programs, so named because of the monocoque design of these structural systems, are dependent upon the specific applications. The commercial aircraft programs focus primarily upon monitoring the health and flight worthiness of aircraft. Arrays of sensors throughout the wings, control surfaces and the fuselage will monitor the structural properties in the context of fatigue cracks and incipient failures. Other types of sensors will monitor the ice build up on wings and control surfaces which can adversely affect aerodynamic performance at take-off and also the controllability of the machine.

While the above programs are largely sensor-based, other efforts are focused on the synthesis of smart structures featuring both sensors and actuators. Some of these programs are directed towards the design of aircraft with non-articulating control surfaces, where the appropriate regions at the edges of the relevant airfoil sections would deform in order to provide the aerodynamic control required. Another program is focused on the development of smart wings whose dynamical response can be automatically adjusted in order to provide smooth flight by tailoring the stiffness and damping properties in discrete sections of the wings.

This philosophy, featuring finite element control segments, incorporates sensors, actuators and microprocessing capabilities. Thus, as the mass of the wing is reduced by the consumption of aviation fuel from the fuel tanks in the wings, the natural frequencies of the wing structure will also change. An application of this approach is schematically presented in Figure 3.24. The sensors will monitor the relevant characteristics of the smart structure, and the signals from the sensors will be fed to the appropriate microprocessors which will evaluate the characteristics of the signals, prior to determining an appropriate control strategy in order to synthesize the desired response characteristics. This will typically be accomplished by controlling the material characteristics of the actuators in the specific finite element segments. The material changes in a typical finite element control segment will in turn alter the global stiffness, and damping characteristics of the smart structure in order to achieve the desired response.

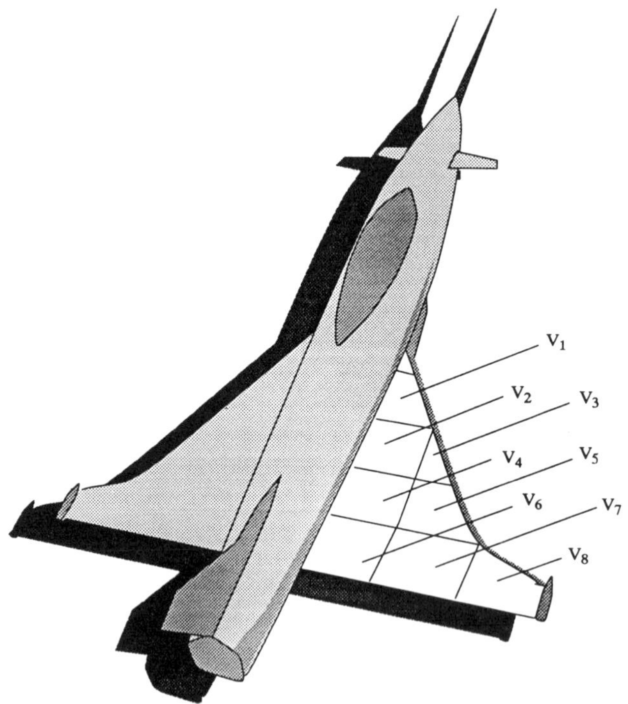

Fig. 3.24 A strategy for actively controlling the dynamic response of an aircraft wing fabricated in smart materials.

The dynamic behavior and performance of state-of-the-art helicopter rotors fabricated in advanced polymeric fiberous composite materials are dependent upon the speed of rotation, the speed of the machine, aerodynamic loads and also the ambient environmental conditions of temperature and relative humidity. The rotor is typically designed for a worst-case scenario, which results in a state of over-design for all other operating conditions because the rotor is a passive system. A consequence of this design philosophy is inferior performance at other operating conditions, because of the attendant design-cascading effects associated with the heavier design.

A rotor designed and fabricated in a smart material would be able to detect changes in temperature and moisture associated with operating in the cold dry Arctic regions, or alternatively in the hot, humid equatorial climates, in addition to detecting dynamic strains associated with wind gusts, for example. These data could be monitored and processed by a microprocessor system in a smart rotor, prior to dictating the appropriate corrective actions to be implemented by a network of embedded actuator systems in the structure, in order to ensure optimal performance under all service conditions.

Consider a smart rotorcraft system which capitalizes on the knowledge-base in the area of smart materials and exploits the unique capability of smart materials to interface with solid-state electronics, by the successful incorporation of intelligent sensor technologies and modern control strategies. This innovative rotorcraft system could have the capacity to truly integrate sensing, processing, and actuator functions in order to respond autonomously to flutter, buffeting, gust fields, and dynamic stall, all of which exhibit large scale hysteresis loops. Furthermore, this system could exploit hybrid optimal control strategies in order to tailor the aerodynamic and elastodynamic performance characteristics of the rotorcraft system, which typically is subjected to variable service conditions while operating in unstructured environments in service.

In such a smart rotorcraft system, the fiber-optic sensing system will instantaneously sense the resulting vibrations and changes in temperature, for example, which will permit the microprocessors to optimally actuate the electro-rheological fluid, magnetostrictive, SMA and piezo-electric domains in the rotor blades as shown in Figure 3.25, thereby tailoring the aero-elasto-dynamic performance of the rotor with a pre-defined performance criterion. The rotorcraft system would typically feature SMA actuators for generating large changes in camber, piezo-electric actuators to provide a capability for changing the surface geometrical characteristics, in order to substantially improve the aerodynamic performance of cambered airfoil section, ER fluids and magnetostrictive materials to control the vibrational response of the structure, for example. Furthermore, SMAs will allow large changes in the geometry of a structure to be readily accomplished. The

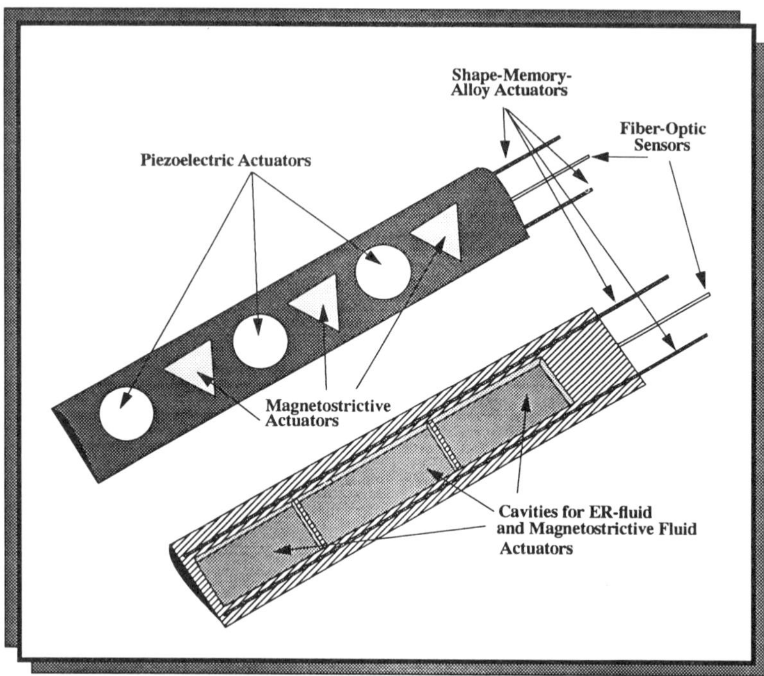

Fig. 3.25 Conceptual rotor blade featuring smart materials.

hybridization of these distinct classes of actuator systems will permit a dramatic alteration in the microstructural characteristics of the smart actuators, which will permit the global geometrical configuration, and surface geometrical characteristics to be changed, and also the global mass, stiffness, dissipative, and aerodynamic characteristics of the ultra-advanced composite rotor structure to be changed. These accomplishments will be achieved by changing the electric field imposed upon the ER fluid and piezoelectric domains, the magnetic field imposed upon the magnetostrictive domains and the thermal field imposed upon the SMA domains.

Military programs for both helicopters and aircraft are not only focused upon the above objectives but also a number of other aspects too. For example, if the aircraft is damaged by enemy action, then what are the consequences, and how does this event affect the original scope of the mission? Smart structures will be developed which will quickly assess the state of the airframe prior to advising the pilot of limitations in the flight envelope and also the degree of severity of the maneuvers that the plane can safely withstand without catastrophic failure, or else automatically adjusting the controls to ensure stable operation of the aircraft. Stealth technologies are also being incorporated in the smart skins of these shell-like aerospace vehicles. The goal is to be able to actively change or reduce the radar signature of the vehicle by controlling the absorptivity of the skin or else changing the local geometry of the skin. The U.S. Air Force claims that the radar image of the 172 foot wide, 370 000 pound, $815 million B-2 stealth bomber is in the insect category when viewed by enemy ground or airborne radar. The generic research associated with these programs which focuses on coupled field phenomena is also relevant to submarines, and their interactions with sonar waves, and also the acoustical noise radiated from military, commercial and domestic products.

The 'smart-materials tool-kit' of diverse classes of actuators, sensors, microprocessors, and data-links available to design engineers can be readily employed to synthesize a new generation of superior products for the sporting goods industries. Typically, the superior performance would be achieved by exploiting the superior dynamic response characteristics of smart materials which would be achieved by actively changing the mass, stiffness, and energy-dissipation characteristics of the smart structure. This technology could revolutionize the sporting goods industry as dramatically as the current generation of equipment fabricated in polymeric advanced composite materials, which have been a major factor in the establishment of numerous records in many different sporting events.

Smart materials could, for example, permit an angler to change the stiffness and energy-dissipation properties of a fishing rod in order to enhance the particular casting technique of the angler, or alternatively, to suit the size and aggression of the fish that has been hooked so as to enhance the pleasure associated with the task of attempting to land this cold-blooded vertebrate animal.

Rackets could be fabricated in smart materials to provide a competitive edge to individuals playing, tennis, badminton, racket ball, squash etc. A schematic diagram of one of these new generation of tennis rackets is presented in Figure 3.26. By controlling the energy dissipation characteristics of the racket, a tennis player will be able to easily return a fierce serve because the frame will absorb the impactive energy of the ball. Furthermore, by controlling the stiffness of the racket to create a very flexible shape, the player will then be able to confuse the opposition by executing a vicious return shot that generally dispatches the tennis ball to the back of the court, but because of the mechanical properties of the racket the ball will barely clear the top of the net prior to landing a couple of yards beyond the net. This shot would, in some sense, replicate the drop shot. Of course other scenarios in racket games can be envisaged.

There are numerous applications in the automotive sector which would benefit from the deployment of smart materials technologies. The development of devices incorporating ER fluids has recently been the subject of

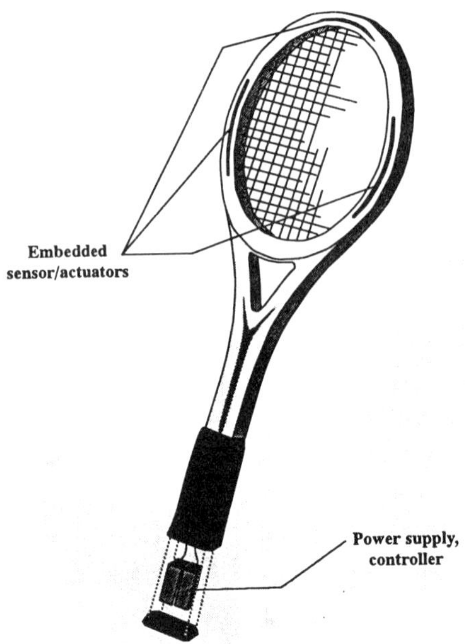

Fig. 3.26 A smart tennis racket.

intense research activities. One of the principal foci has been the development of clutches, as shown in Figure 3.27, and also other transmission components which feature a tailored slippage characteristic controlled by the time-dependent voltage profile imposed upon the fluid.

The change in the bulk viscosity of ER fluids in the presence of externally applied electric fields, can possibly be exploited to both enhance and precisely control the fluid flow around the surfaces of hydrodynamic journal bearings by suitably tailoring electric fields. This innovation would represent a significant advance in the design of lubrication systems which would permit the system to autonomously respond to adventurous departures from steady-state conditions and/or compensate for bearing wear or unstructured changes in operating conditions. Time-varying load patterns, for example, can then be compensated for by transient automated adjustments in the electrical field strength, and distribution precessional motions induced in rotating journal bearings could be corrected by inducing viscous counter forces to dampen off-center drift.

Other research in the automotive industry has focused upon engine mounts incorporating ER fluids and a two chamber design separated by a narrow orifice upon which is imposed a controlled electrical field, as shown in

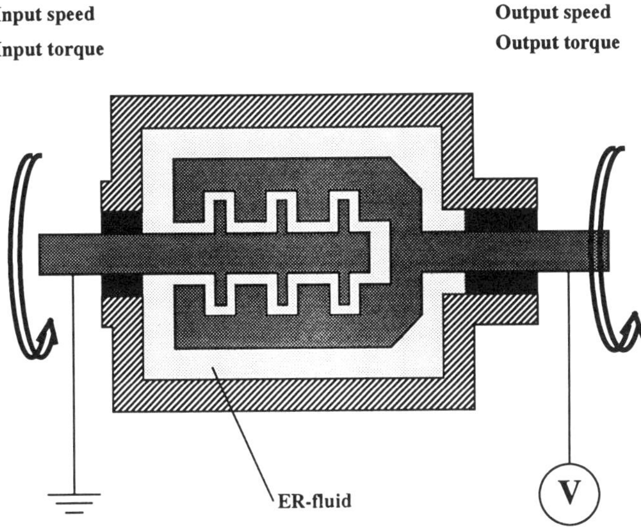

Fig. 3.27 Schematic of an electro-rheological fluid clutch.

Figure 3.28. The objective is to reduce the dynamic loads generated by the engine from entering the automobile structure in order to create enhanced ride characteristics. Such a technology is clearly not limited to vibration isolation in automobiles, but extends to schemes for protecting delicate instruments, schemes for isolating industrial machine systems, and schemes for isolating large-scale civil engineering structures in the earth quake zones, which would feature intrinsic sensing systems and controlled force actuators with the innate capability to neutralize external dynamic excitations.

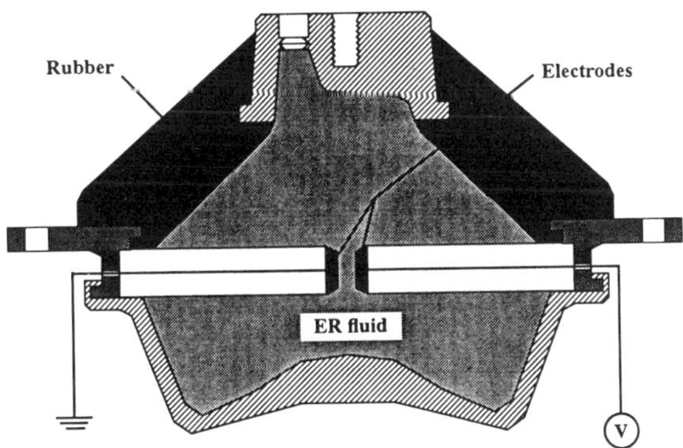

Fig. 3.28 An automobile engine mount featuring an electro-rheological fluid.

The above devices embody narrow orifices through which ER fluids flow. By controlling the voltage field in the neighborhood of these orifices and hence the viscosity of the fluid, the designer can control the rate of flow of these fluids through the orifice. A schematic diagram of this design philosophy is presented in Figure 3.29. This notion has been employed in the design of a flexible fixturing system, schematically illustrated in Figure 3.30, for deployment in flexible manufacturing environments where workpieces of different shapes and sizes must be fixtured by a single fixture.

The notion of automotive vibration isolation readily extends to automotive suspension systems. This work focuses on re-designing the conventional fluid-filled shock absorber to actively control the characteristics of the

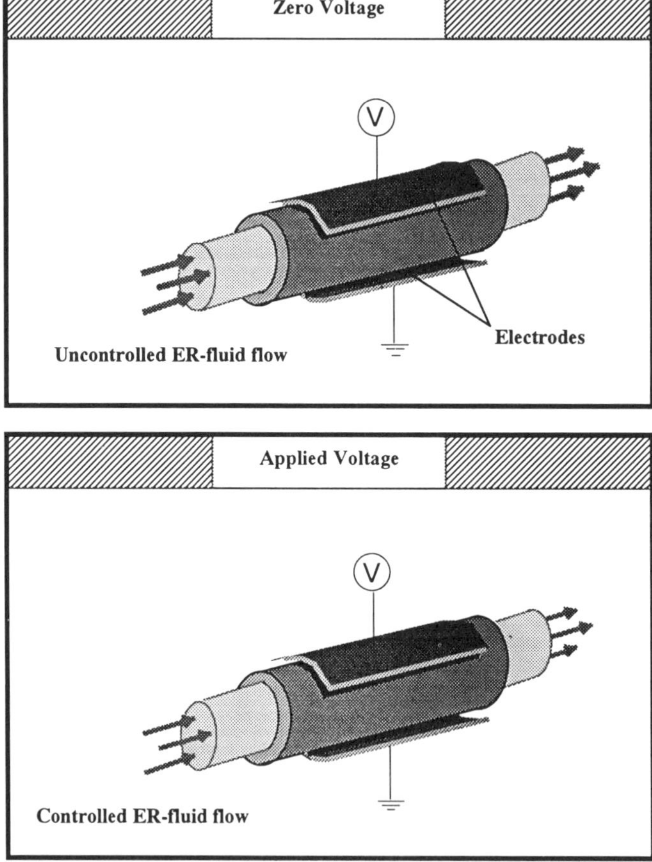

Fig. 3.29 Schematic of a valve design for controlling the flow of electro-rheological fluids.

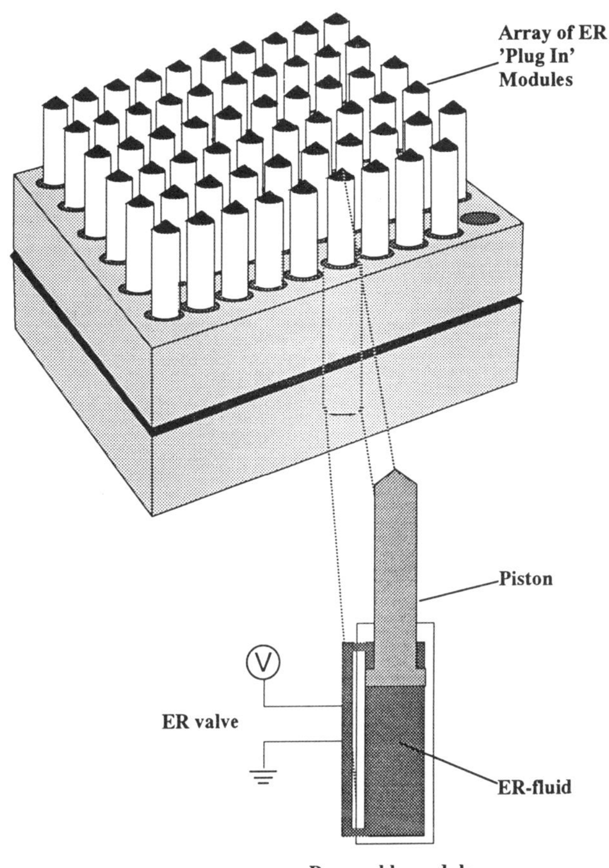

Fig. 3.30 An adaptable fixture featuring an electro-rheological fluid.

product to appropriately respond to the excitation of the road surface, in order to enhance the ride characteristics of the vehicle. Other automotive studies are focused on controlling the dynamic response of the leaf springs by filling them with an ER fluid as depicted in Figure 3.31, hence, the vehicle can be actively controlled by altering the damping properties and stiffness within the beam-shaped members.

The heat and mass transfer properties of electro-rheological and magneto-rheological fluids can be actively controlled by the imposition of appropriate electric and magnetic fields, for enhancing the performance of a broad range of diverse thermal systems such as recuperative heat exchangers, for example. These smart materials systems would typically feature in addition to these actuator-like fluids, the appropriate sensing and microprocessor-based controllers. Systems featuring ER fluids have been developed in order

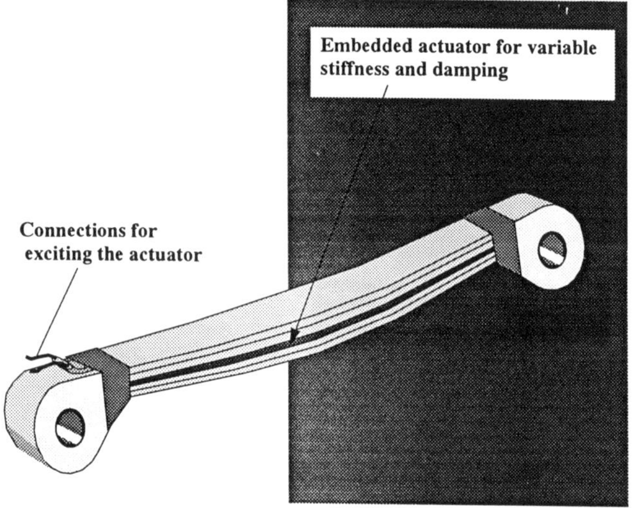

Fig. 3.31 An automobile leaf-spring suspension.

to control the rheological properties as a function of temperature; the thermal conductivity and specific heat as a function of the electrical field strength; the flow rate and hydraulic drag as a function of electrical field; and the convective heat transfer coefficient at a wall as a function of the electrical field because of the dramatic increase in the shearing-rate gradient adjacent to the wall.

Systems featuring magneto-rheological fluids have been developed in order to control the rheological properties as a function of temperature; the conductive heat transfer coefficient as a function of the orientation of the imposed magnetic field; and the convective heat transfer coefficient as a function of the magnetic field strength. These results have broad ramifications in the design of many thermal devices.

Structural members with voids containing ER fluids have been the subject of a number of investigations focused on robotic and mechanism systems. For example, one study has focused on actively controlling the energy-dissipation characteristics of robotic arms upon completing a series of high-speed maneuvers and a schematic diagram of the system is presented in Figure 3.32. The goal of the work was to enhance productivity through the reduction of the vibrational settling-time when the robot motors are not powered upon completion of a series of maneuvers. In other studies the natural frequencies of robot arms containing voids filled with ER fluids have been actively changed as a function of the voltage imposed upon the ER fluid domains, in order to avoid resonance conditions attributed to the excitation frequencies of loads imposed by production tooling at the end

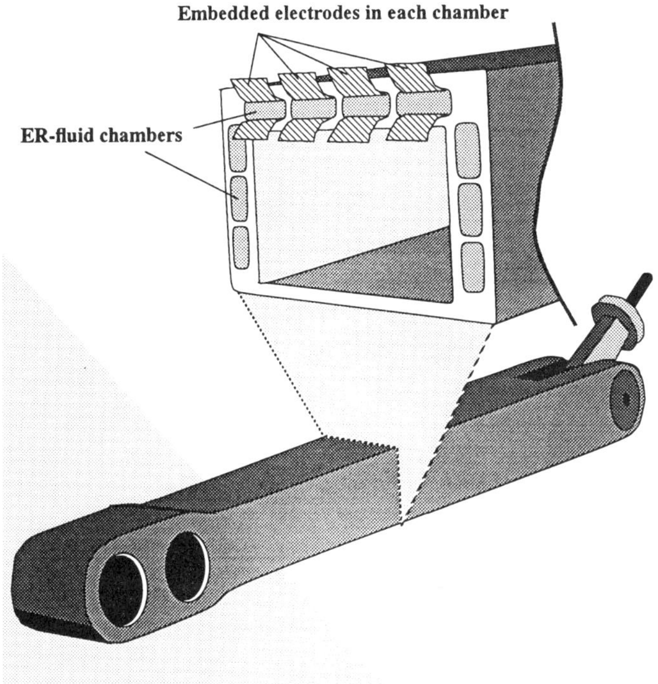

Fig. 3.32 An ER fluid-enhanced robot arm.

effection. Similar studies have been undertaken for linkage mechanisms.

The trend towards flexible automation in the industrial sector has motivated research on more versatile mechanical systems. This versatility can be measured by the ability of a single kinematic chain to generate two or more coupler curves using the same tracer point on the coupler by activating SMA links, as shown in Figure 3.33. Alternatively, the shape of a flexible link, and hence its dynamic response can be changed, by activating a series of embedded piezoelectric actuators in the member. Research has also been prosecuted on experimental micro-robotic systems with SMA actuators, as shown in Figure 3.34.

This trend of flexible automation has catalyzed research on the design of robot end effectors. A typical scenario may require the end effector of a robot to pick up a soft ripe tomato prior to picking up a ball peen hammer and subsequently striking a nail into a piece of wood. *Homo sapiens sapiens* readily accomplishes these tasks since piezoelectric sensors in the finger tips provide feedback information which is coordinated by the brain and relayed to the appropriate muscles via the central nervous system. The goal of mimicking this biological task has been the subject of numerous research groups worldwide. An Italian team has developed a multilayered smart

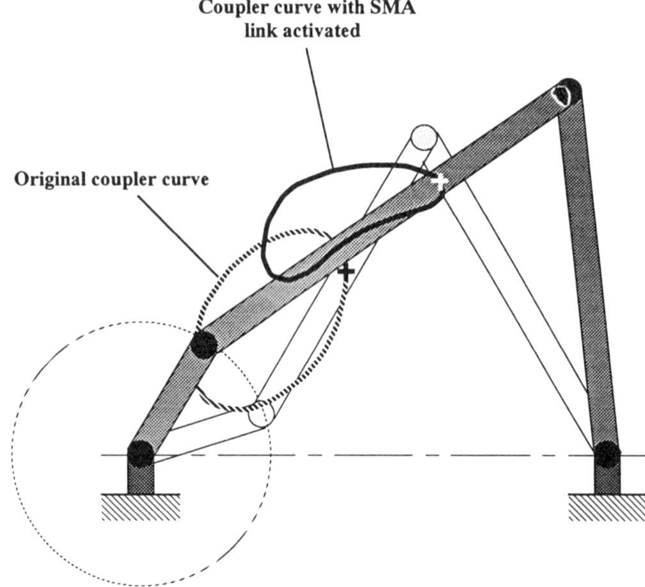

Fig. 3.33 Enhancing the coupler curve characteristics of a four-bar mechanism by utilizing shape-memory-materials.

synthetic skin approximately one millimeter thick to replicate the actions of the human dermis and epidermis. The synthetic dermis comprises an electrically-conducting water-saturated gel sandwiched between flexible electrodes. This artificial dermis is employed to sense the pressure exerted on the gel by the object being grasped by the hand, and consequently the hardness of the object can be evaluated by relating the deformation of the gel to the voltage change across the flexible electrodes. The artificial epidermis is employed to evaluate the surface texture of objects. This is accomplished by utilizing an array of thin flexible piezo-electric disks sandwiched between layers of rubber. The piezo-electric phenomenon relating pressure to electrical charge is exploited in order to permit the smart skin to not only measure forces normal to the surface of the skin but also forces tangential to the surface of the skin, such as those associated with frictional effects, or adhesive tape, for example. While this research has considerable promise, further work is required in order to bring these ideas to the marketplace.

Mundane glass has now gone high-tech. This hard brittle substance, which has been a feature of civilization for the past 2500 years, now possesses smartness through additives which are sensitive to radiation. Thus photochromic glass is obtained by adding copper and silver chloride. The transparency of this type of glass is a function of the intensity of ultra violet

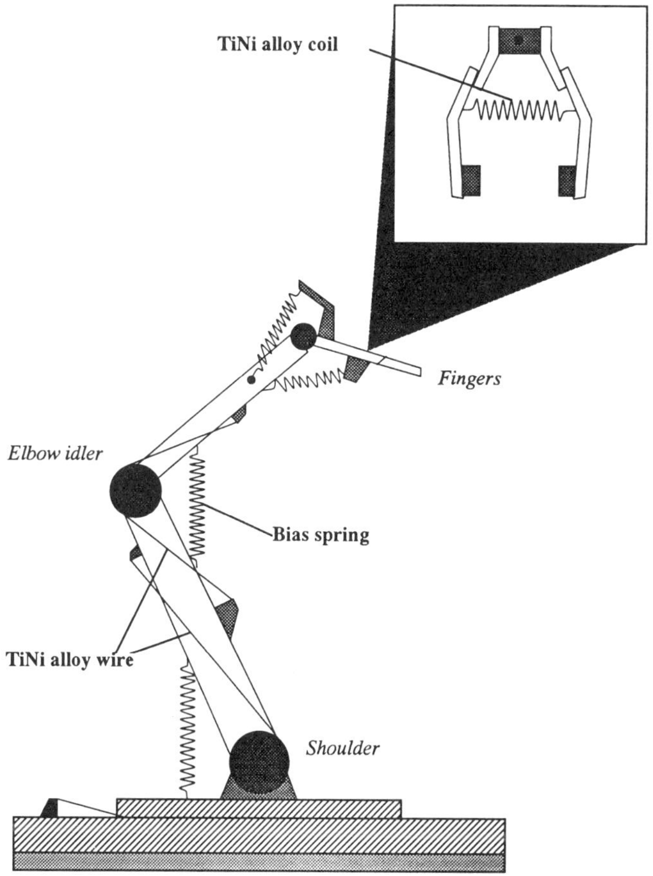

Fig. 3.34 A micro-robot actuated by shape-memory-alloys.

radiation imposed upon the glass object. Typical products are eye glasses, instruments, and windows. Other additives include light filtering dyes and zinc oxide to produce glass that absorbs, filters and scatters light to provide protection against harmful irradiance.

3.9 Smart skins

Aerospace products, such as rockets, satellites, and aircraft, and also naval vessels such as submarines and ships are typically monocoque shell-like structures featuring a thin skin whose structural integrity is generally mission-critical. The safe and optimal operation of the vehicle is dependent upon the environment to which the skin is subjected and also the state of degradation of the skin. It is, therefore, of critical importance that the condition of this membrane be carefully monitored and maintained within specified bounds in order that the system function properly. The critical role of the behavior and structural integrity of the extremities of these vehicles, has triggered the evolution of an embryonic technology which typically would feature actuators, sensors, data-links and microprocessors embedded in the structural material as shown in Figure 3.35 in order to monitor the external loads and the integrity of the structure prior to initiating corrective actions. This class of sophisticated structures has been termed 'smart skins'.

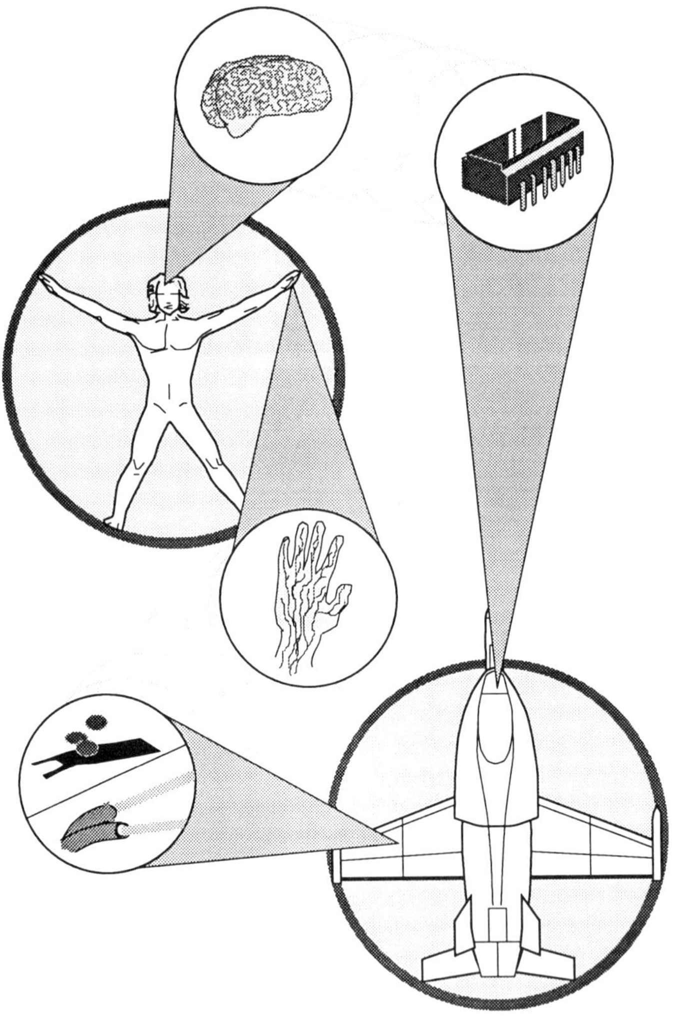

Fig. 3.35 Various smart skin systems.

In the context of space systems, such as the proposed NASA space station or space-based radar, this corrective action may be to actively control the vibrational response of the structure following meteor impact, for example. In the context of military aircraft, however, this corrective action may involve the assessment of battlefield damage on an aeroplane immediately after sustaining a direct hit by ground fire, prior to either advising the pilot of the limitations in his flight envelope and the intensity of the maneuvers that the damaged aircraft can sustain, or else automatically adjusting the flight controls to ensure stable flight. Commercial aircraft can also significantly benefit from a smart skins initiative through the automatic

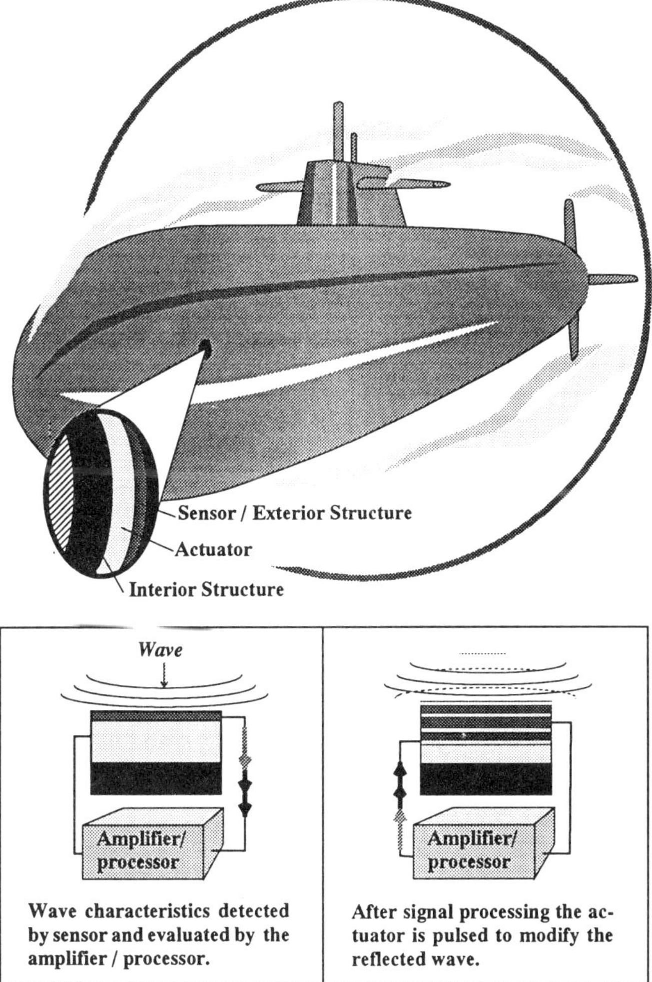

Fig. 3.36 Sonar wave sensitive submarine skin.

monitoring of the airframe structure, which would increase the fleet efficiency because this technology would reduce the time when the aircraft would need to be withdrawn from service for routine maintenance, for example.

The 'smart skins' technology has ramifications for the operation of naval surface vessels and also submarines. For example, the consequences of battlefield damage could be rapidly assessed and the captain advised of his mission options. Alternatively the structural properties of a submarine hull could be tuned to reduce the noise radiated by the vessel from on-board equipment or else change its acoustical signature and become a 'stealth' system in response to enemy sonar, as depicted in Figure 3.36.

Another application of the 'smart skins' philosophy for marine environments focuses on drag-reduction in order to increase the speed of a ship or submarine. This technological application features a compliant shape-controllable skin on the hull of the vessel which would respond in a perturbational fashion to changing hull pressures in order to change the contour of the vessel in contact with the water. The vessel would, therefore, emulate the skin of the dolphin which changes shape in order to achieve minimal levels of drag.

3.9.1 Aircraft with health-monitoring capabilities

The design and operation of both military and commercial aircraft are dictated by stringent requirements which have evolved during the past eighty or so years in order to ensure the structural integrity of these aeronautical systems. However, even when these requirements are satisfied, there have been numerous structural failures in recent years because the fatigue life of the ageing fleet has been exceeded. Figure 3.37 documents the average age of the principal aircraft in the US Air Force. The accidents resulting from these fatigue failures have triggered increasing customer awareness of this engineering phenomenon and two broad mandates have evolved. The first focuses on the safe operation of the current generation of aircraft and is directed towards improved inspection techniques and improved structural analysis techniques for the design of aircraft prior to the retrofitment of superior parts. The second focuses on the evolution of viable techniques for designing, manufacturing, and monitoring the behavior of aircraft structures, while capitalizing upon the knowledge-base associated with the operation of the current fleet of aircraft.

The current knowledge-base may be considered as comprising four domains; design, manufacture, maintenance, and structural operating environment. The domains need to be evaluated in order to distill 'criteria' which adversely impact on the safety and efficiency of the current generation of ageing aircraft. Structural design technologies and structural analysis techniques are in a state of continual evolution as evidenced by the

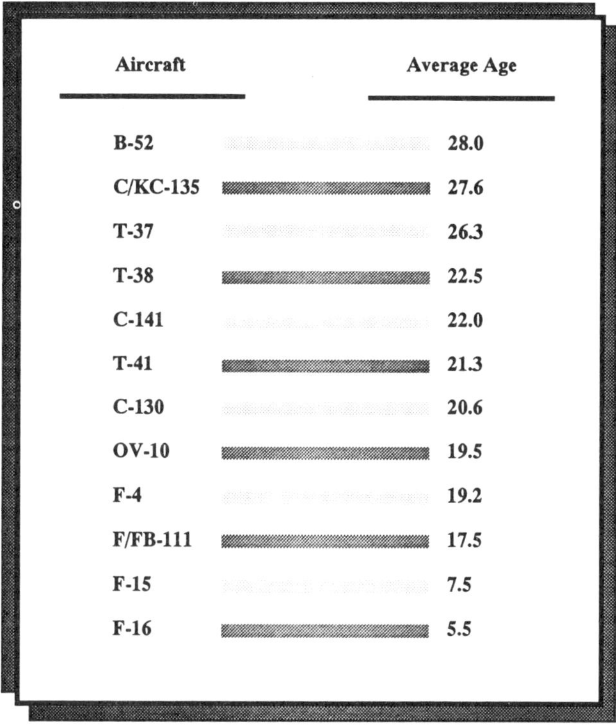

Aircraft	Average Age
B-52	28.0
C/KC-135	27.6
T-37	26.3
T-38	22.5
C-141	22.0
T-41	21.3
C-130	20.6
OV-10	19.5
F-4	19.2
F/FB-111	17.5
F-15	7.5
F-16	5.5

Fig. 3.37 United States Air Force ageing aircraft fleets.

deployment of both advanced fiberous polymeric composite materials and adhesive bonding technologies in aircraft, and also the utilization of sophisticated computer-aided design techniques featuring finite element methods. While there have been significant advances made in the automation and quality control of aircraft parts fabricated in monolithic materials, nevertheless, variations in properties do occur, and this situation has been further exacerbated by the deployment of aircraft fabricated in advanced composite materials. These advanced materials offer high stiffness-to-weight ratios and also high strength-to-weight ratios in these weight-sensitive structures, however, the fabrication of products in these modern materials involves a relatively embryonic field of manufacturing technologies where there is a greater degree of uncertainty in the properties of the final product.

Aircraft maintenance and inspection procedures introduce further parameters into the equation governing the structural integrity of modern aircraft because of the human element. For example, a tool dropped onto a laminated polymeric composite panel may result in delamination of the structure which is not detected by a visual inspection. Furthermore, aircraft must be withdrawn from service in order to perform these routine maintenance and inspection procedures, and this adversely affects the efficiency of operation of the fleet.

The environment in which an aircraft operates, has a major impact upon the fatigue life of the structure. Thus the accumulated structural damage is typically dependent upon the number of cabin pressurization cycles associated with each flight, the number and duration of flights in which violent air turbulence was encountered, operation on rough taxiways and the number of severe impactive landings, for example.

The effects of the structural operating environment and the inadequacies of the maintenance and inspection procedures upon the structural integrity of an aircraft, were recently highlighted by the catastrophic failure of an Aloha Airlines Boeing 737 jetliner on April 28, 1988 when the roof of the forward cabin was ripped off at 24 000 feet while flying from Hilo, Hawaii to Honolulu, Ohau. This nineteen year old plane was making its 89 681st flight. The extent of the damage is clearly shown in Figures 3.38 and 3.39 after the aircraft made an emergency landing on the island of Maui. This accident occurred as the result of the in-flight failure of skin panels in an area which had been the focus of revised Boeing maintenance and inspection procedures in the time period immediately prior to the accident.

Fig. 3.38 Photograph of the Aloha Boeing 737 accident, 28 April 1988.

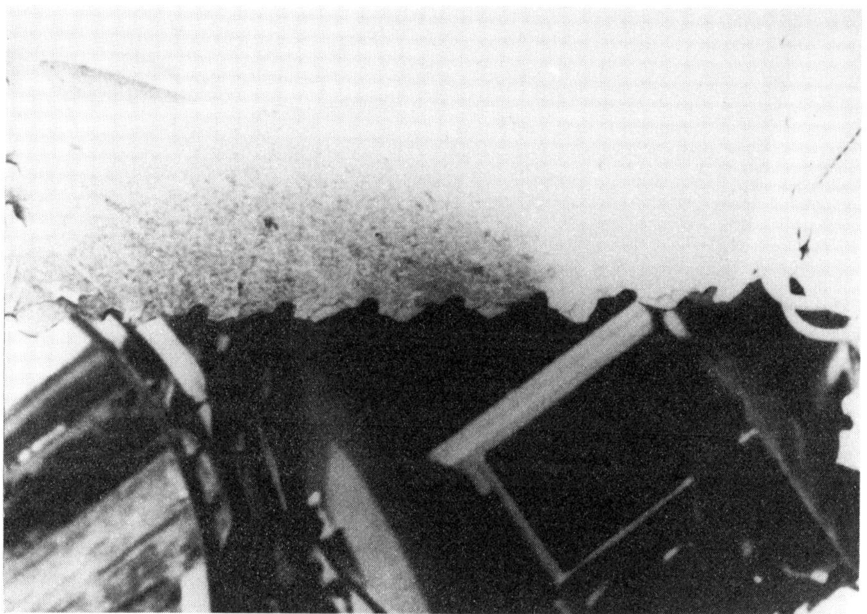

Fig. 3.39 Outer skin panel of the Boeing 737 showing the multiple failures.

Furthermore, the accident occurred despite the release of an FAA airworthiness directive which also was focused on the section of the fuselage in which the failure occurred.

Several key factors were responsible for this catastrophe. Firstly, the aircraft load profiles differed significantly from the norm for Boeing 737 airplanes because the aircraft operated on short flights between the Hawaiian islands. This activity resulted in numerous short-duration fuselage pressurization cycles which contributed to the fatigue failure. Secondly, significant damage had occurred but was not detected prior to the fatal flight. This was evidenced by the comments of a passenger who saw a seven-inch long crack in the fuselage prior to take-off, but did not report it to the airline personnel. Figure 3.40 schematically presents a philosophy for detecting crack propagation in aircraft structures featuring advanced fiberous composite materials, in which embedded optical fibers that sever at the crack tip prevent the light from being measured by the detector. Thirdly, the debonding of a skin panel lap joint resulted in the development of higher stresses at riveted joints, where multiple cracks were initiated from the countersunk rivet holes leading to the catastrophic failure of the fuselage skin.

Thus, it is clearly evident that despite the stringent FAA and Boeing directives focused on inspection and maintenance procedures that human error will continue to plague the safe operation of both commercial and

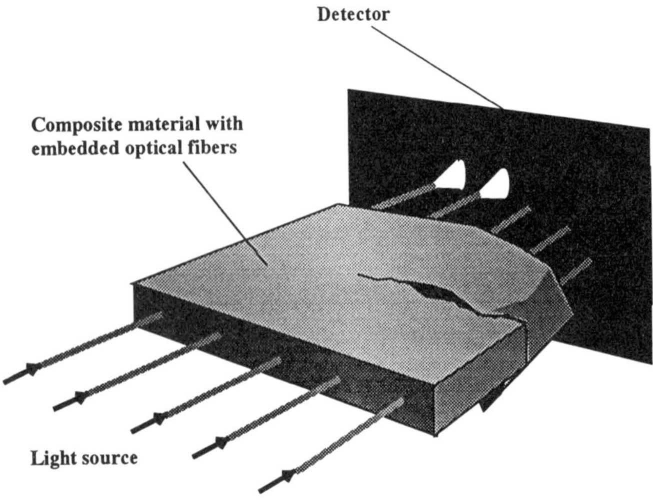

Fig. 3.40 An embedded optic fiber sensing system for damage detection. Embedded optical fibers break at the crack and prevent the light source from being received at the detector.

military aircraft. It is apparent, therefore, that aircraft should incorporate a higher degree of self-diagnostic capabilities in order that the airplanes can safely operate under abnormal conditions such as those experienced by the Aloha Boeing 737.

The notion of automated health monitoring, therefore, emerges as a potentially viable course of action for the next generation of aircraft. Structures would incorporate embedded sensors for providing continuous real-time assessment of structural behavior and integrity when employed in conjunction with artificial intelligence, signal processing, and appropriate computer hardware. Figure 3.41 illustrates a schematic for an aircraft utilizing these capabilities.

3.9.2 Health-monitoring philosophy

Passenger safety is a primary concern of the commercial airlines, but as evidenced by the catastrophic in-flight failure of the Aloha Boeing 737 aircraft, the continuous careful monitoring of an ageing fleet is unable to prevent accidents involving human error. Human error was also the principal factor in the build-up of ice on the wings of the Continental Airlines, Flight 1713, DC-9, aircraft which crashed on take-off from Stapleton International Airport in Denver during the winter of 1987. Similarly ice-build was also responsible for the 1982 crash of Air Florida Flight 90 in the Potomac River adjacent to Washington's National Airport.

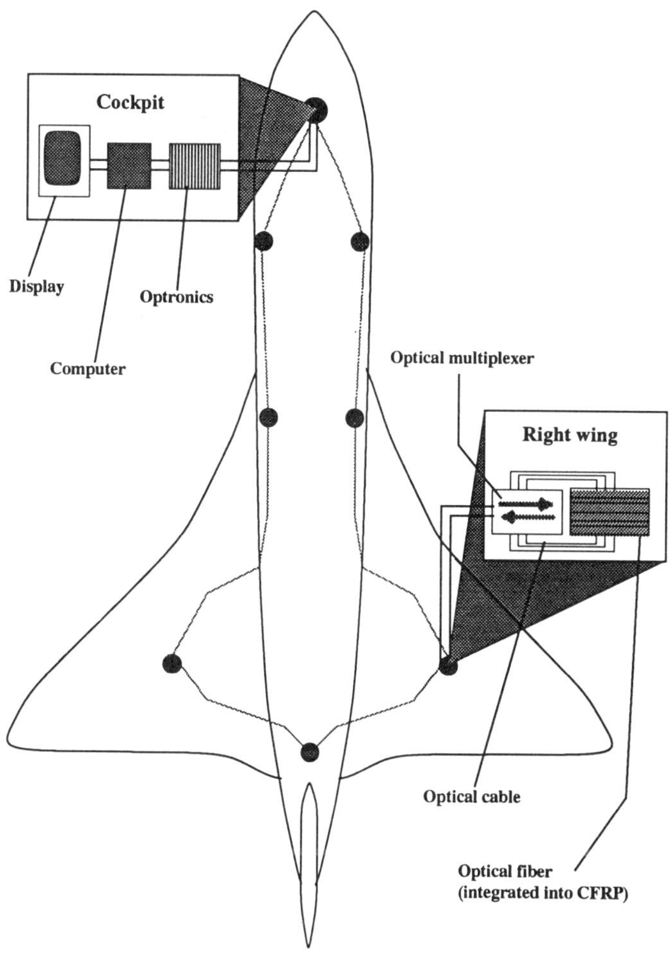

Fig. 3.41 In-flight structure surveillance would require fiber-optic nervous system (FONS) channels in key structural regions of a plane.

An aircraft health-monitoring philosophy would be directed towards preventing these accidents by developing aircraft with ‘smart skins’. These advanced structures would typically incorporate embedded sensors and computer networks which would monitor the flight loading, the environment, and the structural integrity prior to initiating appropriate corrective action as shown in Figure 3.41. This corrective action could be to simply advise the pilot that internal defects or structural failures have occurred in specific sections of the aircraft, or that the ice build-up on the wings exceeds a specified weight threshold, or adversely affects the aerodynamic performance of the aircraft, and the aircraft should return to the departure gate of the airport terminal for further de-icing.

Military aircraft could also feature these capabilities but with the relentless pursuit of superior performance, combat aircraft will feature a more sophisticated degree of health-monitoring capabilities in their arsenal of weapons as pictorially shown in Figure 3.42. A particularly attractive feature of the smart skins philosophy is that the technological features are embedded in the airframe structure without adversely changing the

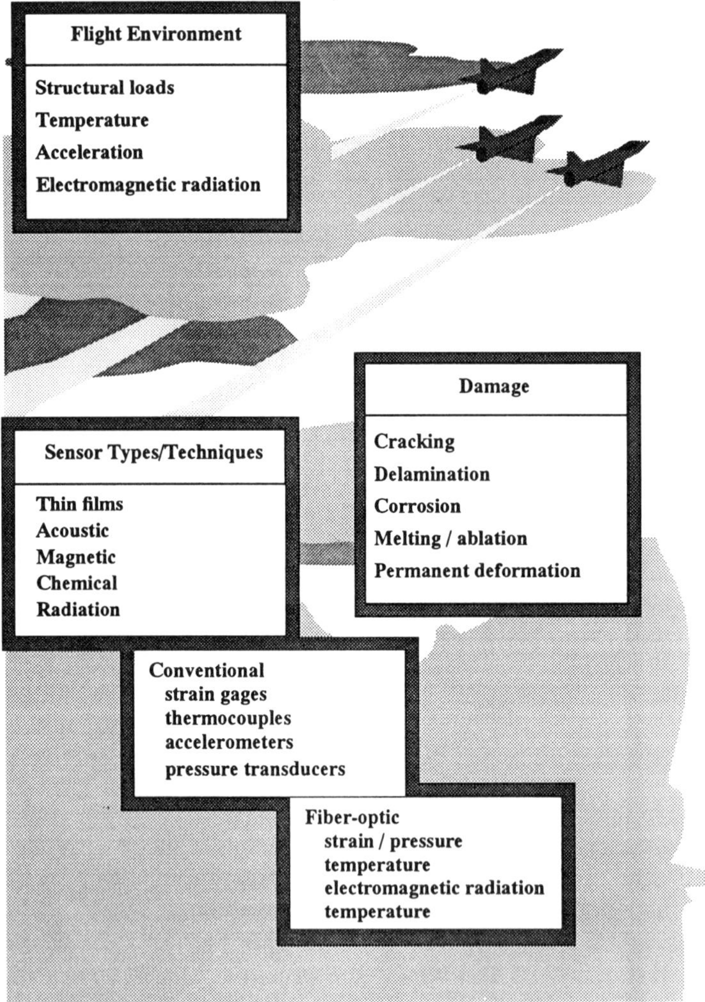

Fig. 3.42 Aerospace vehicle damage mechanisms and potential sensor types/techniques for smart vehicle applications.

aerodynamics of the airplane. This philosophy affords the avionics design engineer great freedom to utilize a larger percentage of the aircraft in contrast to the current philosophy of installing avionics packages in equipment racks in the fuselage. The health-monitoring philosophy is directed towards increasing aircraft performance through the integration of sensors, micro processors, data transmission links and flight control systems to provide a viable dynamic interface with the external flight environment. The sensor arrays will emulate the nervous system of a living organism by monitoring critical locations to detect loading, structural integrity and battlefield damage. The data will be analyzed and employed to ascertain structural integrity, for example, by comparing the current characteristics with the base-line data for the aircraft when it rolled off the assembly line in the factory. This base-line data, at the birth of the product, would be stored in the aircraft computer and employed in a health-monitoring role in addition to predicting the onset of failure, or death of the product.

The smart skins philosophy will involve the embedment of diverse types of sensors in a large portion of the aircraft surfaces which will result in superior radar-coverage to conventional vehicles since sensing and communication at many frequencies will be possible. The anticipated enhancements include improved target identification, superior stealth capabilities and enhanced communication facilities. The kernel of all these cutting-edge technologies is a fundamental understanding of smart materials and structures technologies.

3.10 Aeroelastic tailoring of airfoils

Powered high-lift systems such as rotorcraft and other VTOL (vertical take-off and landing) aircraft cannot generate sufficient wing circulation to facilitate take-off, consequently they rely on high ratios of engine thrust to aircraft weight to augment the basic wing lift. In addition to high engine power, however, significant lift augmentation must be obtained by specialized wing design; and this is usually implemented either through the use of boundary-layer control or by the action of a trailing-edge flap. Several derivative designs have evolved from the basic jet flap illustrated in Figure 3.43(a). Figure 3.43(b) presents a plot of the variations of coefficient of lift, C_L, for the basic jet flap design, for different values of the angle of attack, δ, and the angle of the trailing-edge flap, θ. The deflected flowstream along the flap was found through flow visualization techniques in order to induce an extensive separation bubble on the top surface of the wing, thereby enhancing circulation. The flow visualization produced by the jet flap is depicted in Figure 3.44. It is sufficient to note here that in level flight the flap would of course be realigned to merge with the leading edge of the wing. The contradictory requirements of high lift and low drag which arise

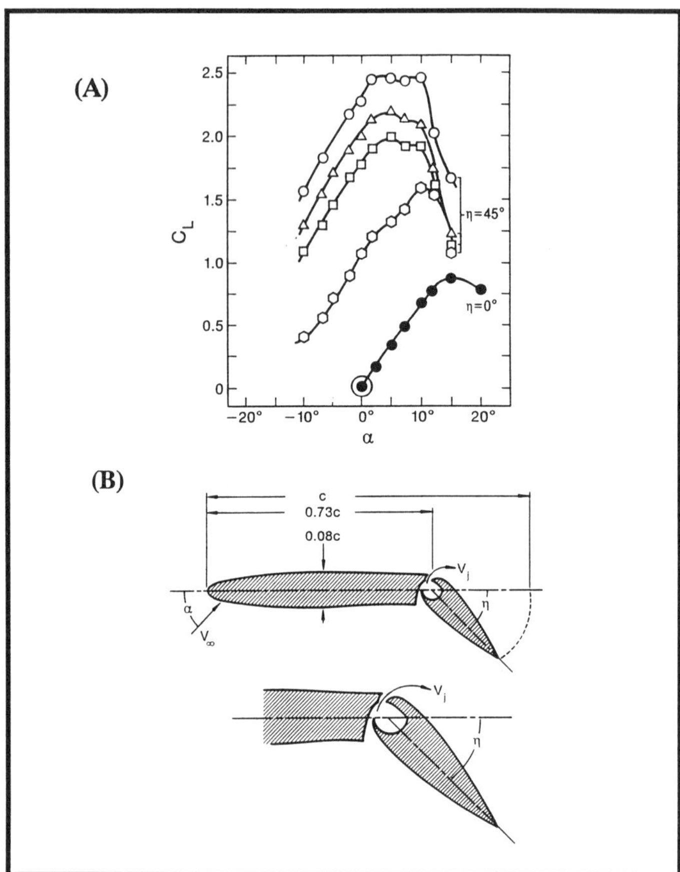

Fig. 3.43 (a) The variations of the coefficient of lift for different values of the angle of attack and the angle of the trailing-edge flap.
(b) A basic design of the jet flap showing the lift augmentation obtained by depressing the trailing-edge flap.

during different stages of flight can be resolved by adjusting the effective camber of the wing. Accomplishing this task will require the development of unique and innovative airfoil designs featuring the appropriate smart materials.

A more subtle adjustment of camber is possible in a lower Reynolds number regime ($\simeq 10^6$) that is more characteristic of ultra-light aircraft. A comparison of this lower Reynolds number regime for ultra-light aircraft is illustrated in the Reynolds number spectrum of flight presented in Figure 3.45. An example of an airfoil which was designed for an ultra-light sailplane of weight 120 kg and 11.5 m span is shown in Figure 3.46.

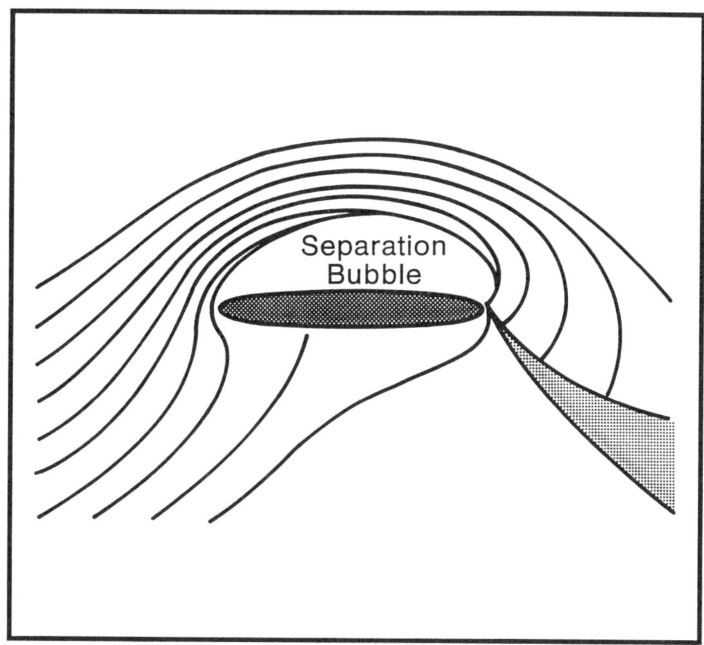

Fig. 3.44 Flow visualization of the separation bubble produced by a jet flap.

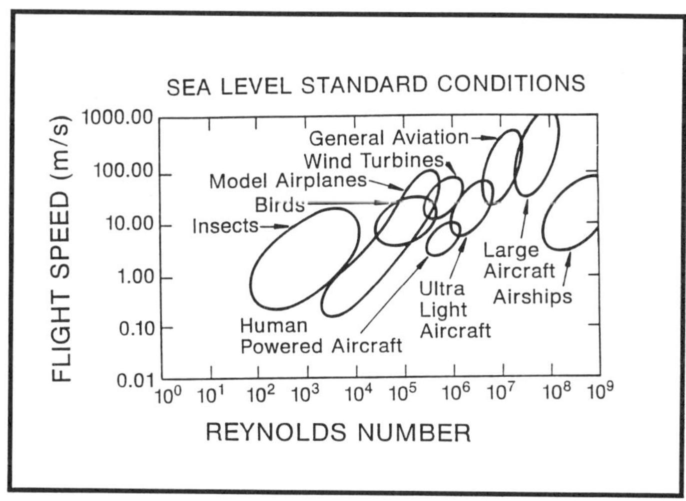

Fig. 3.45 A Reynolds number-spectrum of flight for various aircraft, airships, and animals.

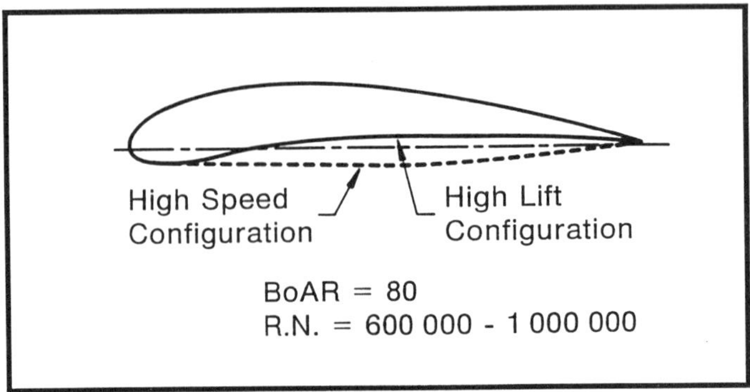

Fig. 3.46 A BoAR 80 airfoil with an adjustable camber.

Sufficient lift could be obtained with this airfoil section only by prescribing an undercamber of the lower surface of the airfoil, but this shape would induce flow separation near the leading edge during high-speed level flight. Thus the full flight-range requirements could not be met by the highly cambered shape needed for high lift. Low drag could be achieved only by dropping the lower surface to increase the wing thickness and thereby prevent flow separation. This airfoil is intended to change the configuration during flight, activated by command much as by a flap. Clearly, considerable fine-tuning and substantial improvement in the aerodynamic performance of cambered airfoils is possible if smart materials are employed to effect this change in shape and hence, the change in boundary-layer characteristics.

During prototypical operation, airfoils either encounter or generate unsteady flow at least part of the time. One aim of the aerodynamicist, therefore, is to reduce the undesirable consequences associated with these situations such as flutter, vibration, buffeting, gust response and dynamic stall. Smart materials could play a crucial role in these situations by actively changing the geometrical characteristic of the airfoil in addition to the stiffness and energy dissipation properties. For low Mach-number applications, incompressibility simplifies matters considerably; then thin-airfoil linear theory, suitably modified by superimposition of individual corrections for camber, angle of attack and large-amplitude deflections, is often employed to evaluate the effect of oscillating surfaces on lift and drag. Second-order perturbation analyses have been employed to investigate the interaction between thick-cambered airfoils and gust fields. Certain ranges of flow conditions, however, cause flow separation leading to effects such as buffeting, flutter and dynamic stall, all of which exhibit large-scale hysteresis loops. When separation is significant, not even the qualitative behavior of the coefficient of lift, C_L, or the coefficient of drag, C_D, at high angle of attack can be reproduced by neglecting the unsteady motion of the

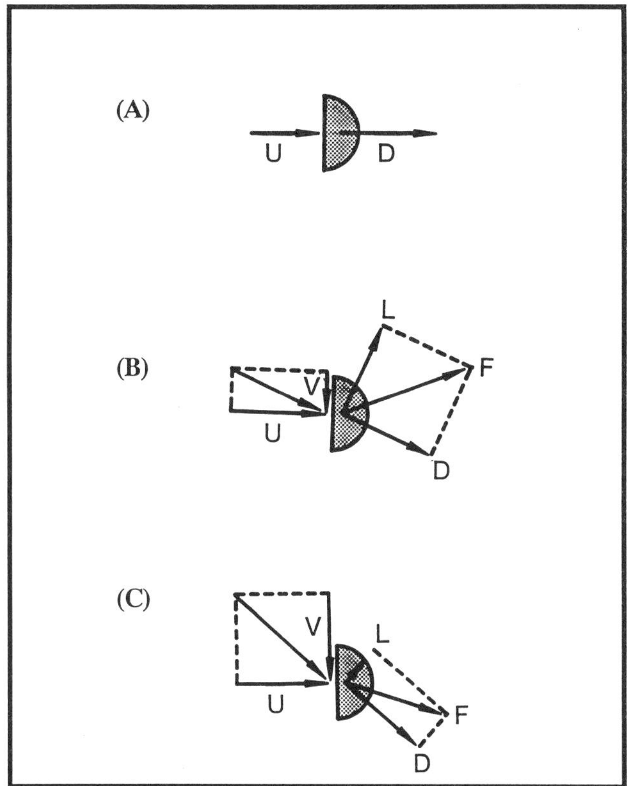

Fig. 3.47 A demonstration of auto-rotation on a D-shaped blade of a Lanchester propeller.

airfoil. This regime of flow-induced vibrations therefore remains largely within the purview of experimentalists.

Autorotation comprises a second category of non-steady flow phenomena pertinent to airfoil design. This phenomenon is exemplified by the Lanchester propeller, illustrated in Figures 3.47(a), (b), and (c) in simplified form. Any slow rotation produces an opposing force, F, that tends to reduce the motion of the propeller, as shown in Figure 3.47(b). If, however, the initial rotation, V, is sufficiently high, the net force may reinforce this motion as shown in Figure 3.47(c).

Under these conditions, the propeller begins to autorotate, and continues to accelerate until a stable equilibrium point is reached, thus, further acceleration reduces the external torque on the blade. This phenomenon is depicted in the 'Riabouchinsky Curve' shown in Figure 3.48. The effect of the Reynolds number on the lift to drag ratios, C_L/C_D, for different surface characteristics of airfoils is illustrated in Figure 3.49. The authors are

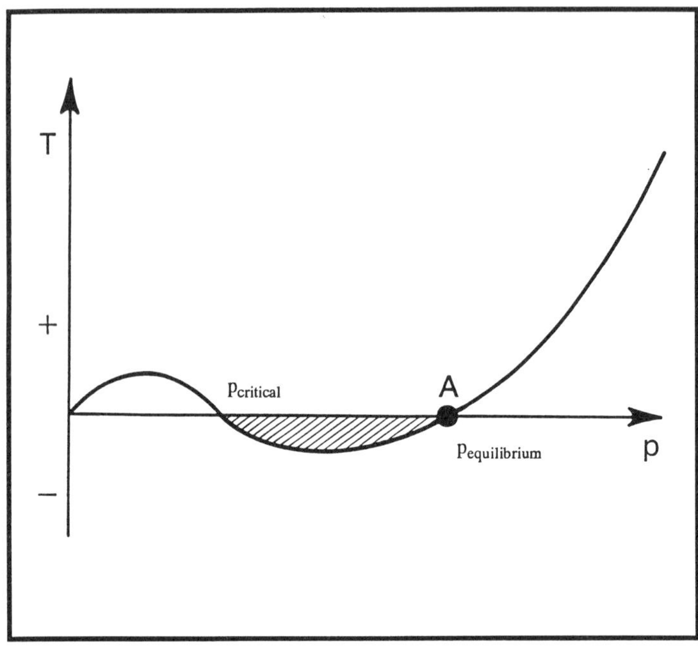

Fig. 3.48 A 'Riabouchinsky Curve' which describes the relation between the initial spin, $p_{critical}$, and the final spin, $p_{equilibrium}$, on the torque, T, induced on the propeller motion.

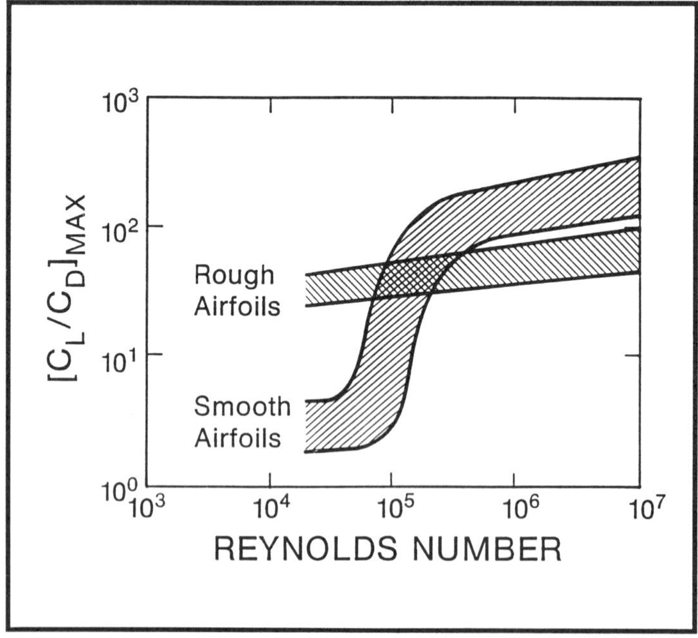

Fig. 3.49 Low Reynolds number airfoil performance.

convinced that significant enhancements in the aerodynamic performance of rotorcraft and aircraft systems will be accomplished by judiciously employing smart materials.

3.11 Synthesis of future smart systems

The future of smart structures will clearly be influenced by advances in such areas as materials science, materials design and manufacture, biomimetics, artificial intelligence, nanotechnology, and neural networks. Consider, for example, the vibrational characteristics of combat helicopter rotors fabricated of advanced composite materials. A present-day optimally designed rotor fabricated of advanced composite materials is passive in the sense that it cannot actively respond to external stimuli in an autonomous fashion. It is clearly evident, therefore, that the vibrational response of the rotor is not optimal for all service conditions except the one for which the rotor was 'optimally designed.' However, if the rotors were fabricated in smart composite materials, their performances could be dynamically tuned to ensure optimal operation under all service conditions and in all unstructured environments.

A helicopter rotor designed and fabricated of smart materials would typically be capable of in-flight structural surveillance, which might include detecting dynamic strains due to changes in payload, in wind gusts, in temperature, and in moisture as well as those caused by battlefield damage. The detection and measurement of changes in the vibrational response characteristics of the rotor, changes in the environment, and the qualitative and quantitative assessment of damage would be undertaken by the sensing network. On the basis of this sensing information, smart structures of the future will be able to initiate appropriate corrective actions almost instantaneously. In a hostile battlefield environment, for example, these activities might involve redistributing loads around highly stressed regions of the rotor structure in order to control the damage that it might sustain.

Another layer of smartness would permit the rotors to 'learn' from past experiences and utilize this knowledge in making decisions for future situations. A more sophisticated degree of smartness associated with the rotor would allow the structure to self-diagnose and assess the extent and quality of damage and then make decisions on the abortion or survivability of the mission. Such a rotor could be fabricated from a smart material that could heal and repair itself, thereby compensating for any battlefield damage that might occur. Alternatively, if an automated assessment of the damage revealed that the mission should be aborted, then the smart structure would have the capability of self-degradation. While this discussion has focused on helicopter rotor systems, clearly these capabilities could also have a profound effect on a diverse range of products in the automotive, defense, construction, and advanced manufacturing industries.

The capabilities that will be demanded of smart structures in the future will require substantial advances in various technologies. Advances in nanotechnology would, for example, permit the design and fabrication of structures with embedded sensors, actuators, and processors to be undertaken in a manner that did not compromise structural integrity. This technology would for the first time allow designers the freedom to manipulate individual atoms and molecules in order to fabricate complex structures one atom at a time. Similarly, advances in biomimetics would allow materials and structures to be engineered with self-repair and self-healing capabilities. These structures would be tailored to exhibit substantial toughness and would incorporate redundancy in order to avoid catastrophic failures. Such capabilities in material design and fabrication would render obsolete conventional engineering philosophies, that relied on substantial overdesign in order to ensure system integrity under a variety of hostile conditions. Advances in neural networks would facilitate the synthesis of materials and structures with exceptional learning capabilities. These smart systems would capitalize on the architecture of a large number of relatively simple processors connected to one another by variable-memory elements. The neural networks would adjust the weights of these elements by experience in order to ensure self-programming capabilities.

4

Electro-rheological fluids

The current philosophy of synthesizing smart materials featuring actuators, sensors, and microprocessors is to embed these subsystems into discrete regions of a structural material such as a fibrous polymeric composite laminate. Thus, for example, the actuator can be a solid, liquid, or gas which, upon activation, changes its characteristics and hence the global characteristics of the smart structure in the neighborhood of the actuator material. The designer has the liberty to engineer the material and/or structure at a macromechanical level by selecting the size, shape and spatial distribution of the actuators throughout the host structural material, in addition to the number of actuators to be activated. Since the actuation mechanism typically involves electrical or magnetic phenomena, there are a wide variety of potential actuator candidates. One of these classes of candidate materials is electro-rheological fluids whose properties are dependent upon the characteristics of the electrical field imposed upon the fluid.

Electro-rheological (ER) fluids belong to a class of colloidal suspensions whose global characteristics can be controlled by the imposition of an appropriate external electrical field upon the fluid domain. The traditional focus of research in this field has been to control the rheological properties of the fluid for discrete hydraulic applications. Other field-effect phenomena include electro-hydrodynamics and also magneto-hydrodynamics, where the focus of attention in these latter disciplines is the interaction between liquids or gases, and electromagnetic fields.

4.1 Suspensions and electro-rheological fluids

When a material, the solute, dissolves in another material, the solvent, to form a *bona-fide* solution, the solute and solvent molecules are generally of comparable size and they are typically distributed uniformly throughout the solution. When these conditions are not satisfied, and the size of the solute particles are much greater than the size of the molecules of the solvent then the system is termed a *colloidal dispersion*. These dispersions have diverse characteristics and they are naturally the subject of study in the field of colloid science, which has, during the past forty years or so, matured from an experiment-orientated discipline to a field which is now supported by rigorous theoretical tools. This theoretical foundation has been established through the efforts of individuals in the diverse subfields of electrostatics, hydrodynamics, polymer science, rheology, synthetic organic chemistry, surface chemistry, statistical mechanics, and thermodynamics, for example.

Colloidal dispersions have a variety of subgroups. These subgroups include *sols* which are dispersions of solids in solids, or solids in liquids; *emulsions* which are dispersions of liquids in liquids; *aerosols* which are dispersions of liquids in gases, or solids in gases; and *foams* which are dispersions of gases in liquids, or gases in solids. ER fluids belong to the first category, but dispersions in the other classifications are also potential candidates for smart material actuator systems.

Colloids may be further classified based on a criterion enunciated in a classical text by *Freundlich*. The criterion involves evaluating whether the colloidal dispersion exhibits solvent attracting characteristics or solvent repelling characteristics. These dispersions are generally termed respectively *lyophilic* and *lyophobic*, but when the dispersing medium is water then the appropriate terms are *hydrophilic* and *hydrophobic* respectively.

A crucial component of any colloidal system is the very large surface area of the micron-sized particulate dispersed phase relative to the surface area of the dispersed phase reconstituted as a single piece of bulk material. Consider, for example, a one centimeter cube of a homogeneous solid material which has a surface area of six square centimeters. If this cube of bulk material is now mechanically broken down into small cubes each with a side of length ten nanometers, for example, then 10^{18} cubes will have been created and their total surface area will be an astonishing six million square centimeters. This example serves not only to demonstrate the relevance of interfacial or surface chemistry to the field of colloid science, but it also motivates comments on the stability of colloidal dispersions.

A lyophilic colloidal suspension is thermodynamically stable because, relative to the bulk material, there is a reduction in the Gibbs free energy function when the solute is mixed with the continuous phase. On the other hand, a lyophobic colloidal suspension is thermodynamically unstable because the Gibbs free-energy function increases when the solute is mixed with the continuous phase. The Gibbs function is a minimum when the dispersed phase is configured in a single lump. Thus a lyophobic colloidal dispersion must feature particles whose surfaces have been treated to cause particulate repulsion. This treatment prevents the coagulation of particles over extended time periods.

A major source of kinetic stability of colloidal particles is the existence of surface electrical charges. These charges are responsible for attracting ions of opposite charges to cluster in the neighborhood thereby forming an ionic atmosphere. This domain comprises two sub-domains of opposite charge, and is frequently termed the electric double layer.

Another important group of substances associated with the stability of colloidal suspensions are the *amphiphiles*. These are molecules comprising two discrete regions: one is oil soluble, sometimes lipophilic or hydrophilic, while the other is water-soluble or hydrophilic. These amphiphiles are strongly surface active because they lower the interfacial tension thereby creating new surfaces easily, and these surfactants are frequently employed as detergents and dispersing agents. Amphiphiles play a critical role in the stability of some classes of ER fluids.

ER fluids are a class of colloidal dispersions which exhibit large reversible changes in their rheological behavior when subjected to external electrical fields. These changes in rheological behavior are typically manifest by the pseudo phase-change of an ER fluid from a liquid state to a solid state, or else a dramatic increase in flow resistance depending upon the flow regime and also the composition of the ER fluid. An ER fluid often has Newtonian characteristics in the absence of an external electric field which become non-Newtonian when the electrical field is imposed upon the fluid domain.

Figure 4.1 graphically presents a simple Bingham-body model for the isothermal constitutive behavior of a typical ER fluid. In the absence of an electrical field the fluid behaves as a Newtonian fluid where a linear

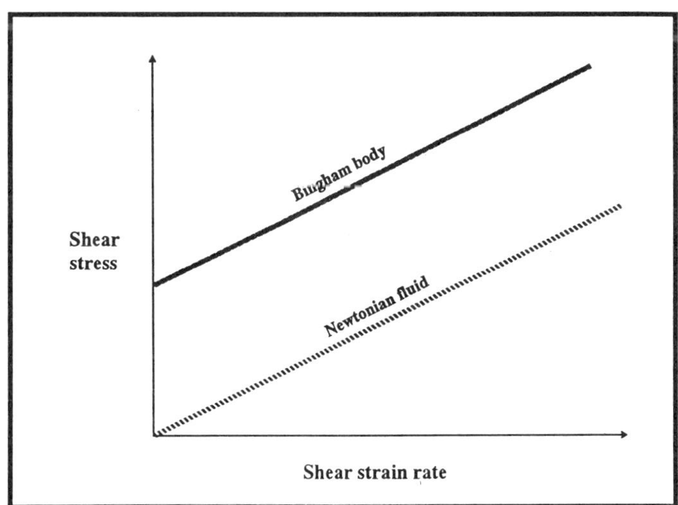

Fig. 4.1 A Bingham-body model for the isothermal constitutive behavior of a typical ER fluid.

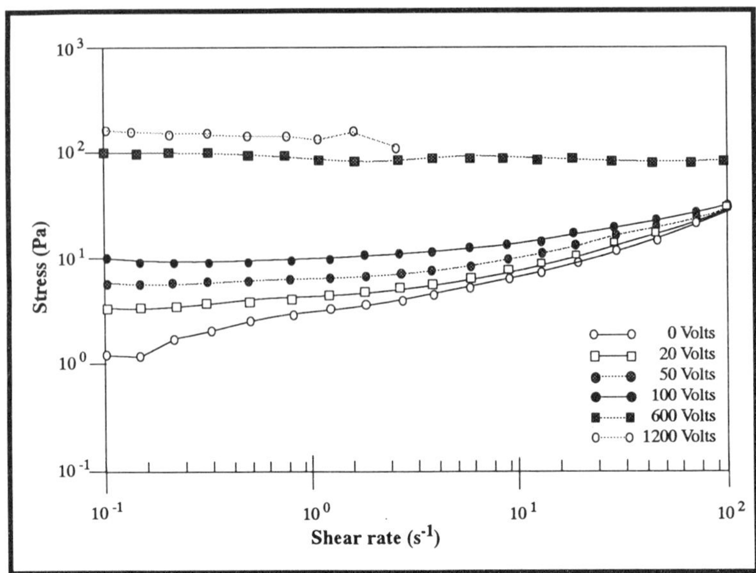

Fig. 4.2 Experimentally determined stress versus shear rate characteristics measured at different voltages for a hydrous electro-rheological fluid comprising 45% corn starch and 55% silicone oil.

relationship exists between the shear stress and the rate of strain. Upon applying an electrical field the ER fluid exhibits a static stress and this shear stress must be overcome prior to initiating fluid flow.

Figures 4.2, 4.3, and 4.4 are log-log graphical presentations of the rheological characteristics of an ER fluid of simple composition comprising 45% corn starch and 55% silicone oil by weight. These experimental results were obtained using a Rheometrics Incorporated RMS-800 mechanical spectrometer which had been upgraded with a high-voltage attachment which could impose a potential difference of 5 kV across the double-walled couette test fixture containing the ER fluid sample under examination. This test fixture comprised two concentric tubes with a radial separation of one millimeter between the two faces which accommodated the fluid sample. The rheological results were obtained by maintaining a prescribed constant potential difference across the test fixture, while varying the strain rate over a prescribed range before selecting a different voltage state and hence different electrical field intensity.

The Bingham-body model, featuring a linear relationship between the shear rate and the shear stress, is clearly demonstrated in Figure 4.2. Furthermore, the increase in the static stress as a function of increasing voltage is also evident from the experimental results. This ability to control the shear stress as a function of the intensity of the electrical field imposed

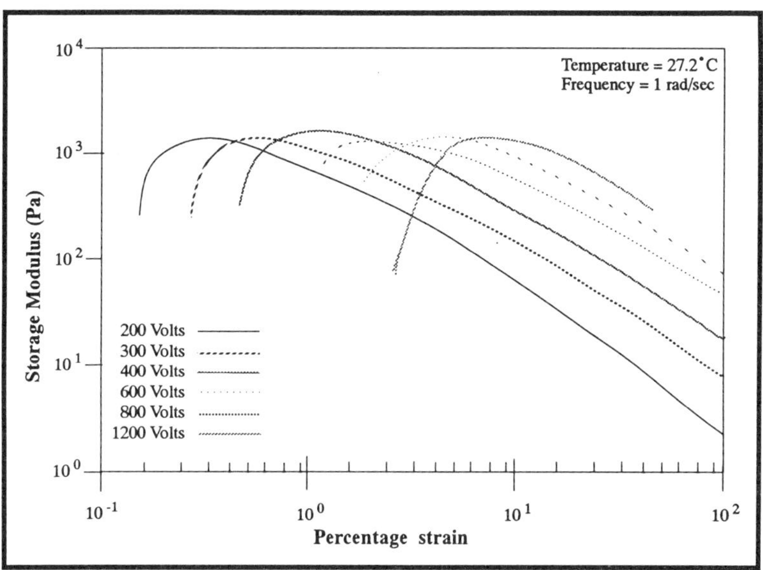

Fig. 4.3 Experimentally determined storage modulus versus percentage strain characteristics measured at different voltages for a hydrous electro-rheological fluid comprising 45% corn starch and 55% silicone oil.

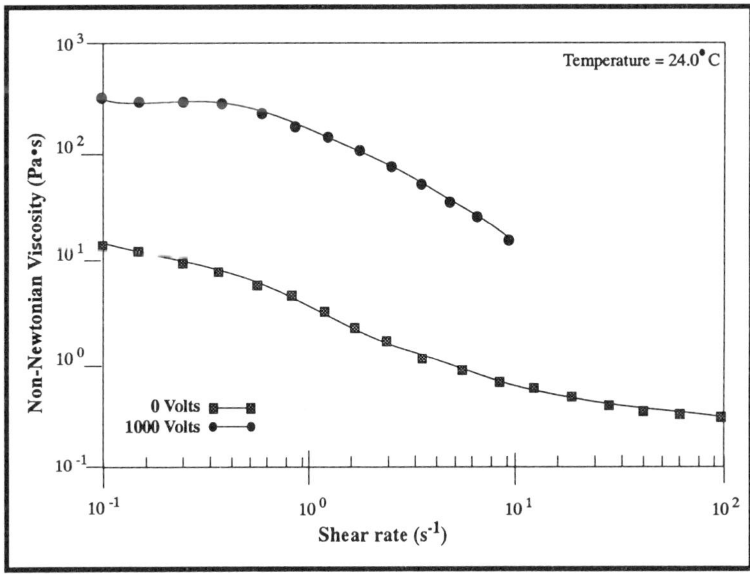

Fig. 4.4 Experimentally determined non-Newtonian viscosity versus shear rate characteristics measured at different voltages for a hydrous electro-rheological fluid comprising 45% corn starch and 55% silicone oil.

upon the fluid domain and also the strain rate are important variables in the design of ER fluid actuators for smart structural materials applications.

Figure 4.3 presents experimental data demonstrating how the storage modulus of the ER fluid sample, which is a measure of the energy-dissipation properties of the suspension, is dependent upon the strain rate and also the applied voltage. Thus by changing these properties appropriately, the system damping can be actively controlled. This is a desirable characteristic in the context of actively controlling the elastodynamic response of smart structures. It is evident from Figure 4.3 that for field intensities between 0.2 kV/mm to 1.2 kV/mm that the storage modulus assumes a value of typically 200 Pa at low strain rates, prior to proceeding through a maximum value of 1000 Pa as the strain rate is monotonically increased, whereupon it continually decreases with any further increase in strain. Furthermore, it is also evident from the experimental results that the profile of the characteristic relating the storage modulus to the strain is of the same shape, but the critical strain increases with increasing voltage.

It is important to note, with reference to the experimental results presented in Figure 4.3, that the maximum displacement of a couette fixture is assumed to be one complete rotation of the outer cup from the initial position. All other strains are expressed as a percentage of this maximum value of strain. Thus the abscissae in Figure 4.3 has units of percentage strain and these data were obtained at a frequency of 1.0 rad/s.

Figure 4.4 presents in a log-log graphical form the relationship between non-Newtonian viscosity and strain rate for the ER fluid comprising corn starch and silicone oil subjected to discrete constant field strengths of 1 kV/mm and 0 kV/mm at a temperature of 24.0°C. When an electrical field is not imposed upon the fluid sample in the couette fixture, the viscosity monotonically decreases as the shear rate increases, thereby exhibiting shear thinning. However, when the ER fluid is subjected to an electrical field intensity of 1 kV/mm at shear rates less than approximately 0.4 per second, the suspension exhibits a Newtonian-like behavior with a viscosity of 300 Pa.s when the shear rate is zero. When the shear rate is greater than 0.4 per second, there is a monotonic decrease in the non-Newtonian viscosity with increasing shear rate.

Furthermore, the non-Newtonian viscosity is significantly greater for the colloidal suspension in the presence of the electrical field than for the fluid in the absence of the electrical field for a prescribed shear rate. For example, when the shear rate is zero the viscosity at 0 kV/mm is 10.5 Pa.s while at 1 kV/mm the viscosity is 300 Pa.s. This significant change in the viscous characteristics of ER fluids as a function of the applied electrical field, has been responsible for innovationally employing these suspensions in a discrete form to a number of engineering applications.

A several orders of magnitude increase in viscosity occurs when electric field strengths of the order of 1–5 kV/mm are imposed upon ER fluids.

These field strengths create particle fibration where the random structure of the suspension shown in Figure 4.5 is transformed into the columnar, chainlike structures parallel to the electrical field generated by the electrodes as shown in Figure 4.6. The dispersion is typically dielectric with non-conducting properties, and until the recent development of anhydrous ER fluids by Block and Kelly, and also Filisko and Armstrong, water was generally added in small amounts.

Table 4.1, which is based upon data presented in a comprehensive survey paper by Block and Kelly, indicates the great diversity of ingredients

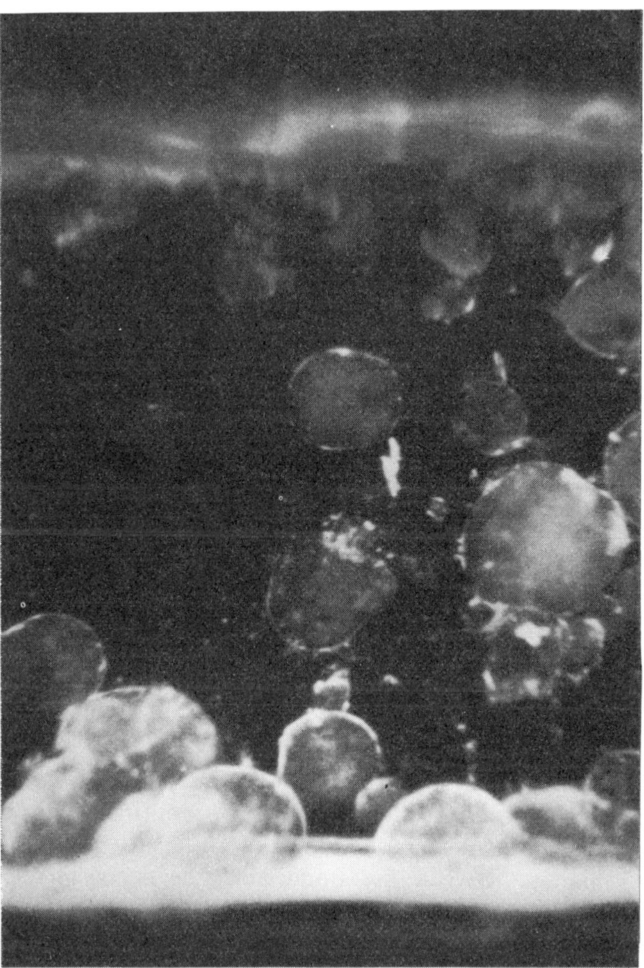

Fig. 4.5 Photomicrograph of the random structure of an electro-rheological fluid, where there is no difference in electrical potential between the electrodes at the upper and lower extremes of the photograph.

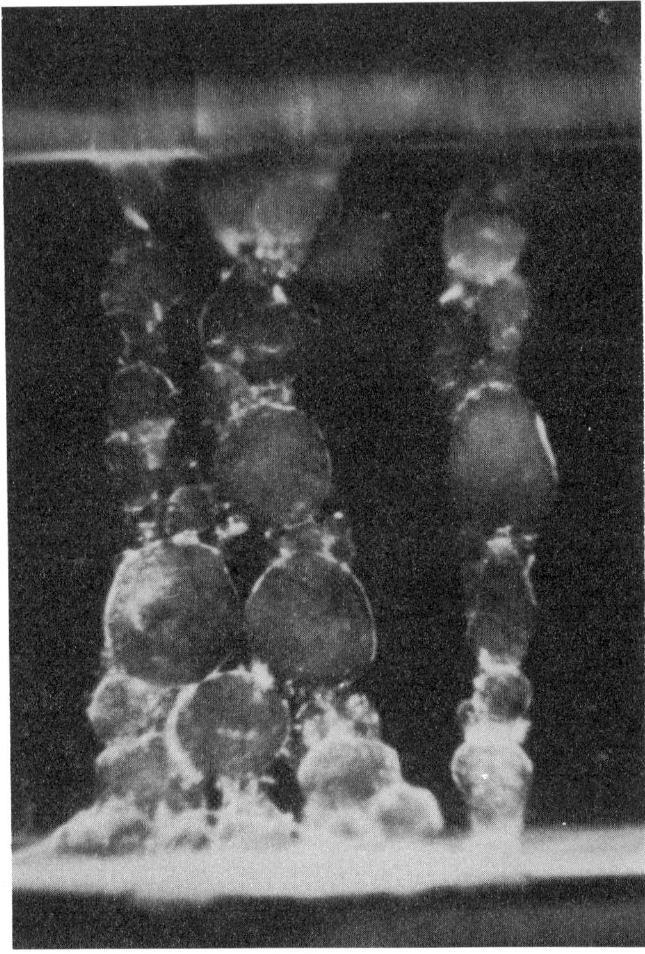

Fig. 4.6 Photomicrograph of the columnar structure of an electro-rheological fluid, when an electrical field of strength of 2 kv/mm is generated between the electrodes at the upper and lower extremes of the photograph.

associated with the synthesis of ER fluids. This diversity is clearly evident in the solvent category and also the solute category but the additive category has a thread requiring small quantities of water to be present in the dispersion. This characteristic has traditionally been considered to be an essential ingredient for the electro-rheological phenomenon. However, recent work on a new generation of fluids by Block and Kelly, and also Filisko and Armstrong, has suggested that this may no longer be true. Nevertheless, Block and Kelly suggest that there are at least seven criteria which must be considered when developing ER fluids, and these are presented in Table 4.2.

Table 4.1 *Typical ingredients for electro-rheological fluids*

Solute	Solvent	Additive
Kerosene	Silica	Water and detergents
Silicone oils	Sodium carboxymethyl cellulose	Water
Olive oil	Gelatine	none
Mineral oil	Aluminum dihydrogen	Water
Transformer oil	Carbon	Water
Dibutyl sebacate	Iron oxide	Water and surfactant
Mineral oil	Lime	none
P-xylene	Piezoceramic	Water and glycerol oleates
Silicone oil	Copper Phthalocyanine	none
Transformer oil	Starch	none
Polychlorinated biphenyls	Sulphopropyl dextran	Water and sorbitan
Hydrocarbon oil	Zeolite	none

The relationship between the shear stress that the fluid can sustain and the applied electrical field is clearly important in the design of any ER-based device or smart structure. Generally a high shear stress is desired for a relatively low field intensity since this constitutive relationship largely governs the power consumption of the system for a prescribed voltage, electrode separation and electrode surface area in contact with the fluid.

A second major parameter affecting the power consumption and the heat dissipation of the system is the electrical conductivity of the fluid. It is desirable for heat dissipation effects to be as low as possible in order to avoid the necessity to design and develop ancillary cooling systems. Furthermore, the ER fluid should be able to operate successfully over a

Table 4.2 *Principal characteristics of electro-rheological fluids.*

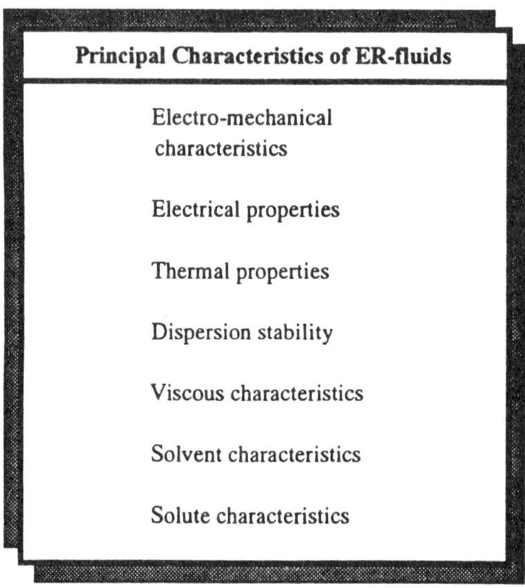

Principal Characteristics of ER-fluids
Electro-mechanical characteristics
Electrical properties
Thermal properties
Dispersion stability
Viscous characteristics
Solvent characteristics
Solute characteristics

broad temperature range. The recent emergence of the new generation of anhydrous fluids has dramatically changed the range of operating temperatures over which ER fluids can be deployed in practice, since the hydrous systems were restricted to operating at ambient temperatures ranging from −20°C to +70°C. The limitations of this thermal domain were attributed to water loss at the higher temperatures, which reduces the potency of the fluid, and at the lower temperatures the ER phenomenon is less evident.

The long term stability of ER fluids in the context of aggregation is an important consideration when developing fluids for commercial applications where ER devices may not be activated for several months. Thus it is advisable to ensure that the density of the particulate phase and the density of the continuous phase be the same in order to minimize sedimentation problems.

It is generally desirable for ER fluids to be characterized by a low viscosity in the absence of an external electrical field and be characterized by a high viscosity when an electrical field is imposed upon the fluid. In the context of the synthesis of ER fluids, a low viscosity generally implies a low particulate volume fraction which necessitates the imposition of higher electrical fields and, hence, greater power requirements, in order to effectively utilize the fluid in practice.

The dispersant employed in ER fluids should have a low viscosity and low volatility while being non-toxic, non-corrosive, and non-flammable. This latter characteristic is clearly important in high voltage applications where the possibility of arcing is always present.

The dispersed phase of an ER fluid should possess the necessary electrical attributes, in addition to being easily atomized from the bulk state, easily dispersed with minimal use of additives, and the solute should be non-abrasive. These heterogeneous dispersions can be considered as comprising four ingredients: the continuous medium or solvent, the particulate medium or solute, organic activator compounds and non-ionic surfactants.

The continuous medium, or solvent, is typically a low viscosity liquid, such as paraffin, silicone oil, or chlorinated hydrocarbons, with viscosities in the range of 0.01 to 10 Pa-s, dielectric constants in the range of 2 to 15, and a high electrical resistivity ranging typically from 1016 to 1010 ohm/meter.

The particulate material of the dispersion includes clays with interstitial moisture, such as kaolimite and diatomite, polysaccharides and silica. The dispersed phase generally comprises solid, semiconducting, or nonconducting materials with dielectric constants in the range of 2 to 40, with particle diameters ranging from one =m to 100 =m, and possessing a surface area of typically 400 m^2/gm. The above particles are generally adsorbents with activator-coated surfaces. These surfaces feature organic compounds which are typically capable of creating hydroxyl-containing materials, aromatic or various amines.

Surfactants are added to dispersions to achieve high concentrations of the dispersed phase and to prevent sedimentation of the dispersed phase. The amount of surfactant added to an ER fluid is quite small, generally being of the order of one to three molecules per square micrometer of the particle surface area.

4.2 The electro-rheological phenomenon

In 1947 Willis Winslow patented a clutch mechanism which exploited the electro-rheological properties of colloidal dispersions of starch or silica gel in mineral oils. This disclosure, and a journal publication by Winslow in 1949 are widely cited in the literature as being pioneering publications, but the electro-rheological phenomenon was previously reported by Koenig in 1885, Duff in 1896, and Quinke in 1897.

Experimental studies of the ER phenomenon have clearly shown that upon applying an external field to a static, non-flowing ER fluid, the random structure of the suspension shown in Figure 4.5 dramatically changes, as the particles rapidly orientate themselves in relatively regular chain-like columnar structures as shown in Figure 4.6. These photomicrographs were obtained using a microscope slide with parallel-sided electrode attachments to impose the electrical field upon the stationary fluid located between the two electrodes. It is evident from Figure 4.6 that the particles align themselves parallel to the electric field, which is in the vertical direction for the two straight, parallel-sided electrodes as evidenced by the black regions at the top and bottom of the photomicrographs.

Figure 4.6 clearly indicates that the global mechanical properties of the suspension are somewhat anisotropic, there has been a local reorganization of the mass distribution, and these columnar structures, if they occur under flow conditions, provide some insight into the change in rheological behavior of the fluid. Experimental work in this area of fluid flow when the ER fluid is subjected to the electrical field is still to be undertaken.

While experimental programs of research on ER fluids have attained some degree of success, there is a distinct dearth of viable theoretical models for predicting the response of this class of colloidal dispersions subjected to coupled field problems involving both fluid mechanics and electromagnetics. While researchers are debating the postulates and assumptions associated with the development of theoretical models for describing the global ER phenomena, they are, nevertheless, in general agreement on the initial conditions. These conditions are that electric dipoles are induced in the particulate phase as a consequence of imposing the electrical field on the dispersion, so that there is interaction between the individual particles and they form columnar structures as schematically presented in Figure 4.7.

4.3 Charge-migration mechanisms for the dispersed phase

The formation of the columnar structures presented in Figure 4.7 is dependent upon the transport mechanism by which these dipoles move and this in turn is dependent upon the nature of the ER fluid. Namely, the porosity of the particles, the characteristics of the surfactant, the hydrous or anhydrous properties of the fluid, and the properties of the chemical activators present on the surface of the particles. Generally the charge transportation mechanisms are characterized as either surface or bulk phenomena, and some of these are succinctly presented in Table 4.3.

These diverse transport mechanisms and also the complex phenomena associated with the interactions between electric and fluidic fields of colloidal systems have presented a major challenge to theoreticians focused on developing viable models for predicting the behavior of ER fluids. However, this dearth of analytical and numerical tools have not significantly hindered engineers in their development of a diverse range of products, which exploit the ability to control the behavior of these fluids by the imposition of an appropriate electrical potential to the fluid domain. This latter feature of controlling the ER fluid phenomenon with electrical signals and circuitry is, of course, an attractive feature when considering the design of these systems.

The ability of engineers to dynamically change, in a controlled manner, the global properties of ER fluids and then exploit the associated phenomena in the evolution of a diverse range of products is clearly evident

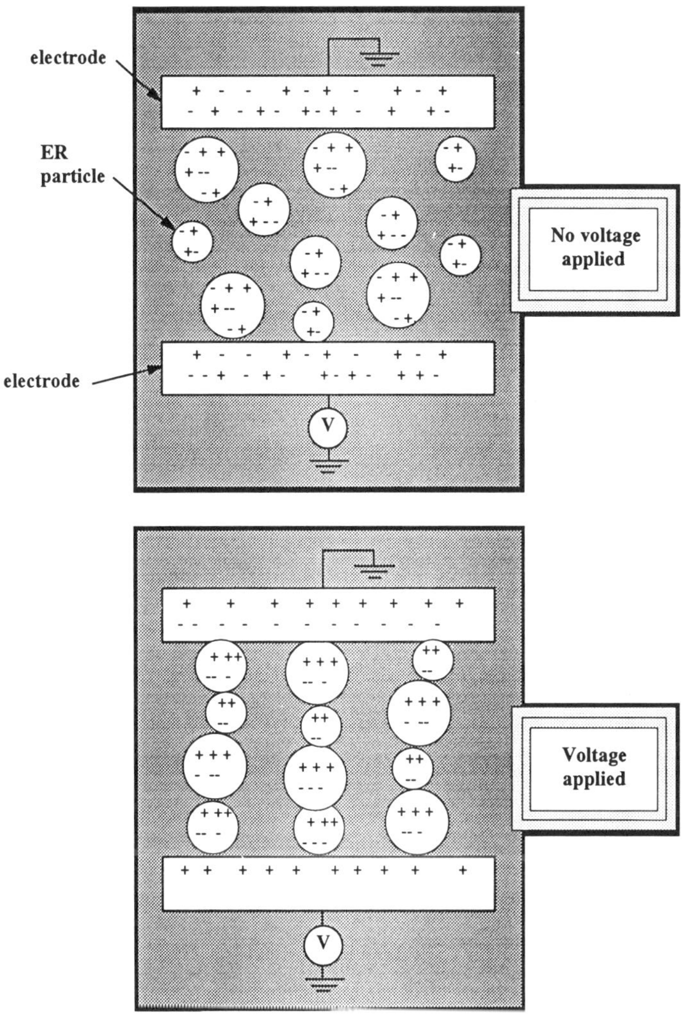

Fig. 4.7 The electro-rheological phenomenon.

upon reviewing the British and Russian patent literature. Typical products include various mechanical devices such as clutches, hydraulic valves, and vibration isolation systems such as engine mounts and shock absorbers. Other work has focused on developing an understanding of the heat and mass transfer phenomena in colloidal systems and this has resulted in the development of double-pipe heat exchangers and recuperative heat exchangers.

Table 4.3 *Charge migration mechanisms in the dispersed state*

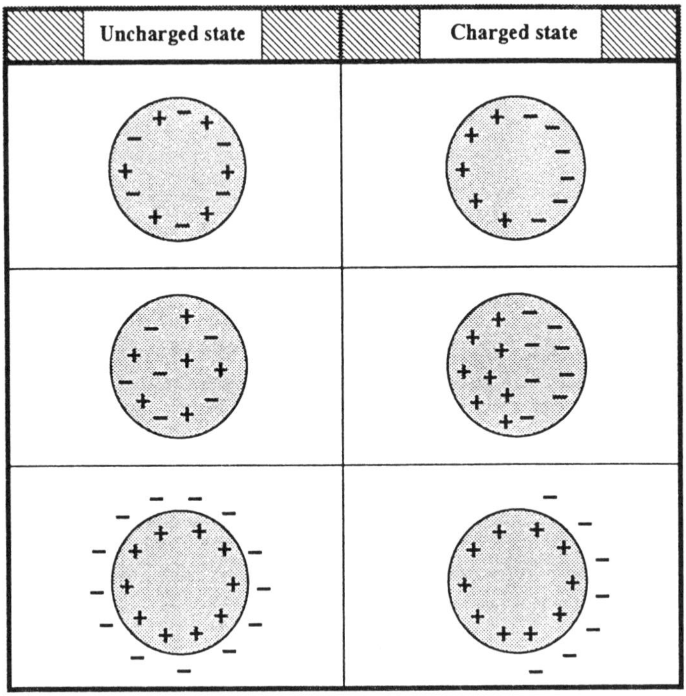

Uncharged state	Charged state

4.4 Electro-rheological fluid actuators

The static response of a given structure subjected to a prescribed loading regime is governed by the stiffness of the structure. The dynamic response, however, is largely governed by the mass, stiffness and energy dissipation characteristics of the structure and also the nature of dynamic excitation. Thus in order to control the static and dynamic behavior of a structure, the structure must in general possess the innate ability to change, upon command, the mass, stiffness and energy-dissipation characteristics. Since ER fluids possess electrically-dependent mechanical characteristics when subjected to static or dynamic electrical signals, then if these ER fluids are embedded within an electrically conductive solid medium, the ability then exists to control the global properties of this material.

The generic idea in the development of ER fluid actuators is presented in Figure 4.8 which presents a general deformable solid in which is embedded a single ER fluid domain. The task for the applied mathematician is to model, and predict the electro-elastodynamic behavior of a system comprising a fluid domain and a solid domain which are subjected to an

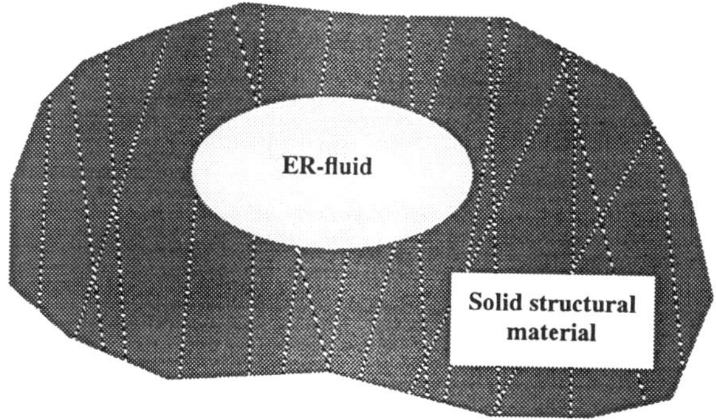

Fig. 4.8 The generic idea of an electro-rheological fluid embedded in a deformable solid.

electrical field and appropriate electrical and mechanical boundary conditions. This is a challenging undertaking when considered in the context of colloidal science, modern advanced anisotropic solid materials, fluid-structural interaction phenomena and electromagnetic field theory. Other non-trivial considerations include control theory, microprocessors and manufacturing issues.

An electric potential must be imposed upon the fluid, and this can be accomplished by several approaches. If the solid is conductive, then upon splitting the solid domain with a plane of incision through the fluid domain, the rebonding of the two solid pieces with an insulator separating the two pieces is one option. Alternatively, if the solid is not an electrical insulator, like a glass-epoxy material, then the surface of the solid adjacent to the fluid can be coated with an electrically conductive material in order to facilitate the imposition of an electrical field on the fluid domain.

The global task confronting the designer is, therefore, how to synthesize an electro-elastodynamic response which satisfies the design specification for this class of smart materials. Clearly since the systems are based on a solid material, a fluid medium and an electromagnetic effect, the designer must carefully consider the stress fields associated with this triumvirate and the interaction between them. This optimal design task clearly involves considerations of material constituents for both the solid domain and the fluid domain and these decisions clearly impact the electro-magnetic phenomenon. Thus, for example, Figure 4.8 presents the simplest case involving a single fluid domain and a single structural material. However, a laminate configuration, as proposed in Figure 4.9, provides a viable alternative involving different ER fluids and also different solids. Thus there are numerous permutations and combinations.

Figure 4.10 presents an algorithm focused upon synthesizing the structural material and the ER actuator material for operation in a dynamic

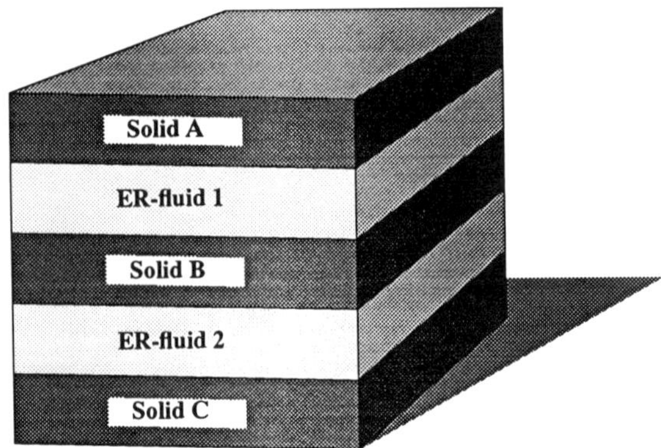

Fig. 4.9 A laminate configuration containing different solids and different electro-rheological fluids.

mechanical environment. When a structural member operates in a dynamic mechanical environment, the transient and forced responses are typically characterized by the natural frequencies and damping characteristics of the structure and also the characteristics of the excitation such as the forcing frequency, for example. While the latter components are excitation-dependent, namely harmonic, stochastic, or nonlinear in behavior, the former properties are dependent upon the inherent mass, stiffness and energy dissipation characteristics of the structure.

These characteristics are largely dictated by the mechanical properties of the structural material and the ER fluid. The selection of appropriate structural materials is a daunting task since the search domain not only encompasses the traditional monolithic materials, such as the steels and the aluminum alloys, but also the advanced engineered polymeric materials. The properties of these engineered materials are exceedingly diverse, not only because of the very large number of fibers and matrix materials available to the materials scientist, but in addition, the engineer has the freedom to select combinations of continuous and discontinuous fibers which govern the global mechanical properties of the structure, and also the synthesis of hybrid materials where, for example, graphite and glass fibers can be unified in a common epoxy matrix material in order to create a material which is different from both a graphite epoxy material or a glass-epoxy material. The stacking sequence, fiber orientation, fiber volume fraction and the manufacturing process are also governing parameters in this synthesis procedure.

The selection of an appropriate structural material must not be considered in isolation, since the solid must interact with an ER fluid which is also part of the smart material. Previous sections have discussed the complexities of synthesizing and analyzing ER fluids and also the diverse ingredients of

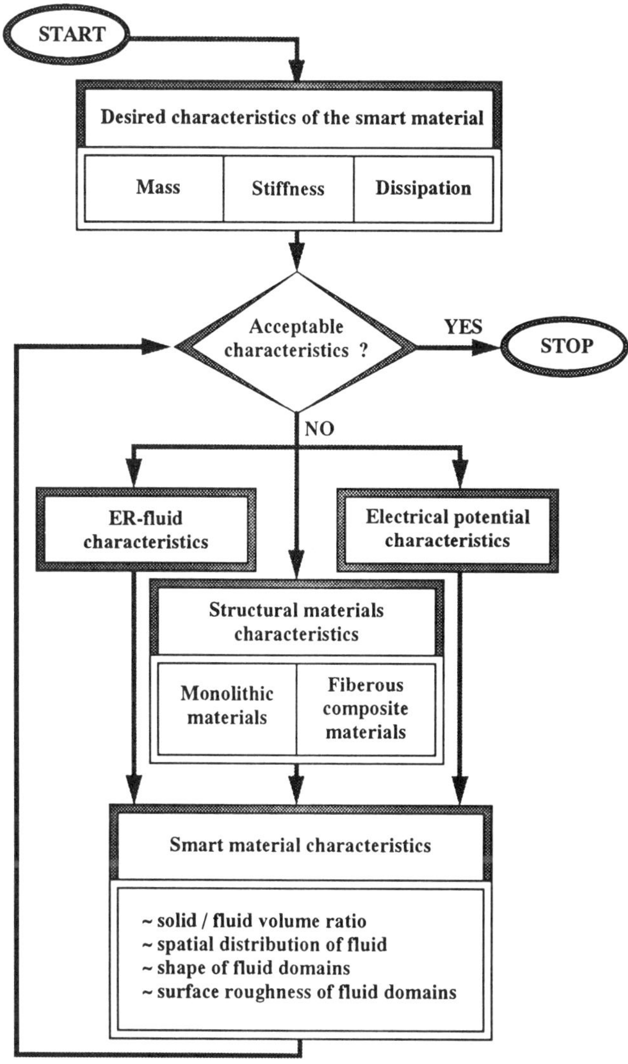

Fig. 4.10 An algorithm for synthesizing the structural material and the ER actuator material for dynamic mechanical applications.

these suspensions. Table 4.4 provides a list of the design parameters for this class of colloidal suspensions.

The incorporation of an ER fluid in cavities within a structural material imposes further constraints on the synthesis of this class of smart materials. For example, the volume of a specific ER fluid necessary to control the response of a prescribed volume of a specific structural material must be ascertained. Furthermore, the spatial distribution of the fluid within the

Table 4.4 *ER-fluid design parameters*

ER-fluid Design Parameters	
cost	toxicity
dielectric properties	viscosity
surfactant	fluid density
additive	hydrous
anhydrous	particulate stability
non-corrosive	power consumption
current density	electrode group
mechanical constitutive properties	thermal stability

solid is important. Should a single cavity be proposed near the neutral plane of a plate, or should the same volume of ER fluid be distributed in multiple voids or even capillaries at different remote locations? The shape and surface roughness of the solid at the fluid-structure interface are also important design parameters since they control the intensity and shape of the electrical field imposed on the ER fluid, and hence ultimately the properties of the smart material, and they also determine the stress concentration characteristics in the region.

Research has been prosecuted on smart beam and plate structures featuring embedded ER fluids, in order to investigate the static response and also the transient and forced responses of smart structures featuring ER fluid domains, subjected to a variety of discrete constant electrical field intensities. Results have not yet been published for the dynamic or static response of these systems, subjected to dynamical electrical fields of various waveforms and frequencies which would result in the structures featuring time-dependent characteristics. Theoretical and experimental work has been undertaken on robotic and machine systems featuring smart links which are subjected to complex loading involving forced, parametric, and inertial excitations. The subsequent pages focus on these experimental research programs.

4.5 A collage of experimental investigations

The objective of the early experimental programs in the field of smart structural materials featuring ER fluids, was to investigate both the static and the transient response characteristics of various cantilevered beam specimens, fabricated in smart ultra-advanced composite materials in a variety of different operating conditions, in order to provide a basis for evaluating the controllability of these structures in real-time. The transient responses were characterized by the fundamental transverse natural frequencies and the damping ratios. Different classes of beams were fabricated featuring different electrode materials, different fluid volume fractions and different geometrical dimensions. In the first set of tests, specimens were fabricated using graphite prepreg tape AS4/3501–6 manufactured by Hercules, Inc. as the face or electrode material, and each electrode comprised three plies with the lay-ups [90/0/90] for specimen classifications A, B, and C, and [0/90/0] for specimen classification D. The

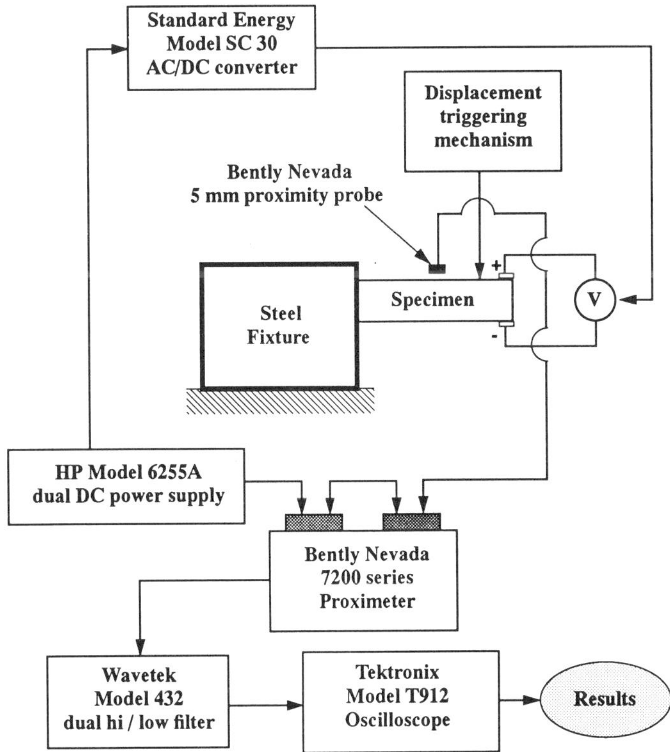

Fig. 4.11 A schematic diagram of the experimental apparatus.

face plates were appropriately configured to provide beams with symmetrical lay-ups and geometry. An RTV silicone rubber adhesive was selected as an insulator in order to provide both excellent bonding between the electrodes and furnish a high dielectric constant.

A schematic diagram of the experimental apparatus employed for evaluating the static and transient responses of cantilevered beams is presented in Figure 4.11. Figure 4.12 presents the static load-deflection

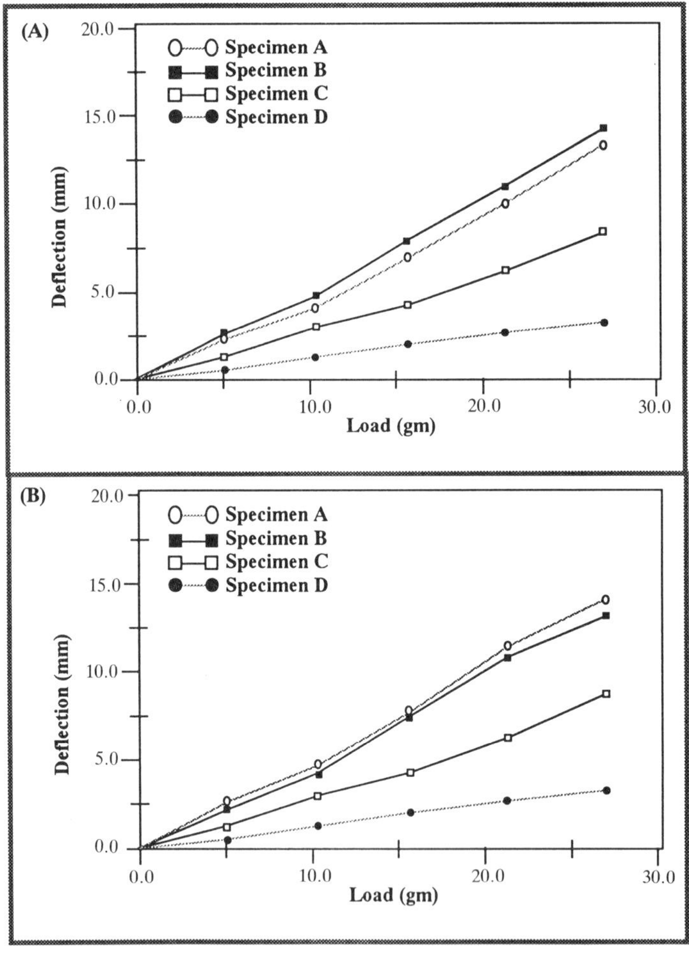

Fig. 4.12 Load deflection characteristics for specimen classification A, B, C, and D, at (a) 0 kV/mm and 24.5°C, and (b) 2 kV/mm and 24.5°C.

characteristics for the four classes of specimens tested. These nominally linear characteristics clearly demonstrate that the type of ER fluid and the lay-up of the AS4/3501–6 electrodes have a profound influence upon the response behavior. Moreover, upon carefully scrutinizing the characteristics in the two reference states, it is evident that when a voltage field of 2 kV/mm is generated, all of the specimens become stiffer relative to their reference state of 0 kV/mm. For specimen classification A, the increase in stiffness is negligible, but for specimen classifications B, C, and D the beam stiffness increases by 7%, 10%, and 6%, respectively. Class B and C specimens contain the same ER fluid, and the same electrode lay-up, but class C specimens have an ER fluid volume fraction of 39.4% rather than 36.1%.

The transient vibrational response characteristics of the beam specimens were obtained by imposing an initial tip displacement prior to removing this constraint and subsequently monitoring the resulting transient response. Figure 4.13 presents photographs of oscilloscope traces of the transient elastodynamic responses of Class B specimens at room temperature. It is evident from these results that the frequency of the response and also the damping ratio of the signal are strongly dependent upon the voltage applied to the beams. Figure 4.14 presents a comparison of the relative frequency and damping ratio increments between Class A and B specimens as a function of the applied voltage. The relative increment in the frequency and damping ratio of each ultra-advanced composite beam specimen in the presence of an electric field, is defined with respect to the corresponding magnitudes in the absence of an electric field, which are employed as datums. Class A and B specimens are identical except for the type of ER fluid filling the beam cavity. Thus, Figure 4.14 clearly indicates that the elastodynamic response of specimens featuring ER fluid IMSL-B-88 are much more sensitive to the applied voltage than specimens containing ER fluid IMSL-A-88.

Figure 4.15 presents a comparison of the relative frequency and damping ratio increments for Class B specimens with a [90/0/90] lay-up and Class D specimens which feature a [0/90/0] lay-up. In the absence of an electrical field, Class D specimens have a higher natural frequency and a lower damping ratio than Class B specimens. This is to be anticipated because of the different lay-ups of the electrodes which results in different stiffnesses and energy-dissipation characteristics of the electrodes.

Figure 4.16 presents the oscilloscope traces of the controlled transient elastodynamic responses of a Class B specimen. The transient response of the specimen for zero applied voltage is presented in the upper photograph, and the transient response of the identical specimen subjected to the same initial conditions for a different electrical field is presented in the lower photograph. This transient response profile was obtained by developing an electrical field corresponding to 0 kV/mm for the first 0.37 seconds of the

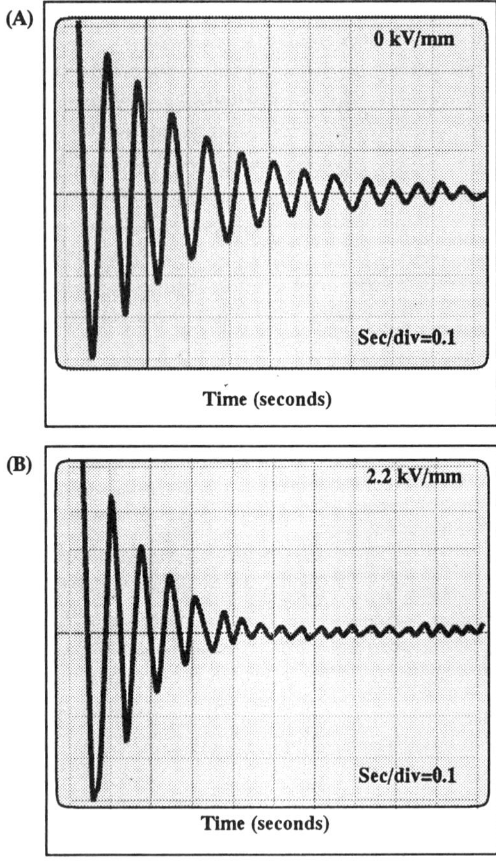

Fig. 4.13 Transient response of a Class B specimen at room temperature (a) with, and (b) without an electric field.

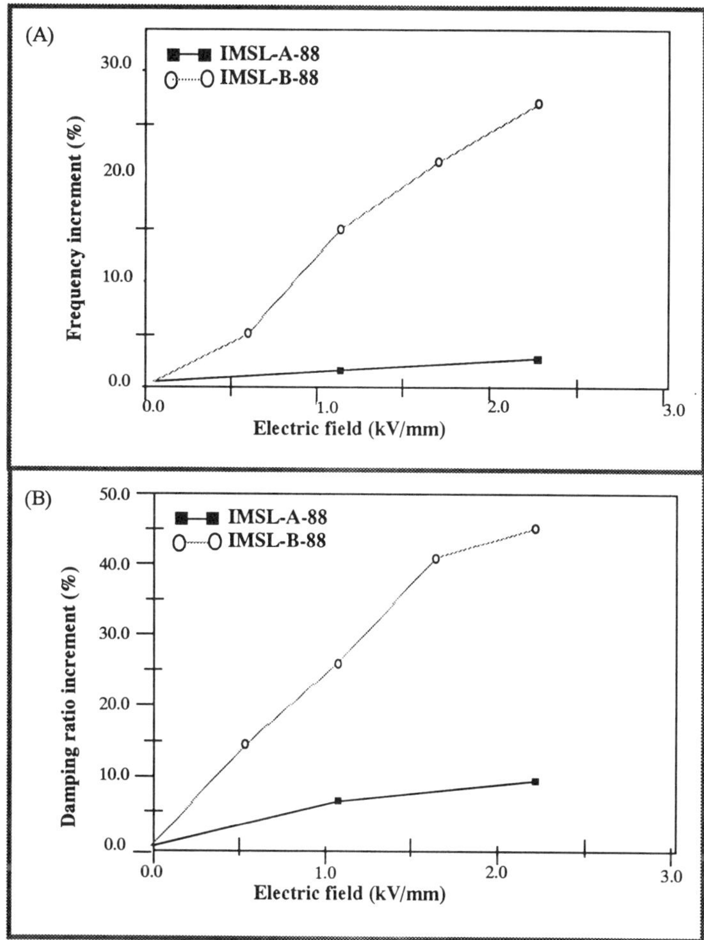

Fig. 4.14 Comparison of the (a) relative frequency-increment, and the (b) relative damping-ratio increment, for specimens A and B featuring different ER fluids at room temperature.

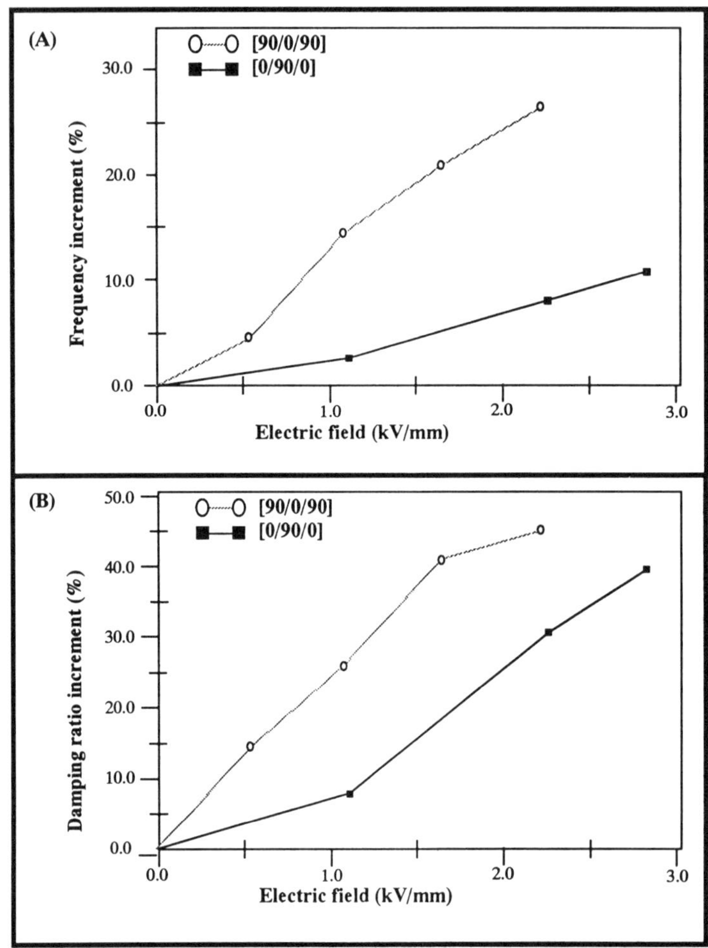

Fig. 4.15 Comparison of the (a) relative frequency-increment, and the (b) relative damping-ratio increment, for two different lay-ups of face material at room temperature.

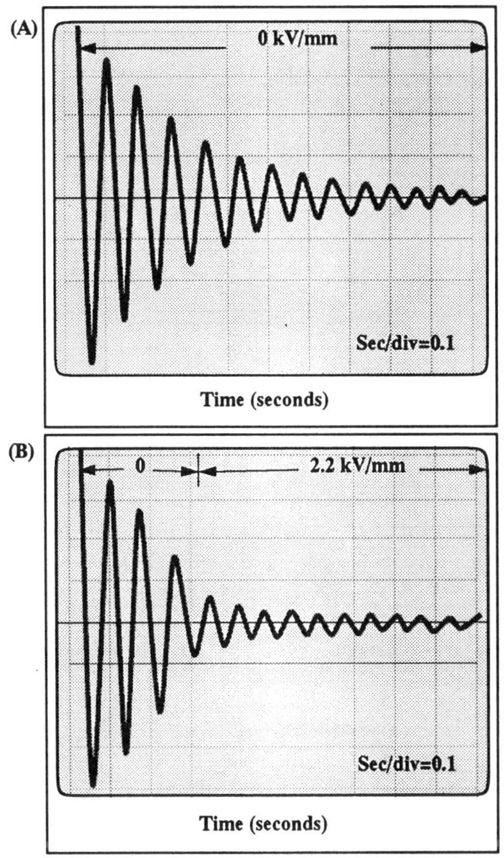

Fig. 4.16 Controlled transient response of a Class B specimen at room temperature, (a) without control, and (b) with control.

response profile, prior to simultaneously imposing a voltage corresponding to a field intensity of 2.5 kV/mm. These two piecewise-constant discrete voltage inputs are active-control inputs based on a bang-bang control strategy. A cursory review of the response profiles presented in Figure 4.16 clearly demonstrates that the amplitude of the vibrational response in the two profiles is identical for the first 0.37 seconds. Subsequently, the amplitude of vibration presented in the lower photograph is substantially attenuated upon generating the electrical field strength of 2.5 kV/mm.

A second set of experiments, focused upon evaluating the transient behavior of smart beams featuring aluminum electrodes and various ER fluids, has been reported in the literature. Figure 4.17 presents frequency increment and damping data as a function of the electrical field strengths

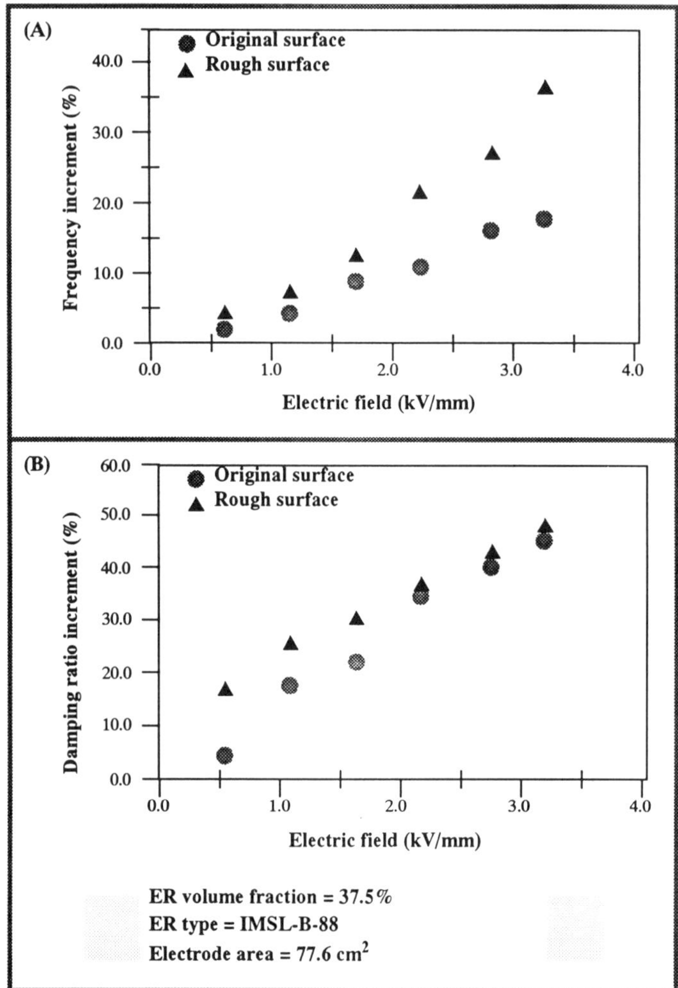

Fig. 4.17 (a) Relative frequency-increment and, (b) relative damping-ratio increment of Class A and C specimens.

defined with respect to the corresponding magnitudes in the absence of an electrical field. The two classes of beam specimens employed in this study differ only in their respective fluid-solid interface conditions. One set of specimens featured a fluid-solid surface that was subjected to abrasion from carbon sandpaper number A235 manufactured by Norton Corporation. The other set of specimens featured a relatively smooth rolled finish. The smart beams featuring the roughened surface yielded larger frequency and damping-ratio increments than the unabraded surfaces at the same voltage.

Figure 4.18 presents schematically the electrode geometries at the fluid-structural interfaces for class D and class E specimens. These specimens only differed in this feature which involves the bonding of circular and rectangular pieces of polyvinyl sheet of the same total surface area. The results of the transient response studies are presented in Figure 4.19 and they indicate that the specimens with the electrodes featuring the rectangular pieces of polyvinyl sheet provided superior response profiles for both frequency and damping than the electrodes featuring the circular polyvinyl disks. Experimental results presented in Figure 4.20 demonstrate that both the frequency increment and also the damping-ratio increment increase when the electrode area in contact with the ER fluid domain increases.

The previous discussions have focused upon evaluating the static and transient dynamic response of smart cantilever beams with embedded ER fluid actuators. Subsequently work has been undertaken on evaluating the dynamic response of these systems subjected to forced excitations. Figure 4.21 presents a schematic diagram of the apparatus and instrumentation employed in these studies. The smart beam is fixtured to the head of an electrodynamic shaker in a cantilever configuration which enables the beam to be dynamically excited in a controlled manner in order to excite flexural vibrations. The response of the beam at a prescribed sinusoidal excitation can be actively controlled by changing the voltage field imposed upon the ER fluid domain in the beam.

Figure 4.22 presents frequency-response results which were obtained by simultaneously exciting the beam with a random noise signal, while the beam was sequentially subjected to two discrete potential differences of 0 kV and 3.42 kV across the upper and lower faces of the beam. In the

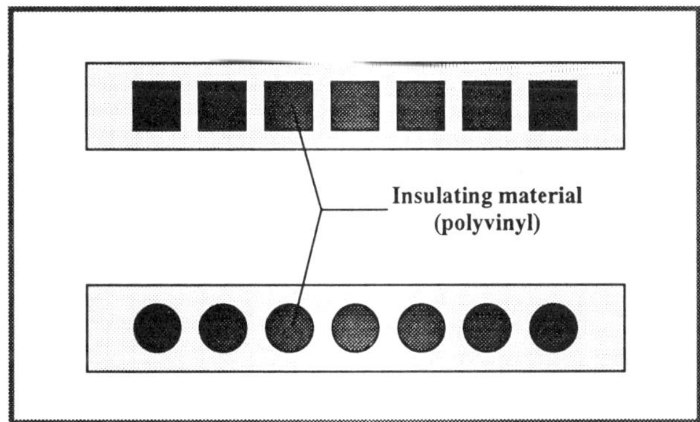

Fig. 4.18 Schematic diagram of different electrode surface geometries for Class D and E specimens.

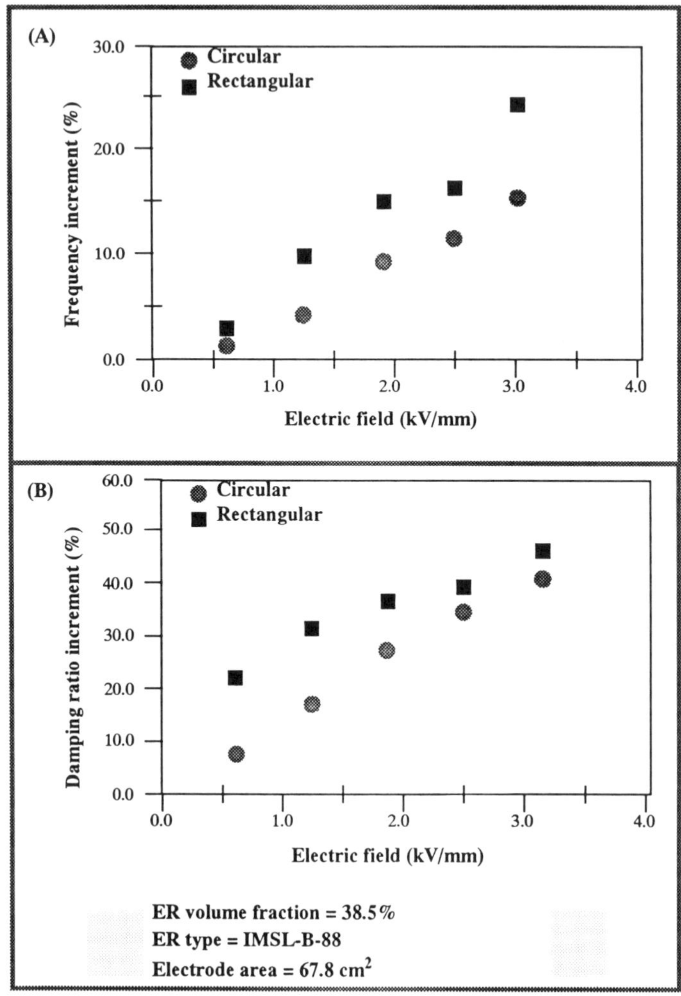

Fig. 4.19 (a) Relative frequency-increment and, (b) relative damping-ratio increment of Class D and E specimens.

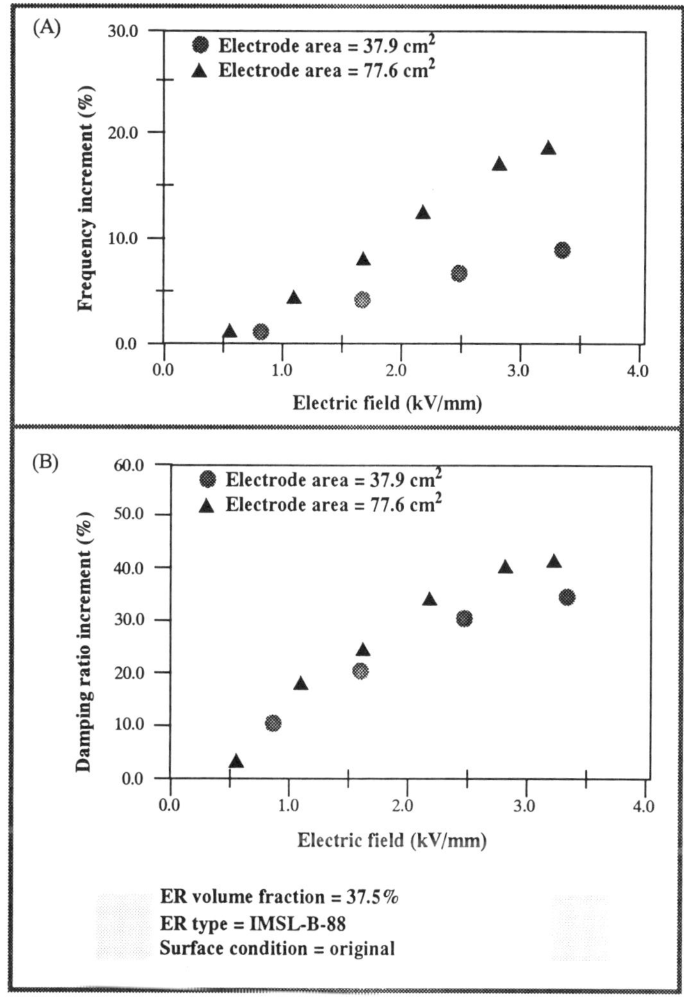

Fig. 4.20 (a) Relative frequency-increment and, (b) relative damping-ratio increment of Class A and F specimens.

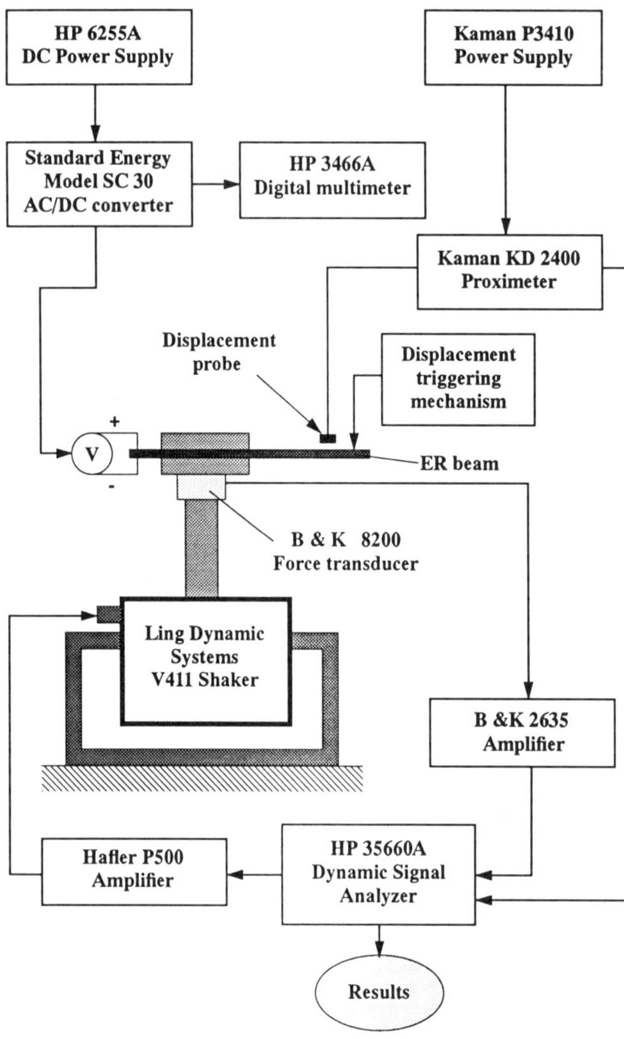

Fig. 4.21 Experimental apparatus for dynamically exciting a smart cantilevered beam.

uncharged state, the first three natural frequencies of the beam in flexure are 20.5 Hz, 79.5 Hz, and 176.0 Hz, while in the charged state of 3.42 kV the first three natural frequencies are 35.0 Hz, 135.5 Hz and a value higher than 176.0 Hz which is not observed on the response curve because the upper cut-off frequency was 215 Hz. It is, therefore, obvious from the experimental results presented in Figure 4.22 that, by controlling the voltage difference between the electrodes of the smart beam, the magnitudes of the

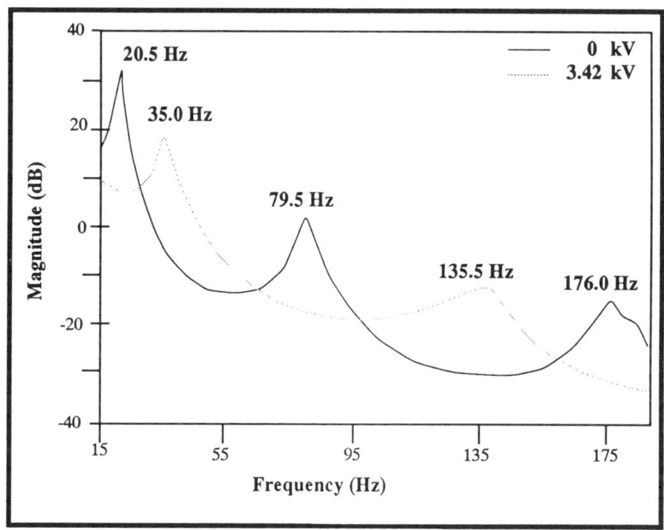

Fig. 4.22 Frequency response of a smart cantilevered beam featuring an ER fluid.

individual natural frequencies can be changed or tailored to assume prescribed values.

Thus, for example, the fundamental natural frequency can be changed from 20.5 Hz to 35.0 Hz by changing the intensity of the electrical field imposed on the ER fluid domain, by changing the potential difference across the beam electrodes from 0 kV to 3.42 kV. This ability to increase the fundamental frequency by 15 Hz, or 70% of the original value has broad ramifications. For example, consider the situation in which this beam, which has a fundamental natural frequency of 20.5 Hz is subjected to a forcing frequency of 20.5 Hz in the flexural mode of vibration. The condition of resonance, with the attendant large and potentially dangerous dynamic stress levels, can be avoided by imposing a voltage on the beam to change the magnitude of this fundamental natural frequency to, say, 26 Hz, thereby avoiding resonance. This capability is readily implemented and observable in the laboratory.

Indeed control algorithms have been developed whereby, for the results prescribed in Figure 4.22, the voltage imposed on the beam can be tailored to ensure that the magnitude of the maximum beam deflections is minimized, by ensuring that the response is below the lower envelope of the continuous and dotted curves presented in Figure 4.22, thereby avoiding the large deflections in the neighborhood of the natural frequencies. Thus for example, if the beam is excited at 79.5 Hz, which is the second fundamental frequency of the uncharged smart beam, then a voltage of 3.42 kV could be imposed on the beam to reduce the response to the magnitude of the chain-dotted line in Figure 4.22; whereas if the beam were subjected to a harmonic excitation of 135.5 Hz and the beam was also being subjected to a potential difference of 3.42 kV, then a 0 kV state would be imposed on the beam in order to drastically reduce the amplitude of the response.

The experimental results presented in Figure 4.22 indicate that the second natural frequency of the smart beam is much more sensitive to the electrical field imposed on the ER fluid domain than the fundamental frequency. Thus the second natural frequency is increased from 79.5 Hz at 0 kV to 176 Hz at 3.42 kV, a change of approximately 121%, while the fundamental natural frequency is increased from 20.5 Hz to 35.0 Hz, a change of 70%, when subjected to the same electrical field intensity. A second controllable vibrational characteristic which may be deduced from Figure 4.22 is the dependence of damping upon the magnitude of the external voltage field imposed upon the ER fluid actuator in the smart beam. This deduction is based on the shape of the curves in the neighborhood of each natural frequency. Thus, the degree of sharpness of response at 20.5 Hz and 0 kV is much more severe, and hence is characterized by smaller inherent damping, than the response of the fundamental frequency at 35 Hz and 3.42 kV. Similarly, the second natural frequency at 79.5 Hz and 0 kV is much sharper and is characterized by less damping than the response at 135.5 Hz and 3.42 kV.

A more comprehensive set of experimental results is presented in Figure 4.23 for a similar smart beam with an embedded ER fluid actuator

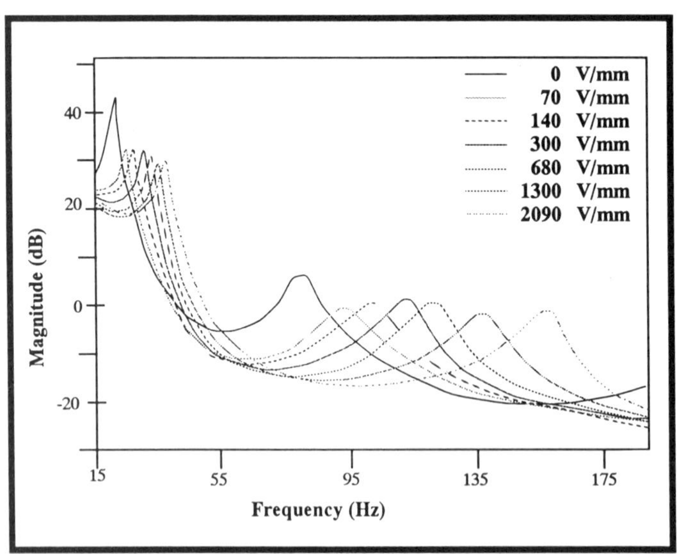

Fig. 4.23 Frequency response of a smart cantilevered beam subjected to discrete voltage states.

subjected to a number of discrete voltage states with field intensities varying from 0 kV/mm of fluid thickness to 2.09 kV/mm. This data emphasizes and illustrates the ability to tailor the dynamic behavior of this class of smart structural materials by controlling the natural frequencies and the energy-dissipation characteristics. Furthermore, the results indicate that the magnitudes of the natural frequencies will increase from the 0 kV/mm datum as the field intensity increases. This ability to increase the magnitude of the natural frequencies may indeed be advantageous in many industrial applications, where the design engineer is required to synthesize a smart structure in which the natural frequencies must be tailored to yield both higher and also lower natural frequencies. This goal may be accomplished by designing a smart structure with a reference state which is characterized by a non-zero electrical potential. Thus, for example, referring to Figure 4.23, an appropriate reference state may be defined as 300 V/mm which enables the magnitudes of the natural frequencies to be either increased or decreased, relative to this value by selecting either larger or smaller electrical field intensities as desired.

Figure 4.24 presents a waterfall plot from an experimental study focused upon controlling the chaotic behavior of a smart beam featuring an embedded ER fluid actuator. The experimental apparatus and relevant

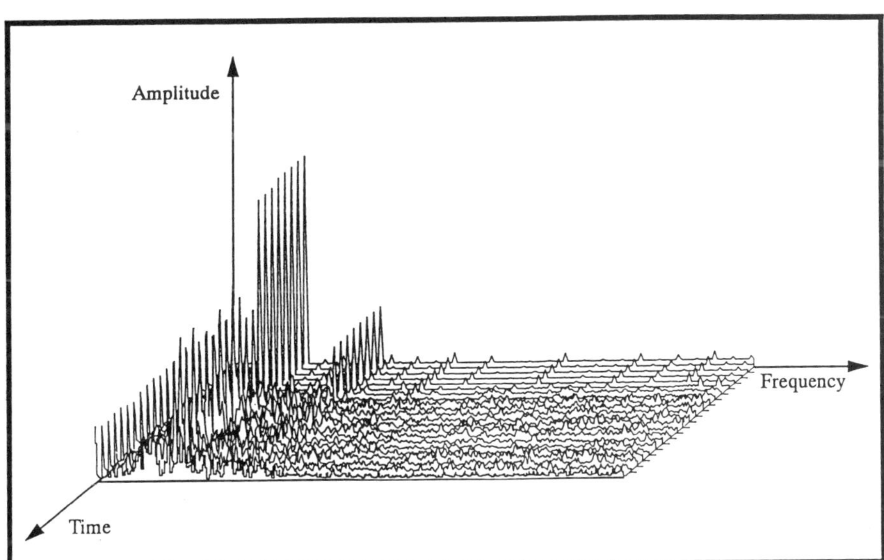

Fig. 4.24 Waterfall plot of the elastodynamic response of a doubly-encastré beam subjected to dynamical excitation, showing that the chaotic response can be controlled by activating the electro-rheological fluid domain within the smart beam.

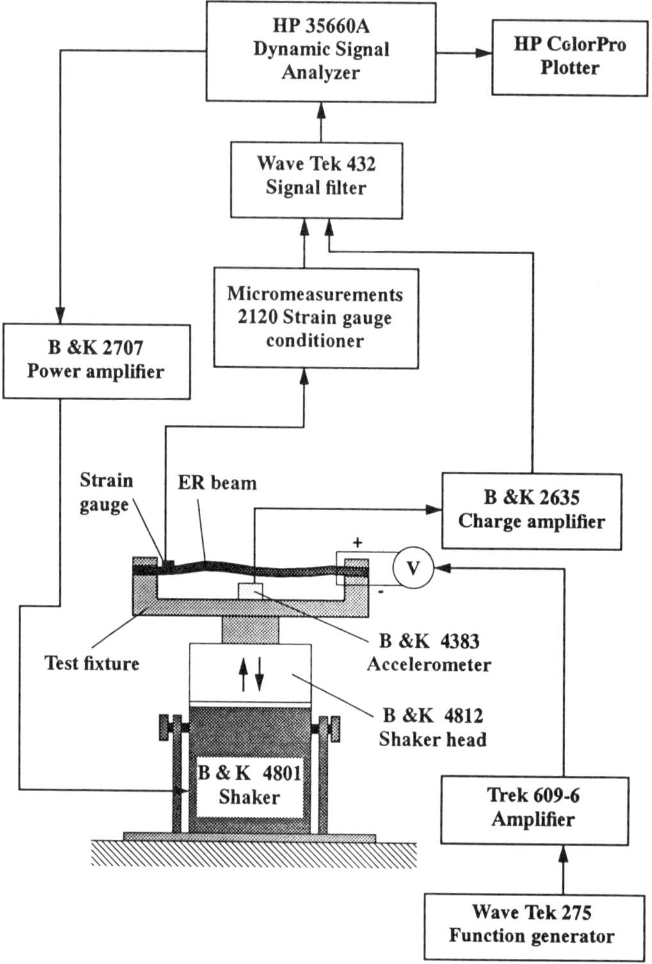

Fig. 4.25 Apparatus for determining the chaotic response of a preloaded, prebuckled doubly-encastré beam subjected to a dynamical excitation.

instrumentation are documented in Figure 4.25, from which it is evident that a test fixture, containing an electrically-insulated doubly-encastré beam, was mounted on the head of an electrodynamic shaker. The construction of the 465 mm long beam, which is shown in Figure 4.26, comprised two aluminum electrodes each of thickness 0.8 mm which were separated by a silicone rubber insulator of thickness 2 mm. The resulting void contained an inexpensive ER fluid with simple constituents featuring 60% corn starch and 40% silicone oil marketed by Dow Corning as 704 Diffusion Pump Fluid. The beam was sinusoidally excited in a direction perpendicular to the face of

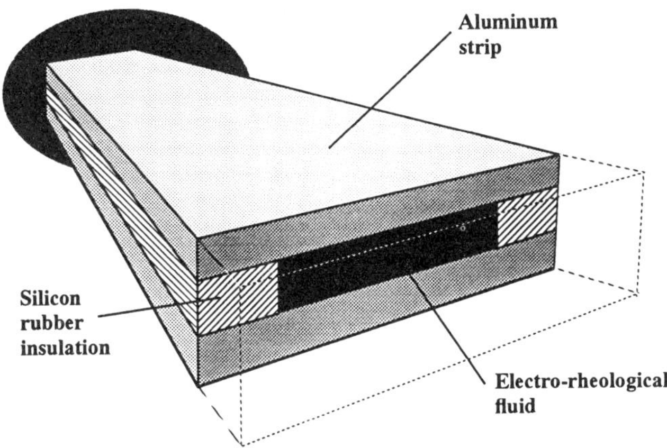

Fig. 4.26 A schematic of a smart beam featuring a void filled with an electro-rheological fluid.

the aluminum plate at different frequencies and accelerations, while imposing different discrete electrical field intensities on the ER fluid domain within the beam. The resulting spectra and time responses curve were studied in order to investigate the ability to control the chaotic responses of the smart beam.

Figure 4.24 presents a time-frequency amplitude plot of the beam demonstrating how a chaotic response regime ensues when the initial voltage state of 1 kV imposed upon the beam is subsequently reduced to 0 kV. It is evident from the experimental data that upon removing the electrical potential from the ER fluid domain, the somewhat deterministic response is overwhelmed by a chaotic regime with no distinct repetition characteristics. This somewhat preliminary study clearly suggests that the chaotic behavior of certain classes of structures deployed in engineering practice, could conceivably be alleviated by fabricating these structures in smart materials featuring ER fluid actuators.

Research has also been prosecuted on controlling the dynamic response of articulating machine systems by employing articulating members fabricated in smart structural materials with embedded ER fluid actuators. Figure 4.27 presents the mid-span transverse deflections of the connecting-rod of the planar slider-crank linkage mechanism presented in Figure 4.28 when operating at 95 rpm. The connecting rod was configured to principally deform in flexure in the plane of articulation of the mechanism. It comprised two aluminum electrodes of 0.5 mm thickness and length 265 mm which were separated by a thin layer of silicone rubber adhesive of thickness 1.5 mm in order to insulate the two electrodes and create a uniform void which accommodated the ER fluid. The flexible smart connecting rod was also insulated from the two sets of metallic multi-element bearings at the

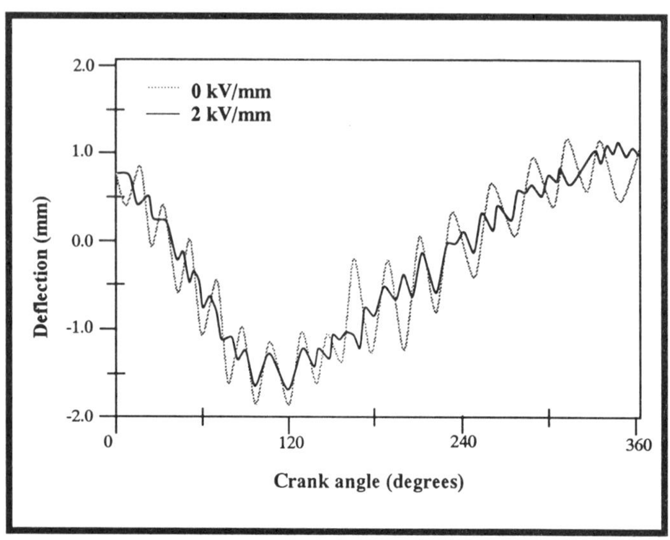

Fig. 4.27 Midspan transverse deflections of the dynamically tunable connecting-rod at an operating speed of 95 RPM.

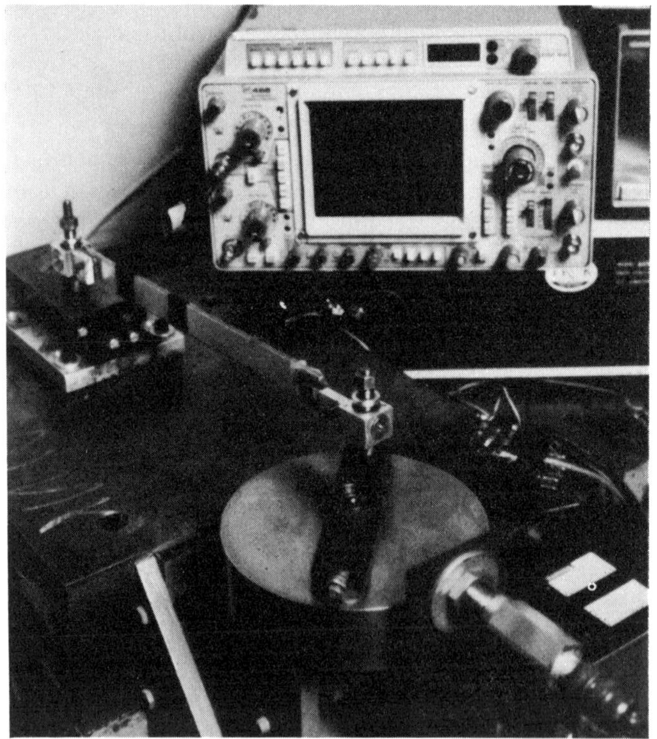

Fig. 4.28 Photograph of an experimental slider-crank mechanism with a smart connecting-rod featuring an embedded electro-rheological fluid domain.

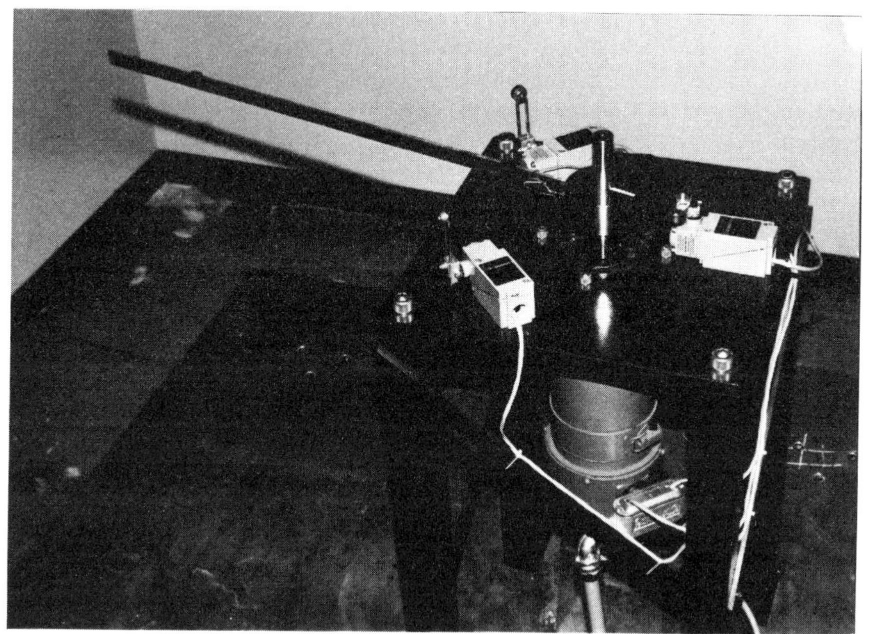

Fig. 4.29 Photograph of a smart robot arm featuring an embedded electro-rheological fluid domain.

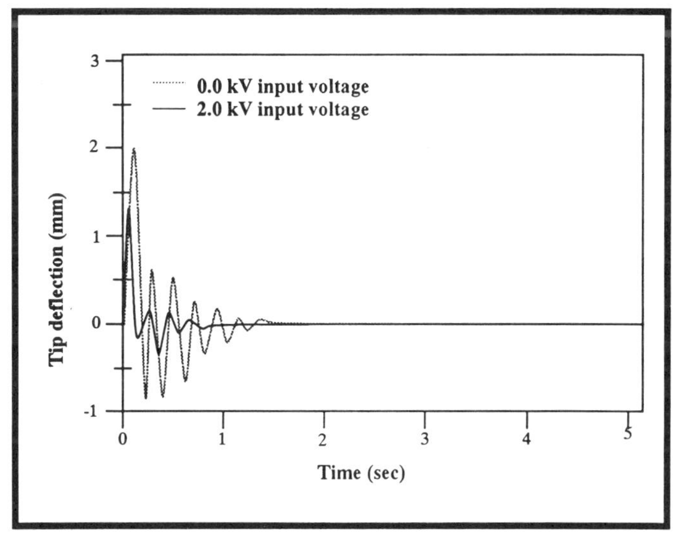

Fig. 4.30 Measured step response with feedback gains of $K_p = 0.5$ and $K_v = 0.25$.

crank-pin and the gudgeon-pin. It is evident from the two sets of experimental results that the amplitude and frequency of the elastodynamic response of this articulating member, which is being subjected to both parametric and forced excitations, can be controlled by adjusting the external voltage imposed upon the ER fluid domain within the smart structural member.

Figure 4.29 presents a photograph of an experimental robot with a smart arm featuring an embedded ER fluid that is driven by an electrical motor at the hub. Figure 4.30 presents the measured step responses to a commanded step angular position of 7.2 degrees with the compensator zero $Z_c = -2$. Feedback gains of $K_p = 0.5$ and $K_v = 0.25$ were employed. The external input voltage of 2.0 kV/mm and 0 kV/mm were continuously imposed on the ER fluid domain throughout the two manouvres. It is clear from the figure that the responses of the robot arm with and without the voltage imposed on the ER fluid domain are dependent upon the magnitude of this voltage. The amplitude of the tip deflection of the robot arm in the absence of the electric field was reduced by approximately 35% when the input voltage of 2.0 kV/mm was imposed on the ER fluid actuator. It is also observed from this response that the settling-time is approximately 1.3 seconds when a voltage is imposed on the fluid domain, while it is approximately 1.6 seconds when an external voltage is not imposed on the ER fluid in the robot arm. Thus the elastodynamic performance of the robotic system can be greatly enhanced by carefully orchestrating the input signal to the electrical motor and also the electrical signal to the electro-rheological fluid actuator embedded within the robot arm.

5

Piezoelectric materials

5.1 Background

Electricity provides the engineer confronted with the task of synthesizing a smart material system with an attractive thread of commonality during the evaluation of candidate sensors, actuators, data transmission links, and microprocessors. The latter two groups are mature fields of scientific endeavor but in the context of materials exhibiting electro-mechanical phenomena for sensing and actuation functions, the field is somewhat embryonic in nature. This chapter opens with a brief review of the basic electro-mechanical properties of some classes of materials prior to discussing piezoelectric materials and applications utilizing these materials.

Piezoelectricity is an electro-deformation phenomenon derived from the Greek word '*piezein*' for '*press*' from which is derived the term 'pressure electricity', that first appeared in the scientific literature in 1880 when Pierre and Paul-Jacques Curie published a paper, describing how various crystals developed an electrical charge on their surface when they were mechanically deformed in certain directions. Their work focused upon crystals of tourmaline, Rochelle salt and quartz. One year later, they discovered the converse effect whereby these crystals changed their shape when they were subjected to an electric field.

This piezoelectric phenomenon is similar to *electrostriction*, which is a property of all dielectrics. The electrostriction phenomenon is evidenced in practice as a small change of geometry of a body when it is subjected to an electrical field. The direction of this small change in geometry does not change if the direction of electrical field is reversed. In sharp contrast to this situation, piezoelectric materials, which are a unique class of nonconducting materials, exhibit a reversal in the direction of geometrical change when the direction of the electrical field is reversed.

Pyroelectricity, which was first recorded in the scientific literature in 1824 and is derived from the Greek for 'heat electricity', is the development of electric polarization in special classes of crystals that are subjected to a temperature change. Typically, the extent of this polarization is proportional to the change in temperature of the crystal.

The pyroelectric phenomenon is only exhibited in crystallized non-conducting substances featuring one or more axes of symmetry that are polar. Thus regions of the crystal featuring the same symmetry develop charges of the same sign, and if a crystal develops a negative charge on one face when the temperature increases, then the same face will develop a positive charge when the temperature decreases. However, it should be noted that the changing inter-ionic forces and changing polarization are a direct consequence of the continuously changing crystal temperature. Consequently, if the crystal is maintained at a constant temperature then the charges will subsequently decay.

The unique characteristics of piezoelectric materials permit them to be employed as actuators or sensors which can be exploited in the synthesis of smart materials utilizing electrical energy for the sensing, communication and actuation functions. Indeed, the deployment of piezoelectric materials in the synthesis of smart materials involves the exploitation of a biomimetics philosophy because the anatomy of *Homo sapiens sapiens* features piezoelectric materials. For example, both skin and bone have piezoelectric properties. Thus, our sensing system at the fingertips involves the generation of an electrical potential at the surface of the skin which is then transmitted to the brain by the nervous system prior to evaluation, interpretation and subsequent action.

The piezoelectric properties of bone play a pivotal role in the biophysical phenomena of bone remodelling. Bone formation and destruction are triggered by the electro-mechanical piezoelectric properties of the bone whereby the electrical signals orchestrate the synthesis of a viable microstructure, of collagen and hydroxyapatite, with the appropriate mechanical properties. This microstructure is dramatically different at different sites in the bone.

Currently, piezoelectric materials are employed in a variety of conventional commercial applications such as phonographic pickup cartridges, where the vibrational motion of the phonograph stylus is converted into a time-varying electrical signal; microphones, where sound pressure waves are converted into dynamic voltages; and devices for controlling the frequency of electrical signals, where the shape of the crystal is carefully shaped in order to ensure that only signals of a specific frequency pass through them.

5.2 Piezoelectricity

Piezoelectricity is a phenomenon that occurs in certain classes of anisotropic crystals subjected to *changes* in mechanical deformation. By applying mechanical deformations to these crystals, electric dipoles are generated and a potential difference develops that is contingent upon the changing deformations. Hence, it must be noted that the electromotive force produced and the associated current developed, in piezoelectric materials, is a function of the continuously changing mechanical deformation, therefore, typical and practical uses are in situations involving dynamical strains of an oscillatory nature.

For the piezoelectric phenomenon to occur in a material, the crystal lattice should have no center of symmetry; thus there must be at least one axis in the lattice along which the atomic structure is different from the atomic structure along the other axes. Consequently, in molecular terms, the piezoelectric phenomenon occurs because of the displacement of ions in the crystal lattice, which couples the displacement of charge associated with the polarization of these materials to the mechanical deformation of the lattice structure.

The manufacture of commercial piezoelectric materials involves exposing the material to high temperatures while imposing a high electric field intensity in a desired direction to create the piezoelectric properties. It is termed 'poling'. Upon completing the poling process, the piezoelectric material exhibits no piezoelectric properties, and it is isotropic because of the random orientation of the dipoles as shown in Figure 5.1(a). Upon developing a poling voltage in the direction of the poling axis in Figure 5.1(b), the dipoles re-orientate and for the condition shown, the cylinder elongates.

The direct and converse piezoelectric phenomena, involving an interaction between the mechanical and electrical behavior of a material, can be usefully modelled by linear constitutive equations involving two mechanical variables and two electrical variables. Thus in matrix form the equations governing the direct piezoelectric effect and the converse piezoelectric effect are written respectively,

$$\{D\} = [e]^T \{S\} + [\alpha^s] \{E\} \tag{5.1}$$

$$\{T\} = [c^E] \{S\} - [e] \{E\} \tag{5.2}$$

where $\{D\}$ is the electric displacement vector; $[e]^T$ the transpose of $[e]$, the dielectric permittivity matrix; $\{S\}$ is the strain vector; $[\alpha^s]$ is the dielectric matrix at constant mechanical strain; $\{E\}$ is the electric field vector; $\{T\}$ is the stress vector and $[c^E]$ is the matrix of elastic coefficients at constant electric field strength.

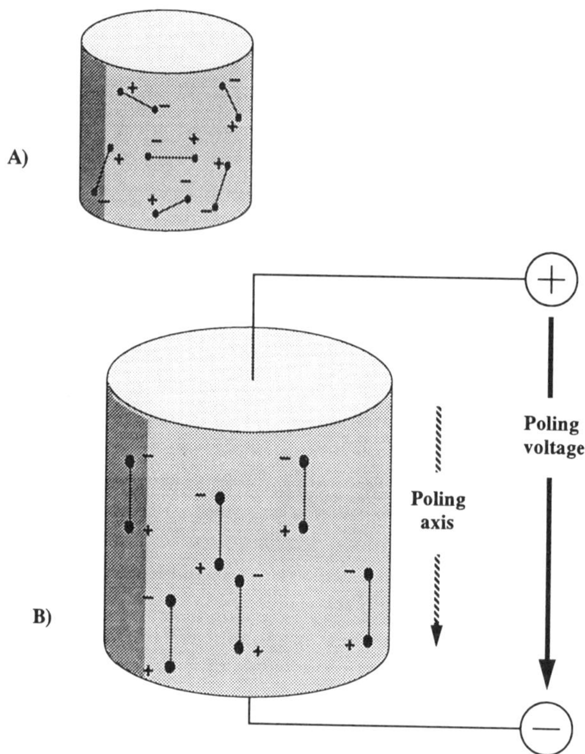

Fig. 5.1 The dipole rearrangement of the piezoelectric effect.

Two basic equations readily distill from these linear constitutive expressions. The first is the electrical expression governing an unstressed material subjected to an electrical field. Since the strain vector contains zeros, equation (5.1) reduces to a relationship relating the field strength to the electric displacement. The second basic equation is the mechanical expression governing the material at zero field strength. Thus since the electrical field vector is only populated by zero elements, equation (5.2) reduces to a relationship relating the stress and strain components of deformation.

Piezoelectric materials possess anisotropic properties. Consequently, from a mathematical perspective, their mechanical and electrical behavior is dependent upon the direction of the external electric field relative to a set of axes fixed in the material. Alternatively, the electrical response of the material is dependent upon the direction of the external mechanical loads, and hence stresses and strains, relative to a set of axes fixed in the material. Thus, design methodologies involving piezoelectric materials must carefully accommodate these anisotropic features. Consequently, with reference to

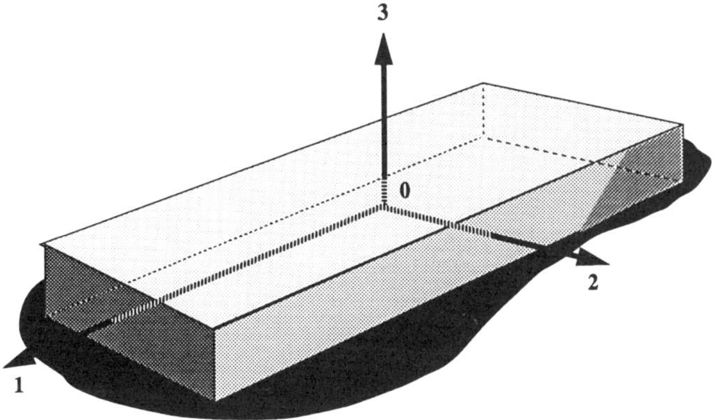

Fig. 5.2 Body axes for defining piezoelectric constants.

equations (5.1) and (5.2), if {D} comprises three elements, and {T} and {S} comprise six elements, then the designer must potentially have access to a comprehensive material data base of these electro-mechanical properties because [e] has 18 elements, $[\alpha^s]$ has nine elements, and $[c^E]$ has 36 elements.

In order to define these electro-mechanical properties relative to prescribed body-fixed axis frames, conventions have been developed for piezoelectric plate-like geometries. Consider a beam of length ℓ, width w and thickness t, the standard convention dictates that subscript 1 corresponds to the length direction, subscript 2 corresponds to the width direction, and subscript 3 corresponds to the thickness direction as shown in Figure 5.2. The coefficients in the matrices featured in equations (5.1) and (5.2) are defined using two subscripts: the first number identifies the axis of the applied electrical field, while the second number identifies the axis of induced mechanical deformation. The axis of polarization is typically in the 03 or thickness direction.

5.3 Industrial piezoelectric materials

The principal commercially available industrial piezoelectric materials are the piezoceramics, such as the lead zirconate titanates (PZT), and the piezopolymers, such as the polyvinylidene fluorides (PVDF). Both classes of materials are available in a broad range of properties to suit diverse applications. An appreciation for the mechanical and electrical properties of these materials may be developed by consulting Figure 5.3 which compares typical piezoelectric polymeric and ceramic materials.

The current generation of smart materials featuring piezoelectric materials

Property	Units	PVDF	PZT	$BzTiO_3$
Density	(10^3) kg/m^3	1.78	7.5	5.7
Relative Permittivity	$\varepsilon / \varepsilon_0$	12	1200	1700
d_{31} Constant	(10^{-12}) m/V	23	110	78
g_{31} Constant	(10^{-3}) V m/N	216	10	5
k_{31} Constant	% @ 1 kHz	12	30	21

Fig. 5.3 Properties of typical commercial piezoelectric materials.

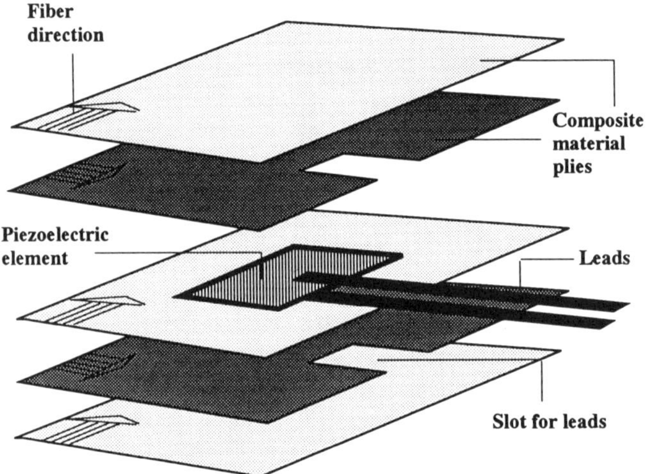

Fig. 5.4 Composite plies with an embedded piezoelectric element.

is generally synthesized with polymeric fiberous composite laminates which readily accommodate embedded piezoelectric actuators and sensors as illustrated in Figure 5.4. The type of fiber and matrix employed for the structural component of this class of smart materials is dependent upon the specific situation in which the smart material is to be deployed, and the properties of the piezoelectric material are also dependent upon this situation. However, the piezoelectric material employed in smart materials is quite thin and is typically utilized in sheet form to maintain and not degrade the structural integrity of the laminate with the introduction of stress raisers and potential delamination sites.

5.3.1 The piezoelectric properties of polyvinylidene fluoride

Polyvinylidene fluoride, which' is a commonly available piezoelectric polymeric material, is a long chain semicrystalline polymer of the repeat unit $CH_2 - CF_2$. The monomer vinylidene fluoride $CH_2 = CF_2$ has a large dipole moment and because these monomer units polymerize in a regular aligned repetitive configuration - $CH_2 - CF_2 - CH_2 - CF_2$ - the polymer is characterized by an extremely high net dipole moment. The magnitude of this dipole moment is responsible for PVDF developing substantially greater piezoelectric characteristics than any other organic material.

Upon completion of the poling process during manufacture, the resulting piezoelectric PVDF film, which is flexible, lightweight and transparent, develops an electrical charge proportional to the *change* in mechanical loading. Thus PVDF is a dynamic material that does not operate under static conditions because of the rapid decay of the induced charge. This class of piezoelectric materials is commercially available in sheet form with a thin layer of nickel or aluminum deposited on the upper and lower faces in order to provide the electrical conductivity for imposing an electrical field upon the polymeric film, or else providing a mechanism for measuring the induced charge associated with the mechanical deformation of the film. Figure 5.5 presents typical properties of piezoelectric PVDF film manufactured by the Pennwalt Corporation.

If a PVDF film is manufactured with a single axis of polarization, such as in the 01 direction in Figure 5.2, then by applying a voltage across the electrodes on the upper and lower faces of the beam in the 02 direction, a resultant strain will develop in the 01 direction. If this experiment is then repeated with a piezoelectric material which is polarized in both the 01 and the 02 directions, the application of a voltage across the upper and lower faces of the plate whose unit normal vectors are in the 03 direction, would produce orthogonal strain components in the 01 and 02 directions. This type of material may be usefully exploited in the design of plate and shell-like structures. The nature of the piezoelectric phenomenon, which is associated with the displacement of ions throughout the crystal lattice of the material, permits this class of materials to be modelled as a distributed parameter actuator when viewed in the context of automatic control theory.

The simplest form of piezoelectric control philosophy involves the imposition of a uniform electric field. However, if the electric field is varied in a spatial manner, then the attendant mechanical strain will vary spatially too, thereby providing two variables in the control scheme. These variables are time and the spatial field distribution for appropriately configuring the piezoelectric elements. These controlled deformation fields permit controlled forces and bending moments to be imposed on structures.

Properties of Piezoelectric Film		
	Compressive Strength	60 Mpa (10^6 N / m^2)
	Tensile Strength MD TD	160 - 300 MPa (10^6 N /m^2) 30 - 55 MPa (10^6 N /m^2)
	Temperature Range	-40 °C to 80 °C
	Water Absorption	0.02 % H_2O
	Maximum Operating Voltage	750 V / mil = 30 V /μm
	Breakdown Voltage	2000 V /mil = 100 V /μm
C	Capacitance	380 pF /cm^2 for 28 μm film $\varepsilon/\varepsilon_0$ = 12
c_v	Speed of Sound	(1.5 - 2.2) x 10^3 m /s (transverse-thickness)
d_{31}	Piezoelectric Strain Constant	23 x 10^{-12} m /V
d_{33}	Piezoelectric Strain Constant	-33 x 10^{-12} m /V
E	Young's Modulus	2 x 10^9 N /m^2
g_{31}	Piezoelectric Stress Constant	216 x 10^{-3} V m /N
g_{33}	Piezoelectric Stress Constant	-339 x 10^{-3} V m /N
k_{31}	Electromechanical Coupling Factor	12 % (@ 1 kHz)
k_{33}	Electromechanical Coupling Factor	19 % (@ 1 kHz)
t	Thickness	9, 28, 52, 110, 220, 800 μm (10^{-6} m)
ε	Permittivity	(106 - 113) x 10^{12} F /m
$\varepsilon/\varepsilon_0$	Relative Permittivity	12 - 13
ρ_m	Mass Density	1.78 x 10^3 kg /m^3
ρ_e	Volume Resistivity	10^{13} ohm-meters
tan δ_e	Loss Tangent	0.015 - 0.2 (@ 10^1-10^4 Hz)

Fig. 5.5 Typical properties of piezoelectric film.

5.3.2 Discussion

Piezoceramics are brittle materials which are manufactured as small simple-shaped parts with a relatively small surface area. These brittle piezoelectric elements are typically manufactured in 2 inch by 3 inch sheets in thicknesses varying from 4 x 10^{-3} to 12 x 10^{-3} inches. In sharp contrast to the mechanical characteristics of piezoceramics, PVDF materials are flexible plastic materials that are manufactured in thin sheets which can be readily cut into a wide variety of different shapes and sizes. This tough lightweight plastic material is manufactured in various thicknesses from 0.4 x 10^{-3} to 30 x 10^{-3} inches, and it is easily shaped prior to bonding to a substrate with commercial adhesives.

The significantly different mechanical properties of these two classes of piezoelectric materials can be exploited in different applications. Thus, for example, the high rigidity of ceramic elements ensures that electrical energy is efficiently converted into mechanical energy which ensures good actuation capabilities. On the other hand, PVDF films are characterized by a low charge characteristic, thereby rendering PVDF plastic materials to be weak electro-mechanically compared with the piezoceramic elements in low frequency applications and near resonant frequencies. However, PVDF elements are more sensitive to mechanical loads over a wider range of loading conditions than piezoceramics, consequently they are good candidates for sensor applications. For example, the diverse sensitivity of polymeric films to mechanical deformation has permitted PVDF materials to measure the impact of micron-sized particles in space, and also the stresses developed in ballistic tests by armor-piercing shells. The pressure sensitivity of these polymeric materials ranges from micro-torr to mega-bar.

Polyvinylidene fluoride (PVDF) films have wide-band frequency dependent characteristics from near D.C. to frequencies in the GHz range depending upon the thickness of the film. For example, a 28 μm sheet has a fundamental resonant frequency near 40 MHz; very thin films have resonant frequencies in the GHz range. The relative dielectric permittivity of PVDF's is typically of the order of 12 while piezoceramics are of the order of 1200 or higher. Consequently the piezoelectric stress constant g_{31} is approximately 216 x 10^{-3} (V/m)/(N/m^2), which is typically 20 times greater than that of a piezoceramic material.

PVDF materials possess dielectric strengths of the order of 40 V/μm while the dielectric strengths of typical piezoceramics are of the order of 2 V/μm. Thus polymeric materials can be exposed to much higher electric fields than their piezoceramic counterparts. PVDF films are approximately one-third the weight of piezoceramic materials. Furthermore, because they are extremely compliant, PVDF film sensors can be bonded to lightweight flexible structures undergoing large dynamic deflections without significantly distorting the elastodynamic characteristics of the original structure. However, because PVDF materials are so compliant they are not as effective as the piezoceramics in the role of electromechanical actuators for smart material applications, especially at low frequencies.

Polyvinylidene fluoride films are extremely robust when subjected to diverse chemical environments with high humidity, solvents, acids, and ultra-violet radiation, but, the material is quite sensitive to electromagnetic radiation, mandating appropriate shielding in certain applications. In addition, PVDF films are temperature sensitive, resulting in decreasing piezoelectric performance with increasing temperature. A typical limiting temperature for PVDF applications is 100°C, however, piezoceramic elements may generally be employed in environments twice this temperature.

5.4 Smart materials featuring piezoelectric elements

A cursory review of the technical literature reveals that there are numerous companies marketing polymeric and ceramic piezoelectric elements for diverse classes of commercial products. It is evident upon reading this technical literature that polymeric and ceramic piezoelectric elements fabricated in a thin-sheet form can be utilized as electro-mechanical transducers for microphones, relays, strain gauges and thermistor-based devices for example, depending upon the shape of the element, the direction of polarization of the element, and whether the element is electrically or mechanically excited.

Consider, for example, two piezoceramic sheets in strip form bonded together and fixed at one end to form a cantilever beam structure as shown in Figure 5.6(a), where the arrow associated with P denotes the direction of

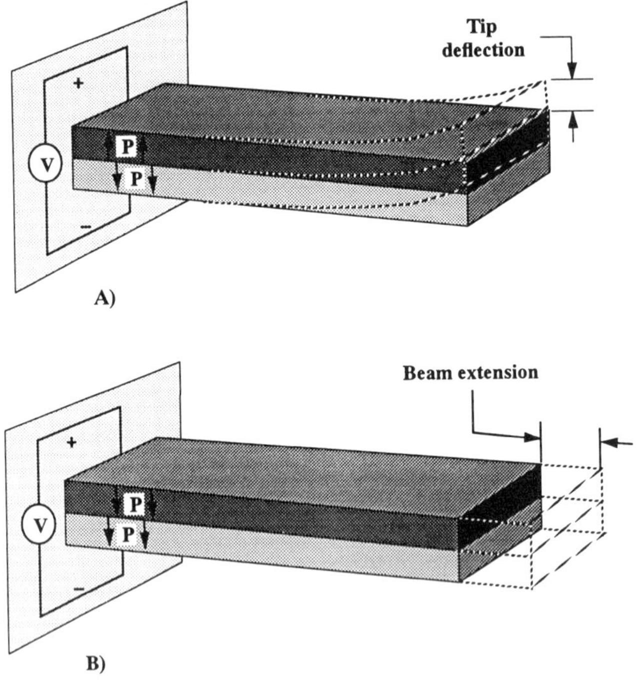

Fig. 5.6 Piezoelectric cantilever beam configurations upon application of an external voltage.

polarization. When configured in this way, the application of a potential difference across the piezoelectric sheets results in the axial contraction of one sheet and the axial extension of the other sheet. The outcome of this electrical excitation is a mechanical deformation field that generates a transverse beam deflection at the tip, almost an order of magnitude greater than the axial deformation.

If the direction of polarization of the upper piezoelectric element is reversed, as shown in Figure 5.6(b), then the application of the same potential difference across the beam will result in an axial extension of the beam. Conversely, if a negative potential difference is generated, then the beam will contract.

The two illustrative examples of piezoelectric beams presented in Figure 5.6 provide the nucleus for synthesizing smart structures featuring piezoelectric actuators. Consider, for example, an array of several piezoceramic thin-sheet actuators which are bonded to the upper and lower faces of a vibrating beam. These actuators can be employed to exert restoring forces and bending moments to the structure so that the vibrational response characteristics may be tailored to comply with specified performance characteristics. The positioning of the piezoelectric elements on the upper and lower faces of the beam permit the maximum bending moment to be exerted on the structure for prescribed element properties. Furthermore, these lightweight actuators do not adversely affect the mass or stiffness properties of the original structure, consequently the original dynamic response characteristics of the structure are unchanged when the piezoelectric elements are not activated.

The generic idea presented in Figure 5.6 of electrically actuating piezoelectric elements in order to develop a desired mechanical deformation of the elements, can be extended to the synthesis of smart structures with controllable dynamic response characteristics. This is accomplished by simultaneously considering the mechanical properties of the load-bearing structure, the electro-mechanical properties of the piezoelectric materials employed for the actuators and sensors and their spatial distribution in the smart structure, the mechanical properties of the bond at the interface between the piezoelectric elements and the substrate, the control algorithm for the system, and the desired response characteristics.

Figure 5.7 presents schematic diagrams of a smart material featuring embedded piezoelectric elements in the material at locations removed from the axis of symmetry, or neutral axis. Typically this is accomplished with polymeric composite laminates where the manual manufacturing process of hand laying up each ply individually, prior to debulking and autoclaving, permits piezoelectric actuators or sensors to be sandwiched between the plies, or alternatively, these thin piezoelectric sheet-like elements may be embedded in recesses cut in the composite prepreg material. The thickness of the prepreg tape is typically of the order 3×10^{-3} to 10×10^{-3} inches

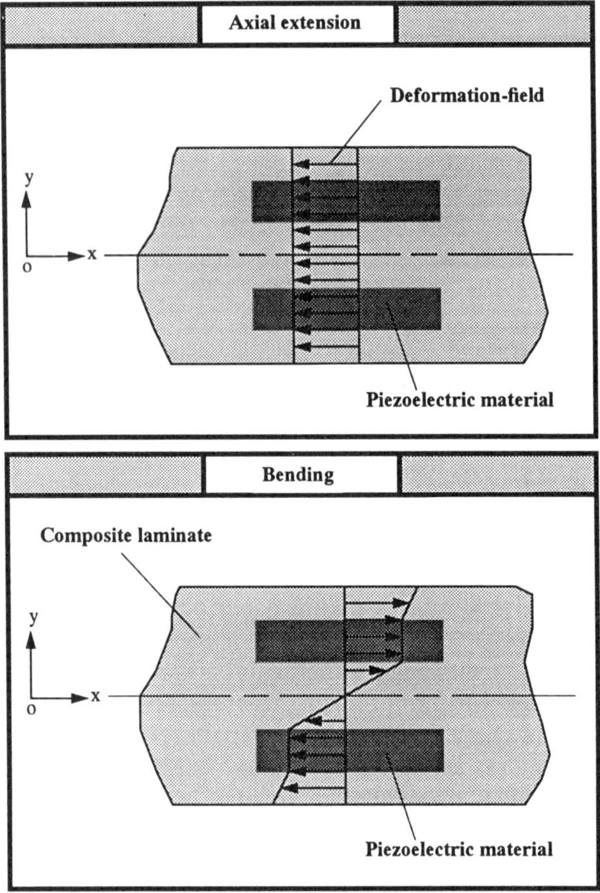

Fig. 5.7 Smart composite laminate with embedded piezoelectric actuators.

which is comparable to that of the piezoelectric elements which range from 4×10^{-3} to 10×10^{-3} inches. There are, of course, concerns associated with the structural integrity of this class of smart materials due to stress raisers caused by the embedment of the small piezoelectric actuators and also the electrical wiring which typically connects the piezoelectric elements to a microprocessor outside the smart material. This field is the subject of ongoing research.

By appropriately configuring the direction of polarization of the piezoelectric elements relative to the plane of symmetry of the material and also the excitation voltage, the actuators can be employed to generate axially time-dependent contractions or extensions of smart structures in the 01 direction, or alternatively, the actuators can be employed to generate time-dependent bending moments and hence flexural deformations of the smart structure. By coupling the individual piezoelectric actuator to appropriate control algorithms and sensors, the basis then exists for creating a smart structure capable of controlling a variety of dynamic performance characteristics. In all applications, issues of bonding must be addressed in addition to the tailoring of the stiffness of the piezoelectric materials and the stiffness of the substructure or load-bearing domain of the smart material. Ideally, the bond between the piezo-electric material should be perfect and the stiffness of the actuator material should be greater than the substructure, in order that the piezoelectric actuator material can generate relatively large strains in the substructure and hence modify, in a controlled fashion, the global response of the smart material.

In one extreme, if a piezoelectric actuator is bonded to the surface of a thin compliant beam then the beam and the actuator would deform in unison and the actuator would dictate the mechanical behavior of the beam. However, in the other extreme, if the same piezoelectric actuator were bonded to the surface of a very thick beam with a high modulus of elasticity relative to that of the actuator, then the beam would not experience significant deformation and the piezoelectric element would therefore have a minimal influence on the dynamic behavior of the beam.

Smart materials featuring piezoelectric elements embedded in fiberous composite laminates are currently manufactured by a hand-layup technique, in which the prepreg tape is appropriately orientated and stacked, and piezoelectric elements and the associated leads are embedded prior to being subjected to a prescribed pressure and thermal sequence in an autoclave to cure the resin matrix system. This thermal sequence typically subjects the laminate to temperatures above 250°F for the commercial resin systems. It is clearly important, therefore, to ensure that the Curie temperature of the embedded piezoelectric element is greater than the maximum temperature experienced by the smart laminate during manufacture. Since the Curie temperature of a piezoelectric material defines the thermal limit of the piezoelectric properties of the material, if this temperature is exceeded during the manufacturing cure cycle of the composite laminate, then the material will lose its piezoelectric properties and consequently the smart material will lose its actuation or sensing capabilities. This thermal limitation is a major concern when embedding PVDF films in laminates because the upper operating temperature of these materials is typically 200°F.

Piezoelectric actuators and sensors require an electrical potential to be developed across the electrodes on the upper and lower faces of the PVDF or PZT film in order for them to operate successfully. Since graphite-epoxy laminates are high-performance fiberous composite materials which are ideal candidate materials for this class of smart structure, the designer must ensure that the electrical conductivity of the graphite epoxy substructure does not result in the piezoelectric elements being short-circuited by direct contact between the substructure and the element electrodes or the connecting leads. This situation, which does not occur with glass-epoxy systems, is typically overcome by insulating the piezoelectric element and the associated leads from the substructure by wrapping them in a very thin insulating material which is subsequently bonded in place. The selection of the insulating medium and the bonding of this medium to the piezoelectric actuator, or sensor, and the substructure must be performed carefully in order not to reduce the effectiveness of the element.

The philosophy of employing piezoelectric elements as actuators and sensors in smart structures has been widely reported in the literature. Some

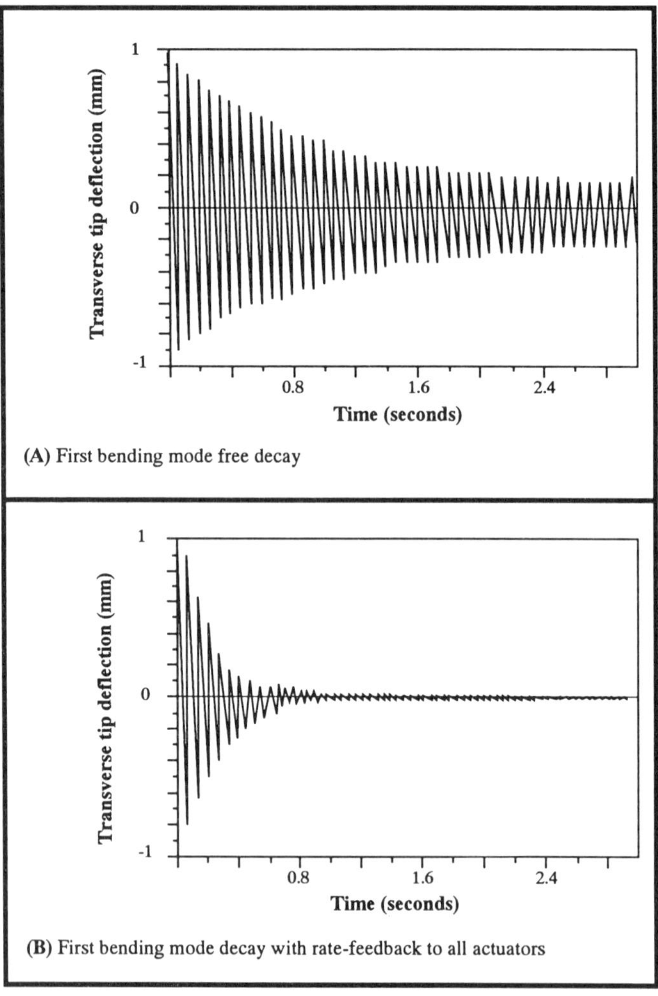

Fig. 5.8 Control of a smart beam featuring embedded piezoelectric actuators.

investigators have reported on studies involving the embedding of piezoelectric elements in fiberous polymeric composite laminated plates and beams while others have undertaken studies where the piezoelectric elements have been bonded to the surface of the structure. Figure 5.8 illustrates experimental results of the transverse tip deflection of an aluminum cantilevered beam featuring four sets of piezoceramic actuators. A simple rate-feedback controller was employed to attenuate the response in the first bending mode. This was accomplished by first integrating the

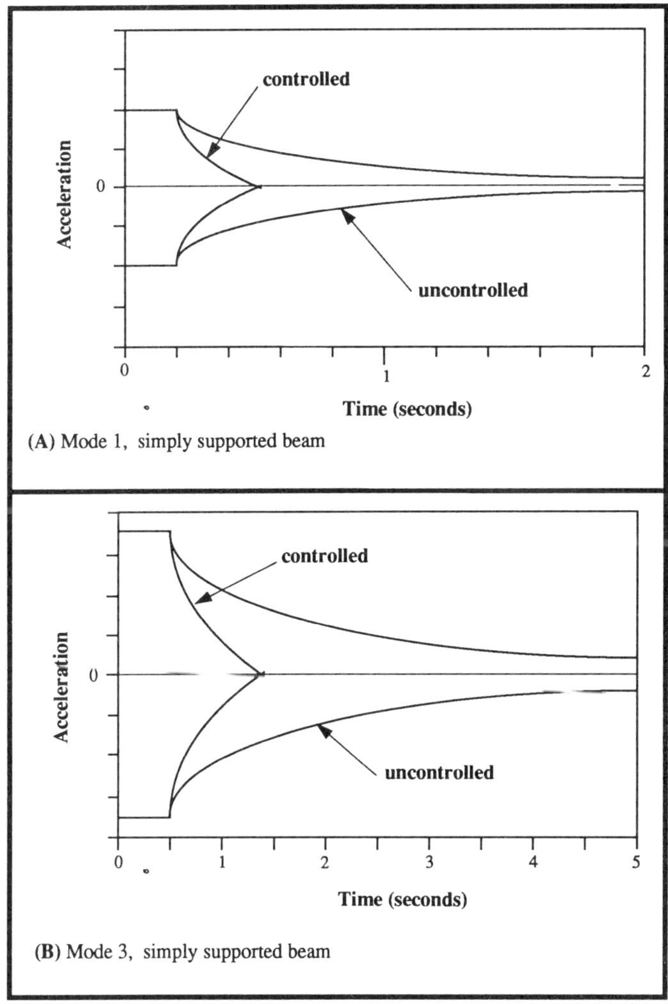

Fig. 5.9 Acceleration response envelope of a smart beam featuring embedded piezoelectric actuators.

signal from an accelerometer positioned at the tip of the beam prior to amplification and activating the piezoceramic elements in order to develop an order of magnitude increase in the attenuation of the signal. A similar profile is presented in Figure 5.9 where the focus of attention is the acceleration characteristics of a simply supported beam.

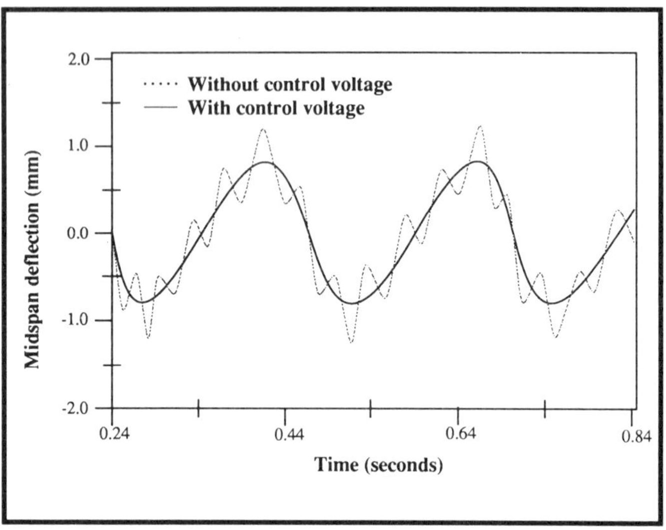

Fig. 5.10 Midspan transverse deflection of the connecting-rod at an operating speed of 250 rpm.

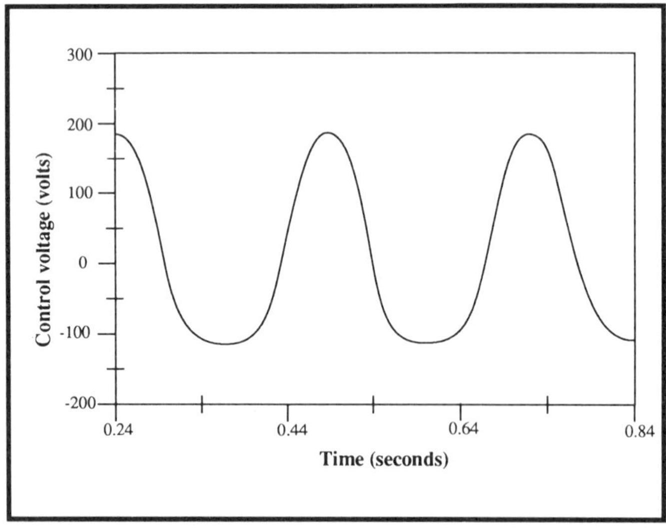

Fig. 5.11 Control voltage characteristics for activating the piezoelectric actuators bonded to the connecting-rod.

Figure 5.10 presents theoretical results for the midspan transverse deflection of a flexible connecting-rod of a slider-crank mechanism operating at 250 rpm. The flexible articulating aluminum member featured two sets of surface bonded piezoceramic actuators. The dotted line presents the typical elastodynamic response of the articulating member which comprises a low frequency quasi-static response at a frequency the same as the operating speed of the mechanism, and a higher frequency component near the fundamental natural frequency of the flexible connecting rod in flexure. A linear optimal controller was employed in conjunction with modal analysis to generate the continuous oscillatory line shown in Figure 5.10 by employing the control voltage depicted in Figure 5.11 to activate the piezoceramic actuators bonded to the surface of the aluminum beam.

Clearly these three examples are not all-embracing illustrations of the field of smart structures featuring piezoelectric elements. However, they do serve to illustrate the potential for this class of smart piezoelectric systems to control the elastodynamic response of both structures and also members of articulating mechanical systems such as mechanisms, machines, and robotic systems.

6

Shape-memory materials

6.1 Background on shape-memory alloys

The shape-memory phenomenon manifests itself when a shape-memory-alloy (SMA) is plastically deformed in the low temperature martensitic condition, and upon removal of the external loads regains its original shape when heated. The exact mechanism by which the shape recovery takes place is not very well understood, however, the process of regaining the original shape is known to be associated with a reverse transformation of the deformed martensitic phase to the higher temperature austenitic phase.

Typical materials which exhibit the shape-memory effect include the copper alloy systems of Cu-Zn, Cu-Zn-Al, Cu-Zn-Ga, Cu-Zn-Sn, Cu-Zn-Si, Cu-Zn-Ni, Cu-Au-Zn, Cu-Sn, and the alloys of Au-Cd, Ni-Al, and Fe-Pt. Nitinol, a nickel-titanium alloy, is the most common of the SMAs or transformation metals.

Nickel-titanium alloys (Nitinol, NiTi) acquire their name from Ni (Nickel)-Ti(Titanium)-NOL (Naval Ordnance Laboratory). NiTi alloys, featuring a near-equiatomic composition, can be plastically deformed in their low temperature martensitic phase and then be restored to the original shape by heating them above the characteristic transition temperature. Typically plastic strains as high as 6% to 8% can be completely recovered by heating the Nitinol in order to transform it to the austenitic phase, and constraining it from regaining the memory shape can result in stresses of 100 000 psi, for example. The typical material characteristics for Nitinol are presented in Figure 6.1 from which it is evident that the yield strength of martensitic Nitinol is approximately 12 000 psi.

There is a substantial database on the thermal, electrical, magnetic, and mechanical characteristics of Nitinol, however, the influence of residual stresses and high temperatures associated with the fabrication and processing of Nitinol-based composites, for example, is not well understood.

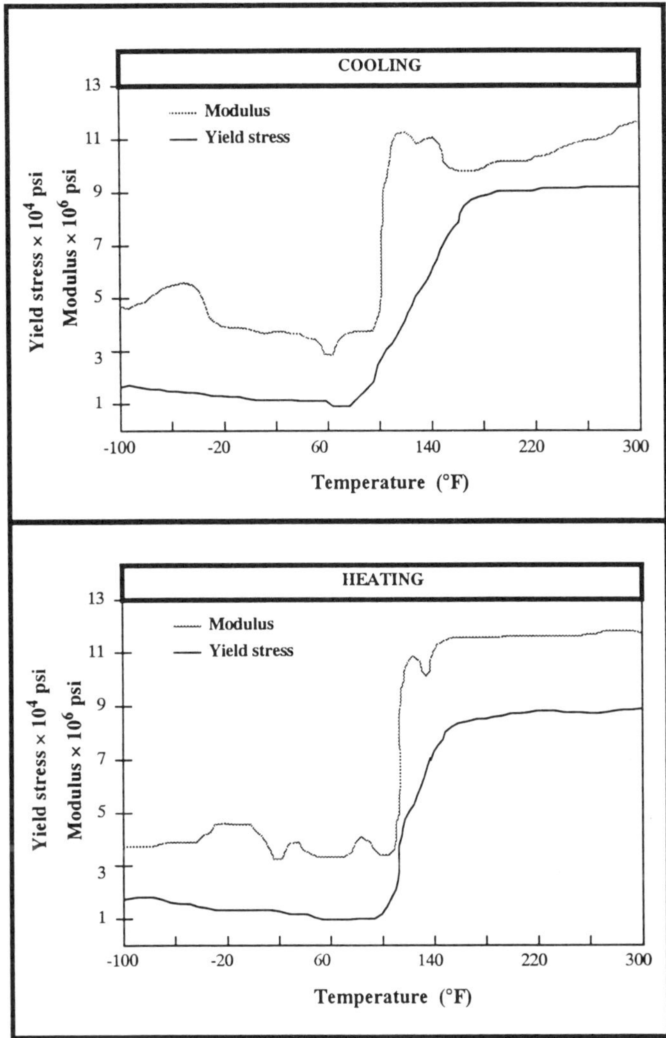

Fig. 6.1 Typical material characteristics for Nitinol.

Furthermore, the extent, duration and repeatability of the shape-memory effect as well as the dynamic actuator and sensing characteristics of Nitinol are also not understood. Shape-memory effects in alloys and metals are characterized by deformation mechanisms which are associated with shape changes due to martensitic transformations. Therefore a thorough understanding of martensitic transformations is an important prerequisite to the understanding of the shape-memory effect in alloys and metals.

The transformation of steel in a high-temperature austenitic phase to a low temperature martensitic phase was first observed by the German metallurgist Adolf Martens. The extremely fine structure observed in the martensitic phase results from a lattice transformation without atomic diffusion. The face-centered cubic austenite transforms into body-centered cubic lattices or body-centered tetragonal lattices. These diffusion-free martensitic transformations manifest themselves in a variety of alloys and metals.

Martensitic transformations fundamentally involve a lattice transformation featuring shear deformation and a coordinated atomic movement, which maintains a one-to-one lattice correspondence between the lattice points in the parent phase and the transformed phase. The martensitic phase is a substitutional or interstitial solid solution. The transformation is diffusion-free which yields the same concentration of solute atoms dissolved in the martensitic phase as in the parent phase. Martensitic transformations are typically characterized by well-defined shape changes or surface reliefs.

Figure 6.2 shows the formation of a typical surface relief due to the

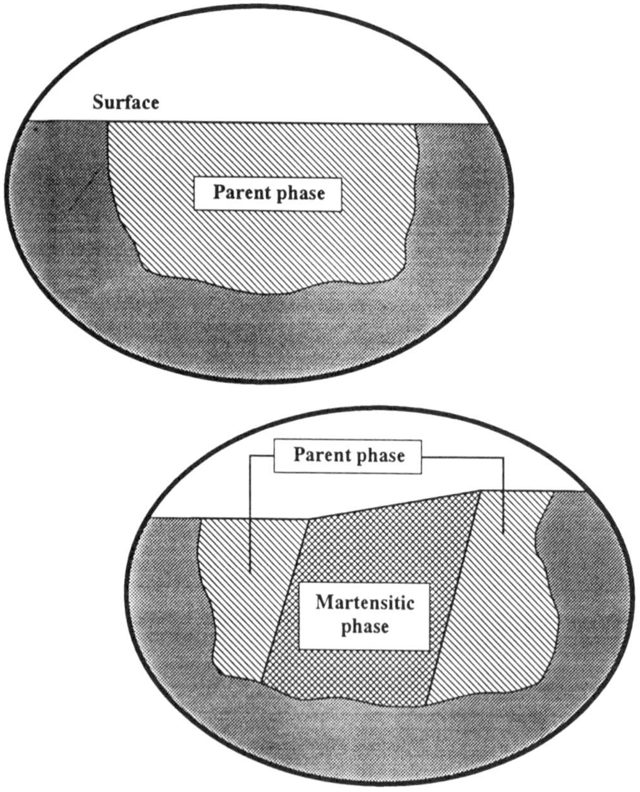

Fig. 6.2 Surface relief characterizing martensitic transformations.

manifestation of a martensitic phase in some zones of the surface due to the phase transformation. Furthermore scratch lines etched on the surface of a typical specimen in the parent phase become distorted at the boundaries between the phases as a consequence of martensitic transformation. The orientation of the surface relief and the distortion of scratch lines are very well-defined through quantitative descriptions which are strongly dependent upon the crystal orientation of the parent phase. The formation of surface reliefs and distortion of scratch lines clearly demonstrates that shearing deformation is involved in the transformation mechanism.

Martensitic crystals are necessarily characterized by lattice defects due to the complementary slip and twinning deformations which are induced by the shearing deformations which change the parent phase lattice. These complementary deformations are characterized by the lattice invariant strain, and dislocation stacking and twinning faults associated with these complementary deformations have been experimentally observed through electron microscopes. The twinning faults play a significant role in the shape-memory effect.

The transformation from the parent phase to the martensitic phase can be induced only when the chemical free energy of the martensitic phase is lower than that of the parent phase. Furthermore, the difference between the chemical free energies of the two phases must be greater than the nonchemical free energy, such as strain energy or interface energy, in order to provide the excess of non-chemical free energy which is the essential driving force for the transformation.

The equilibrium temperature, which is defined as the temperature at which the chemical free energy of the martensitic and parent phases are equal, plays a critical role in the transformation thermodynamics. The material must be suitably cooled to a specific temperature below the equilibrium temperature in order to ensure a sufficient excess of chemical free energy which subsequently initiates the transformation as shown in Figure 6.3. Martensitic transformations can advance only if the temperature is below a critical value for the initiation of the transformation which is denoted by M_I in Figure 6.3. In ferrous alloys this athermal nature of the transformation does not permit the martensitic crystals to continue to grow after they are formed. The nucleation of new martensitic crystals in the remaining solid parent phase moves the transformation forward with the crystals growing at approximately one-third the speed of elastic waves in this class of solids.

In addition to the athermal transformation discussed above, it is possible to initiate transformation after an incubation period at a constant temperature. This mechanism for isothermal transformation can be triggered if the material is maintained at a temperature above M_I, or if the material is super-cooled below M_I and this temperature maintained after an athermal transformation has been induced. The extent of transformation

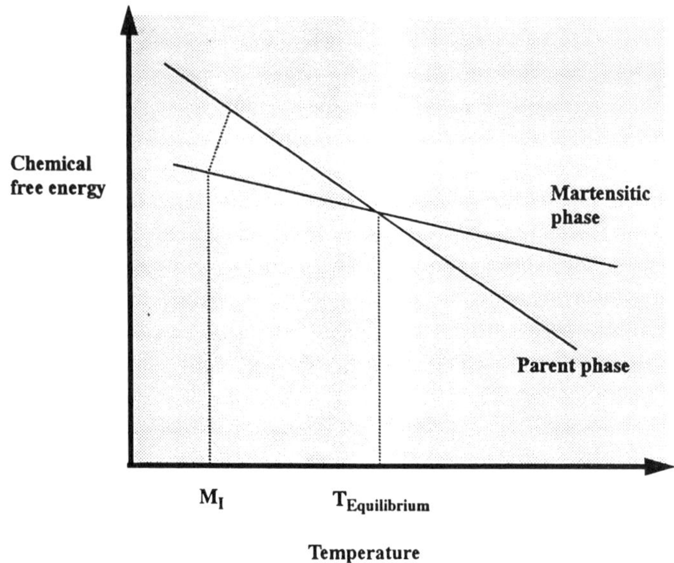

Fig. 6.3 The effect of temperature on the chemical free energies in the martensitic transformations.

increases with time, and new martensitic crystals are created and grow at very high speeds.

Both the athermal and isothermal transformations are typically characterized by the appearance of individual martensitic crystals, which grow rapidly to their final size. No additional growth in these crystals is stimulated either by a rapid reduction in temperature or the passage of time, which is typical of non-thermoelastic transformations. In sharp contrast to this scenario, thermoelastic transformations are characterized by the nucleation of the martensitic crystals, which grow at a velocity proportional to the cooling rate during falling temperatures, and result in crystal growth rates which may even be visible to the naked eye. Similarly, input of additional heat energy results in the volumetric shrinking of the crystals. These thermoelastic martensitic transformations are critical for the manifestation of the shape-memory effect. Similar growth and shrinkage of crystals can also be observed when external forces are applied.

Thermoelastic martensitic transformations can occur only when the interface energy and the energy required for plastic deformation is negligible. These conditions are fulfilled if the transformation involves very small structural and volume changes, while manifesting a good coherence between the parent phase lattice and the martensitic phase lattice. Typically, if the parent phase has an ordered structure these conditions are satisfied. Most of the alloys which permit thermoelastic martensitic transformations have super-lattice structures where the fundamental lattices are body-centered cubic (BCC). Even in alloys, such as InTl, FePc, and MnCu, which feature disordered lattices, the parent phases are exceptionally face-centered cubic (FCC). Barring a few exceptions, almost all SMAs, which admit thermoelastic martensitic transformations feature super-lattices with BCC structures and are classified as Beta-phase alloys.

Martensitic transformations are characterized by a reverse transformation which is absent in the deformation fields associated with slip or twinning, therefore the deformation behavior of alloys undergoing martensitic transformations is significantly different from those pertaining to ordinary metals and alloys. The fundamental characteristic of the reversible shape-memory effect is that regardless of the severity of the deformation of the low-temperature martensitic phase, heating induces a reverse transformation which allows the specimen to return to the original shape corresponding to the high temperature austenitic phase. If this shape does not change upon cooling the specimen and transforming it to a martensitic phase, the phenomenon is referred to as a 'one-way' shape-memory effect. Under certain conditions a partial reversion of shape may occur due to a 'one-way' or 'partially reversible' shape-memory effect. Generally, even when the specimen does not recover 100% of its shape, the shape-memory effect is still referred to as 'reversible.' A reversible shape-memory effect can be induced in SMAs by employing the following techniques:

a) subject the martensite to substantial deformation beyond a prescribed limit;
b) subject the parent phase to deformation, and cool the specimen under suitable constraints, and maintain this state under stress for a prolonged period of time;
c) deform the martensitic phase, and heat the specimen in order to induce the reverse transformation;
d) deform the specimen after creating minute precipitation in the parent phase.

All these techniques create sites of internal stress through some mechanism within the high-temperature parent phase, which is essentially homogeneous and has the ability to recover its shape in the reverse transformation upon heating. Furthermore, these sites of internal stress control the martensitic transformation which is initiated by cooling. Therefore by employing techniques a, b, and c discussed above, the sites of internal stress which are created, are irreversible defects such as deformation-induced dislocations. Technique d results in sites which consist of stable stress-induced martensite, which does not undergo any reverse transformation upon heating.

Actuator and sensor materials for smart materials applications are typically characterized by a structural re-organization at the molecular or mesoscopic levels. In order to fully exploit the capabilities of these smart materials, a thorough analytical understanding of these molecular re-structuring phenomena must be developed. The entire subject of martensitic phase transformations (MPTs) is one of considerable technological and scientific interest. In recent years, theoretical understanding of the basic physical properties of these systems has developed by applying finite element techniques to solve continuum elastic models for cubic to tetragonal phase transitions, and also by considering simple atomistic model Hamiltonians which incorporate the essential ingredients of a martensitic system. There have also been some attempts to construct phenomenological Hamiltonians which incorporate the basic underlying physical characteristics of MPTs and solve these models using Monte Carlo techniques and molecular dynamics techniques. In fact it is now believed that the MPTs are basically first order displacive structural phase transitions. Such phase transitions have been of great interest to physicists working in the area of structural phase transitions in solids.

Since the underlying physics of martensitic phase transitions can be understood through studies of lattice displacements of atomic length scales, coupled anharmonic oscillator systems serve as the simplest models in order to study the dynamics of MPTs. No exact analytical studies of the dynamics of coupled anharmonic oscillators are currently available. Self-consistent phonon theories, while adequate for describing short-time dynamics are clearly inadequate for long-time dynamics whose knowledge is essential for the practical use of shape-memory systems, for example. Molecular dynamics simulations can provide valuable information pertaining to the dynamics but not for very long time-scales due to the limitations of length and time-scales in simulations.

Recently, an exact approach, the Recurrence Relation Method (RRM) has been developed for solving the dynamics of systems described by hermitian Hamiltonians. This method has been extremely successful in deriving several new results and recovering old ones for relaxation functions of both classical quantum and many-body systems. Current efforts in this area are found on studying a variety of different model Hamiltonians, which can be utilized to probe the thermodynamic properties and pseudo-elastic response characteristics of martensite systems.

Efforts are being currently focused on establishing constant stress molecular dynamics simulations of simple martensite systems using phenomenological, two-body effective potentials fitted to available electronic structure calculations which incorporate the finite wave vector phonon anomalies. In addition to calculating the MPT temperatures as functions of external stress and different softening parameters the relevant correlation functions could also be calculated and employed as Landau-Devonshire parameters in a long wavelength description of the system. The Landau-

Ginzberg theory, which allows for the macroscopic inhomogeneities, could then be solved to study the effect of finite size and other factors. This may require a knowledge of the surface effects on the Landau-Devonshire parameters which could be obtained from finite size simulations. This work could then be extended to investigate the role of impurities on the MPT through their effects on the Landau-Devonshire parameters.

6.2 Applications of shape-memory-alloys

There are more than ten basic alloy systems which exhibit the shape-memory effect. By combining these alloy systems, or by adding other elements to these systems, the number of alloy systems increases by an order of magnitude. For practical engineering applications, most of these alloys are impractical either because of their high cost or because several alloys can be used exclusively in their single crystal form. Therefore, for practical applications TiNi and CuZnAl and their combinations with small quantities of other elements are the only viable SMAs currently available in the marketplace.

TiNi and CuZnAl alloys feature significantly different properties due to the inherently different constituents from which they are comprised. Furthermore, the manufacturing process and the technology associated with both of these classes of SMAs are also quite different. It is clearly evident from a comparison of the properties of these two alloys that the performance characteristics of TiNi SMAs are superior to those of CuZnAl SMAs. Therefore, in applications where a highly reliable product with a long fatigue life is desired, TiNi alloys are the exclusive materials of choice. Typical applications of this kind include electric switches and actuators. However, if high-performance is not mandated and cost considerations are important then the use of CuZnAl SMAs can be recommended. Typical applications of this kind include safety devices, temperature fuses, fire alarms, etc.

Shape-memory-alloys are unique in the sense that when deformed at low temperatures they revert back to their original shape upon heating. However, some permanent deformation may remain in the alloy. In order to maintain a satisfactory shape recovery the deformation strain must be constrained from exceeding a critical value which is a function of a variety of processing parameters and the service environment. The critical strain, of course, depends upon the geometrical attributes of the product, the qualitative and quantitative nature of the loads, and the number of loading cycles that the product is designed to withstand. Although a precise determination of the critical strain value is rather difficult, for a smaller number of load cycles, the strain must be restricted to under 6% in TiNi and 2% in CuZnAl. Furthermore, when the product is designed for a very high number of load cycles, the strain must have an upper bound of 2% for TiNi and 0.5% for TiNi and CuZnAl respectively. Similarly, heating a SMA product to a temperature above some critical temperature is not recommended. When SMA products are subjected to such elevated temperatures for prolonged periods of time, the shape at the high temperature is partially memorized, which inhibits the product from accurately reverting to a previously memorized shape. This scenario suggests that prolonged exposure of SMA products to high temperatures beyond some critical temperature has a deleterious effect on the memory of the alloy in the sense that a fairly accurate memory is rendered 'blurry' or 'fuzzy'. The critical temperature for TiNi is approximately 250°C, and for CuZnAl the critical temperature is approximately 90°C. Extended exposure to thermal environments with temperatures above these critical temperatures results in an impaired memory function regardless of the magnitude of the load.

Another technical consideration in the practical application of SMAs pertains to fastening/joining protocols when SMAs are fastened and/or joined to other conventional materials. These significant issues arise due to the fact that SMAs undergo expansions and contractions which are as high as 2% to 6% under routine service scenarios. These changes in dimensions contrast sharply with the magnitude of expansions and contractions encountered in typical traditional materials which have an upper bound of approximately 0.5%. Therefore, if SMAs are welded or soldered to other materials they can easily fail at the joint when subjected to repeated loading. TiNi and CuZnAl can also be brazed using silver fillers, however, the brazed region can fail due to cyclic loading of only a few hundred cycles even when the strains are very small. It is therefore desirable to devise some other mechanism for joining SMAs to other materials as shown in Figure 6.4. It is evident from the figure that the SMA is in contact with the conventional material over a very small region, and furthermore, it is also free to move relative to the other material. For similar reasons SMAs cannot be plated or painted. However, in some cases, CuZnAl is plated or painted to avoid corrosion or stress corrosion cracking.

6.3 Continuum applications: structures and machine systems

Shape-memory-alloy-based smart structures are typically fabricated with reinforced composite materials. Fibers of SMA are embedded into either a matrix material, such as an epoxy resin system, or in a laminated fiber-reinforced composite material as shown in Figure 6.5. By changing the temperature beyond the phase-transition point, the shape and mechanical characteristics of the SMA, and hence the shape and global mechanical characteristics of the smart structure are changed too. In typical beam, plate and shell-like structures the SMA fibers or films are embedded in symmetric

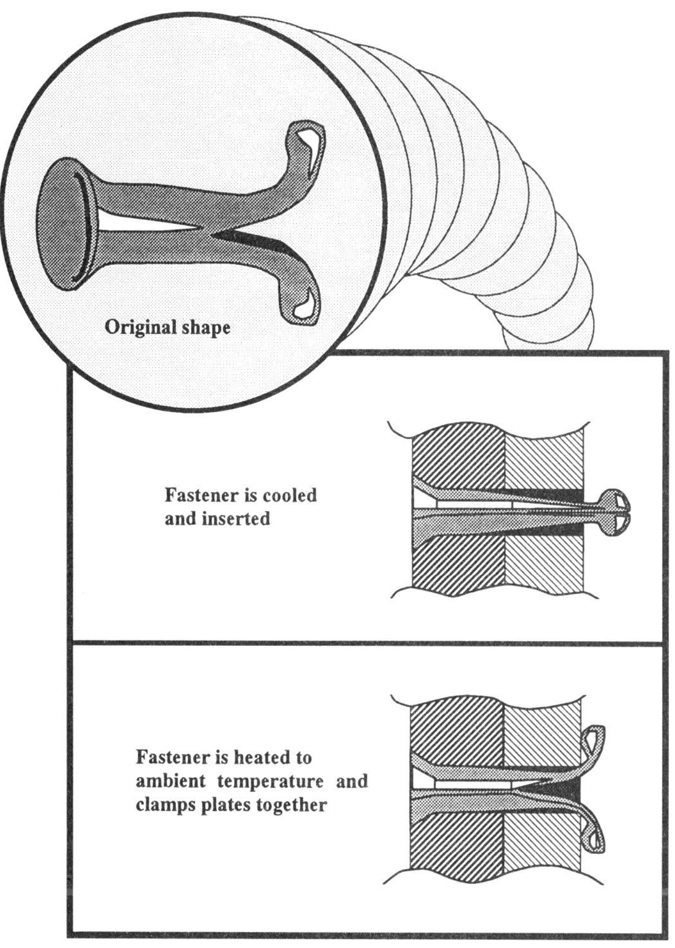

Fig. 6.4 A shape-memory-alloy fastener.

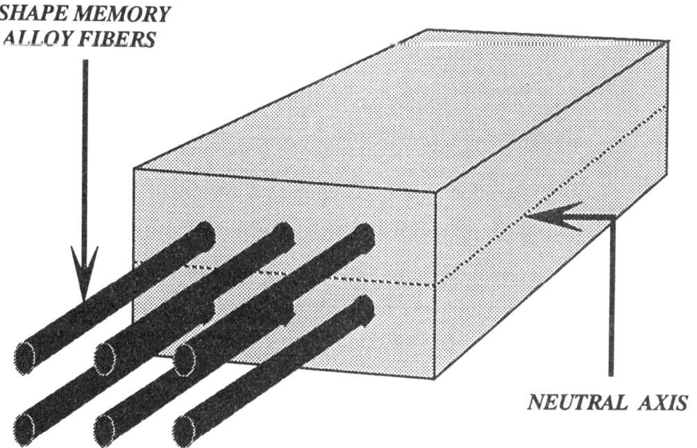

Fig. 6.5 Shape-memory-alloy-based smart structure.

pairs off the neutral axis/surface in order to control the geometry of the structure. Prior to the fabrication of the smart structure the SMA fibers are plastically deformed and are constrained in a configuration which is different from their memory configuration at the end of the curing cycle. Typically, in service, the fibers are electrically heated and once the phase-transition temperature is achieved the SMA fibers have a tendency to regain their memory configuration. In this process a distributed shear load is generated along the length of the fibers because of the off neutral-axis location, which results in the controlled application of a bending moment to the smart structure with the attendant change in configuration. Sleeves can be created within the composite laminates in order to accommodate the plastically-elongated SMA prior to clamping the SMA to both ends. When the SMA is heated, if the fibers were not clamped, they would contract. When one end of the beam is free, the fibers in a sleeve will exert a concentrated force on the ends of the structure in a direction that is always tangential to the structure at the point where the fibers are clamped to the structure. However, when both ends of the beam are fixed, heating the SMA results in fibers with a significantly increased stiffness and applied tension that will resist any transverse motion.

Steady-state vibration control which may also be used for structural acoustic control can be accomplished with SMA reinforced composites by employing 'Active Modal Modification.' The modal response of a structure can be tuned by heating the SMA fibers to change the stiffness of all or portions of the structure. For example, when Nitinol is heated to cause the material transformation from the martensitic phase to the austenitic phase, the Young's modulus changes by a factor of approximately four as shown in Figure 6.1. The stiffness is increased by a factor of four and the yield strength increases by a factor of ten. These changes in the material properties occur due to a phase transformation, and they do not result in the generation of any appreciable force and they do not need to be initiated by any plastic deformation.

The SMA fibers may be placed in or on the structure in such a way that when activated there are no resulting deflections but instead the structure is placed in a residual state of strain. The resulting stored strain energy changes the energy balance of the structure and modifies the elastodynamic response. This is referred to as 'Active Strain Energy Tuning.'

Active strain energy tuning utilizes both the embedded fiber and sleeve methods described earlier. The difference between the embedded fibers and the fibers in a sleeve is that in the first case the force of the SMA is distributed over the length of the fiber and in the latter case the force is concentrated at the end of the structure or is used to resist transverse motion. Both of the design concepts described above have been incorporated into prototypes and their potential demonstrated on a limited scale. Typical results for plate deflection under a uniform pressure load employing active

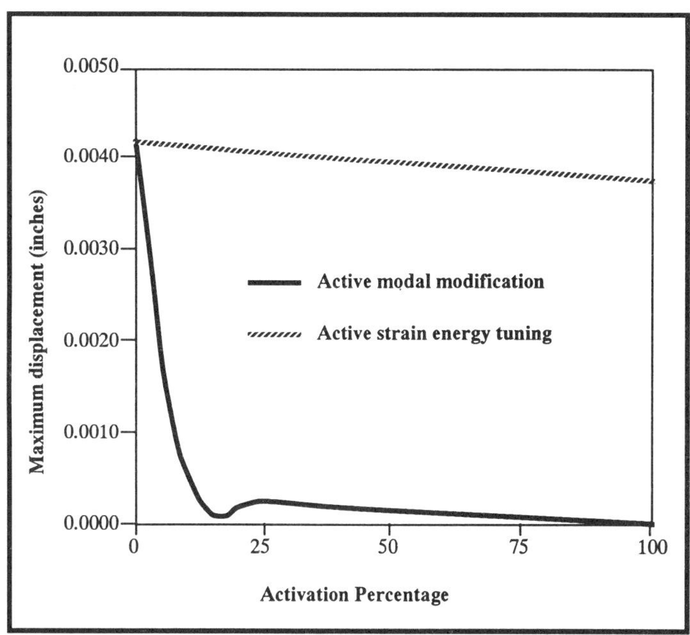

Fig. 6.6 Maximum plate deflection under a uniform pressure load using active shape-memory-alloy (SMA) control.

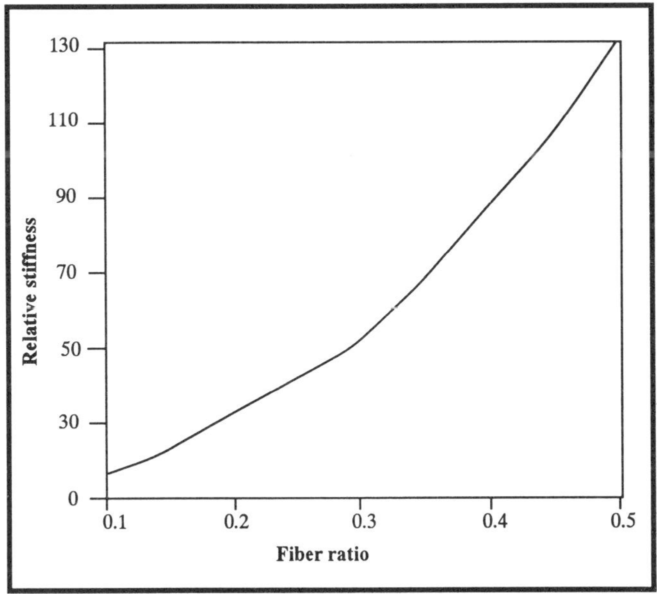

Fig. 6.7 Flexural stiffness tuning of quasi-isotropic plates using active strain energy tuning.

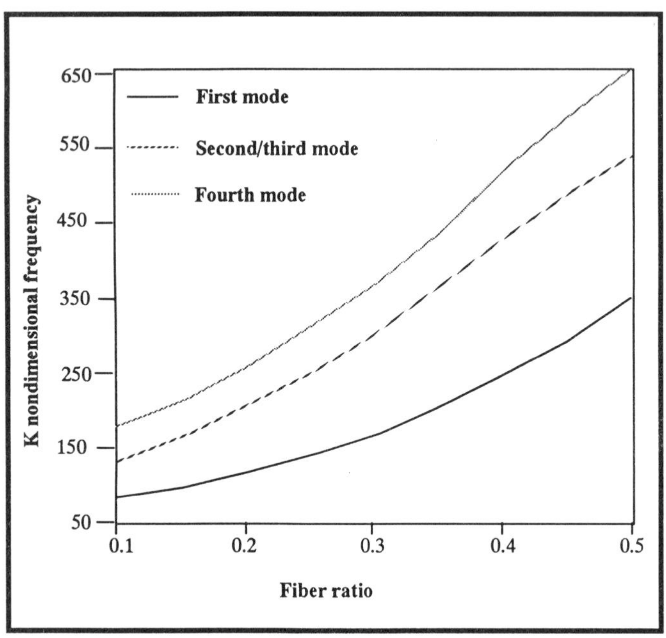

Fig. 6.8 Variation of natural frequencies as a function of fiber volume fraction.

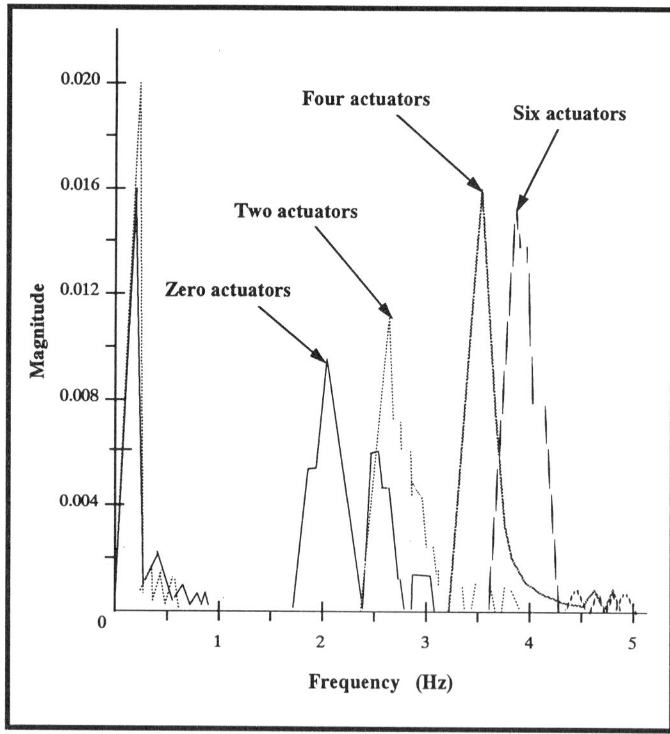

Fig. 6.9 Natural frequencies of a SMA composite beam with zero, two, four, and six actuators.

control of the shape memory actuators are presented in Figure 6.6. Results for the flexural-stiffness tuning of quasi-isotropic plates are shown in Figure 6.7, and the variation of natural frequencies as a function of fiber volume fraction and the number of actuators is presented in Figure 6.8 and Figure 6.9 respectively.

6.4 Discrete applications

Historically, SMA based devices have been discrete in nature with typical applications in switching and actuation. A typical application of this kind involving a temperature fuse is schematically presented in Figure 6.10. Shape-memory-actuators can be moved back and forth and rotated by heating or by employing other forms of control. Electrical energy is typically employed for heating purposes, however, hot water, hot air, high-frequency induction heating, microwave heating, laser light, infrared rays, etc. can also be employed for heating. Shape-memory-alloy actuators have the distinct advantage of small size, because they are constructed as a single unit of SMA with a simple internal material-based mechanism for actuation. This scenario is in sharp contrast with traditional actuators which employ more than one component and rely on more complex actuation mechanisms. Furthermore, SMA actuators are not influenced by other environmental factors such as humidity and vacuum, for example. Under such conditions, components of traditional actuators such as seals associated with motors and hydraulic cylinders, for example, encounter substantial difficulties in delivering optimum performance. Shape-memory-alloy actuators can therefore be employed as positioning devices for scanning electron microscopes. Other application areas include reaction vessels, nuclear reactors, chemical plants, etc.

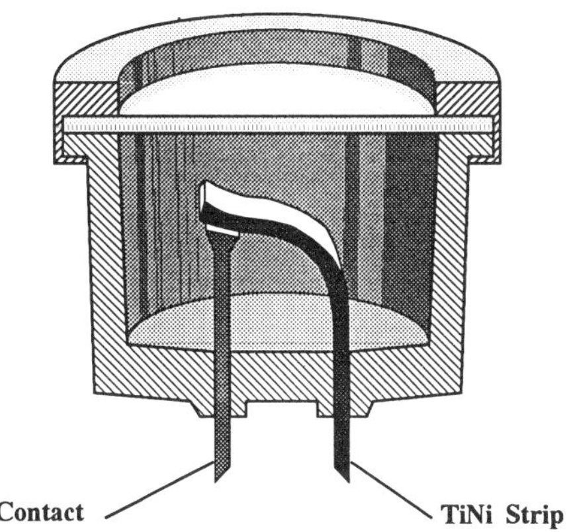

Fig. 6.10 A temperature fuse employing a shape-memory-alloy.

The Japanese have recently focussed their efforts on employing SMAs in the field of robotics. Two main types of SMA actuators for robots have been developed, namely, biased and differential. Biasing uses a coil spring to generate the bias force that opposes the unidirectional force of the SMA. In the differential type, the spring is replaced with another SMA and the opposing forces control the actuation. The assembly of a micro-robot actuated by SMAs is schematically shown in Figure 6.11. This micro-robot features five degrees of freedom corresponding to the capabilities of the human fingers, wrist, elbow, and shoulder, for example. The variety of robotic maneuvers and operations are coordinated by activating the TiNi coils in the fingers and the wrist in addition to contraction and expansion of straight TiNi wires in the elbow and shoulders. Digital control techniques exploiting a pulse current, in which the current is modulated with pulse

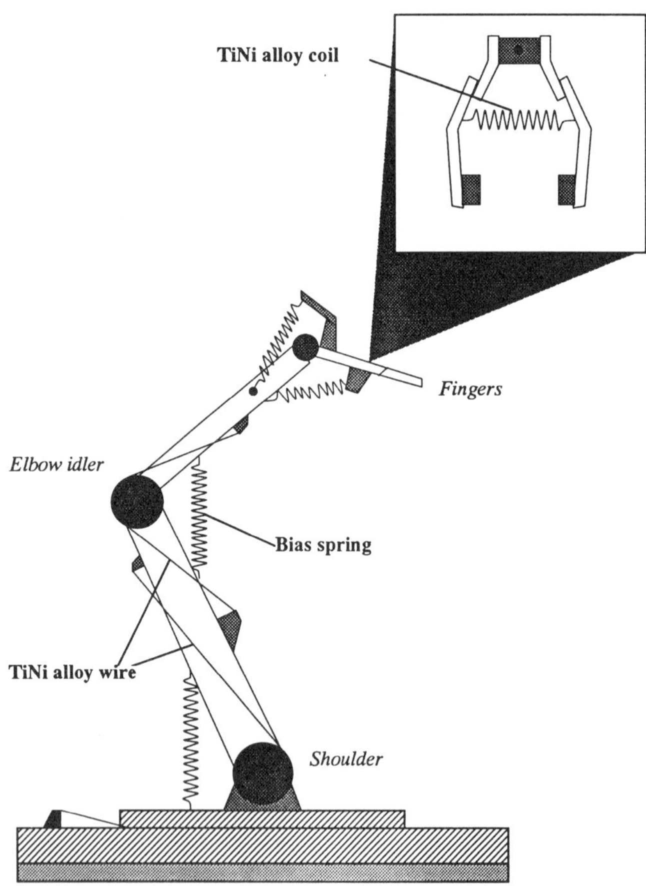

Fig. 6.11 A micro-robot actuated by shape-memory-alloys.

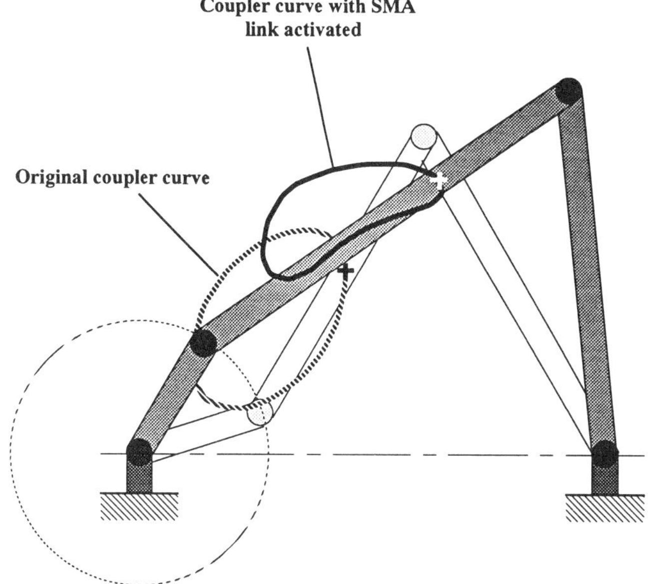

Fig. 6.12 Enhancing the coupler curve characteristics of a four-bar mechanism by utilizing shape-memory-alloys.

width modulation, are employed in all of the components in order to control their spatial positions and speeds of operation. Recently, researchers at the Intelligent Materials and Structures Laboratory at Michigan State University, have developed a new class of machine systems incorporating SMA actuators which have the capability to deliver variable output characteristics as shown in Figure 6.12.

TiNi sheets and rods twisted into cones and volute springs have been investigated for deployment in satellite antenna applications since the early Sixties. Once the satellite is launched and a stable orbit is established, the antenna is warmed by either employing electrical heating, or alternatively incident sunlight is utilized for heating purposes. Upon heating the SMA components of the antenna, the antenna then unfolds in space in a manner similar to the unfolding of an umbrella, as schematically shown in Figure 6.13.

In the Seventies, SMA based pipe couplings were developed. In these applications a TiNiFe alloy with a transformation temperature of −150°C was fabricated as a tube featuring an inner diameter, which was 4% smaller than the nominal outer diameter of the pipe which had to be joined. During the joining process the coupling was maintained at a low temperature by using liquid nitrogen. A tapered plug of suitable dimensions was then forced into the coupling in order to increase the diameter of the coupling by

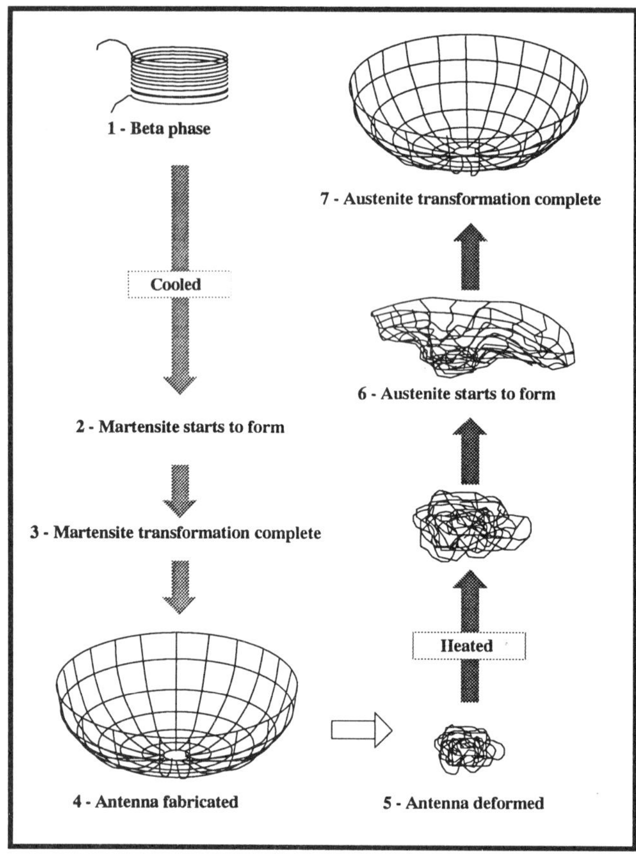

Fig. 6.13 The shape-memory-alloy memorization process.

approximately 8%. The two pipes to be joined were then inserted at the two ends of the coupling. As the coupling's temperature rose to the room temperature it had a tendency to shrink in order to return its inner diameter, in this case, to the dimension before the forced expansion. The ends of the two pipes are therefore held together simply, in service, with a high level of reliability. Figure 6.4 presents a schematic diagram of the pipe coupling. Thousands of such couplings have been employed in nuclear submarines, warships, and pipes on the ocean floor. Indeed, pipes as large as 150 mm in diameter have been connected with such couplings. F-14 fighter jets have employed such couplings without leaking or reliability problems. In addition to reliability, SMA couplings have significant advantages as compared to traditional couplings. Furthermore, since high temperatures are not involved with the assembly process there is minimal temperature-induced damage to the surrounding parts. This is in sharp contrast to the potential for thermal damage due to high temperatures associated with welding, for example. Furthermore, SMA couplings simplify overhauls since they can be disassembled by substantially reducing the temperature.

Shape-memory-fasteners are typically employed in situations where one side of the part to be fastened is not easily accessible. The SMA fasteners are cooled in liquid nitrogen where they feature a cylindrical configuration similar to a traditional bolt or rivet. The SMA fastener is inserted into the hole in the part. Upon reaching room temperature the end of the fastener in the inaccessible region returns to the memorized position in which the end of the fastener is spread apart as shown in Figure 6.4. In this configuration the part is adequately fastened. Shape-memory-alloys are also employed as electrical connectors, clamps and seals.

Several applications of SMAs have been explored in the medical field. In

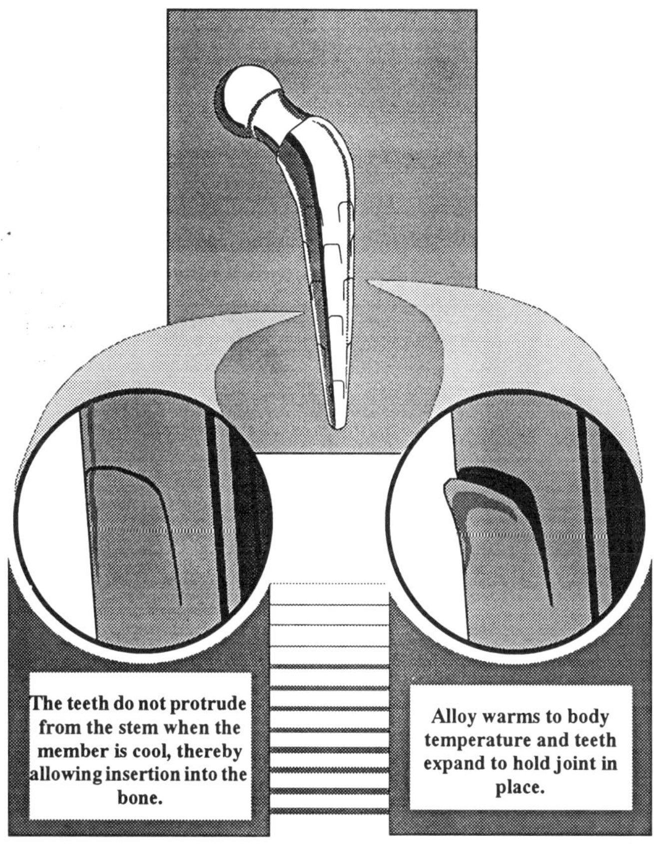

Fig. 6.14 A shape-memory-alloy artificial hip joint.

medical applications, in addition to mechanical characteristics, highly reliable biological and chemical characteristics are also very important. The material must not be vulnerable to *in vivo* degradation, decomposition, dissolution, or corrosion. Furthermore, the material must be biocompatible and not result in cytotoxicities, such as carcinogenicity and genetic abnormalities, for example. Therefore, TiNi alloys are the only SMAs that are being exclusively considered for medical applications. However, because of limited studies of TiNi for medical applications the biocompatibility of TiNi SMAs has not yet been fully established.

TiNi SMAs have been employed in the development of artificial joints including femoral heads and sockets. A typical application of SMAs in artificial hip joints is illustrated in Figure 6.14. TiNi SMAs are employed for bone plates and marrow pins for healing fractures in the femur, tibia, and the intestrocharter bones, and superelastic TiNi wires are employed for

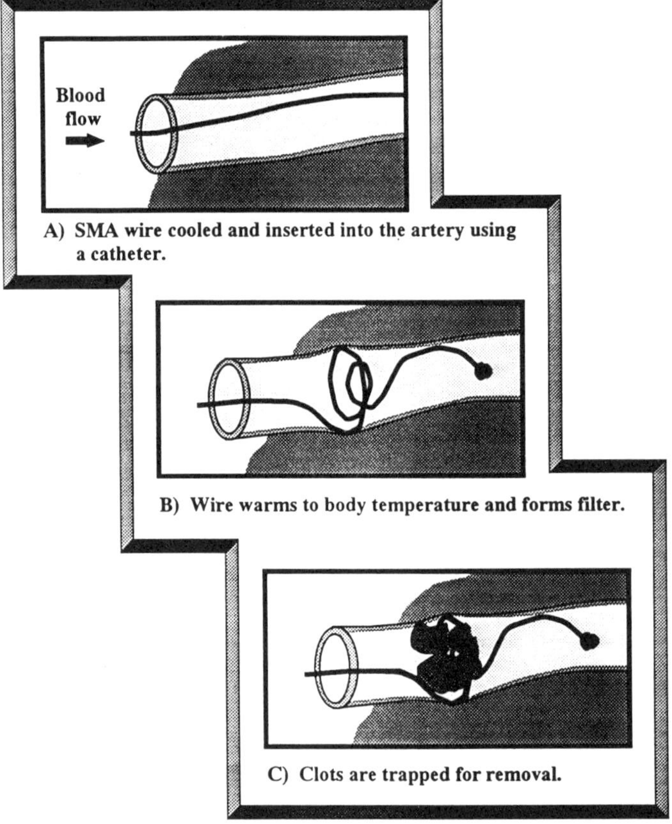

Fig. 6.15 Shape-memory-alloy blood clot filter.

connecting broken bones. An artificial heart featuring TiNi wires as the drive elements has also been developed. Prototype filters employing TiNi wires for trapping blood clots moving through the vena cava have been developed for an exploratory procedure. These wires feature a memorized configuration which permits them to act as filters and trap blood clots. The reference configuration corresponds to a straight wire, which is guided into the vena cava through a catheter. Once inside the vena cava, the TiNi wire is heated to the temperature of the blood in the biological tubular vessel. This change in temperature triggers the reversion of the wire to a complicated memorized shape for a blood filter as shown in Figure 6.15. Several other applications such as orthodontic wires and dental castings have also been considered in the medical field. Shape-memory-alloys have tremendous potential for applications in a variety of areas. Some of the impediments to accelerating the application of SMAs in such applications are discussed in the next section.

6.5 Impediments to applications of shape-memory-alloys

The following issues are perceived to be impediments for the commercialization of several SMA applications discussed in the previous section.

Slow response-time

The response-time of smart materials featuring SMAs is greater than that for the competing technologies due to the inherent thermal time-constants associated with this class of material.

Hysteresis

The hysteresis characteristics of SMAs must be carefully evaluated and clearly understood in order to ensure the desired performance and repeatability under variable service conditions.

Energy

The energy required to activate the shape-memory phenomenon is significantly larger than that required with other approaches for certain classes of problems because of the attendant thermal losses primarily by conduction.

Temperature constraint

The shape-memory phenomenon occurs at a single specific temperature for a prescribed class of materials. Hence, in order to activate the material, the SMA must be heated to the prescribed transition temperature at the boundary of the martensitic and austenitic domains of the phase diagram.

This single actuation temperature could be problematical if the material is required to operate in unstructured thermal environments.

Stress concentration

The embedding of metallic wires in a host material for continuum applications will result in stress concentrations at the interface between the two constituents when the SMA wires are activated. These stress raisers will clearly have an undesirable effect upon the structural integrity of the part and its fatigue life.

Certification

Smart materials based on SMAs must satisfy the stringent FAA regulations in aircraft applications, and space qualification criteria for space applications. Automotive applications of this technology must comply with NHTSA requirements.

Interaction

The chemical reaction between the SMA and the host structural material must be carefully addressed in order to guarantee viable performance under variable service conditions.

Continuum Applications

Traditional applications of SMAs have been restricted to discrete applications. The knowledge-base for continuum applications, where layers of SMAs are embedded in structural materials, is embryonic in nature. Additional research and development is required in order to establish design rules to synthesize viable cost-effective structures with prescribed performance characteristics.

Control Issues

Control strategies for this class of smart materials and structures are still in the embryonic stage of development. Several theoretical, software and hardware related issues must be addressed prior to commercialization of SMA-based smart materials technologies.

6.6 Shape-memory plastics

A variety of shape-memory-plastics are currently being developed. Norsorex, for example, is the trade name of the polynorbornene polymer manufactured by Norsler in France. Norsorex has excellent shape recovery properties at shape memory temperatures of over 35°C. However, Norsorex has several characteristics which have impeded its commercialization in engineering applications. Norsorex is characterized by a super high molecular weight of over 3 000 000. Furthermore, the shape recovery temperature of Norsorex cannot be changed easily. For commercial applications shape-memory-plastics must be easily processible and feature a wide temperature range in which the shape recovery temperature can be changed. Other considerations, such as strength, durability, environmental resistance, etc. are also significant.

Zeon Shable is a relatively new shape-memory-plastic which addresses several requirements for commercialization. Zeon Shable yields itself to easy processing by injection molding, extrusion, and blow molding, for example. The plastic is available in five different grades which feature five different shape recovery temperatures ranging from 40°C to 80°C, which are 10°C apart for each grade of the material. Zeon Shable features very good strength properties, and good environmental and chemical resistance.

In the injection molding process, the plastic is initially heated, melted and injected into a mold, which is maintained at a specified temperature. Upon cooling, a hard molded plastic is obtained as a result of the primary molding operation. When the molded plastic is heated up to the shape recovery temperature the product manifests itself in a soft rubbery state. In this soft state, the product can be easily subjected to large deformations by the application of appropriate force fields. The deformed product in the rubbery state can be cooled in order to retain the same geometric configuration even in the absence of the applied force field. This process is usually referred to as secondary molding. When it is heated again to the shape recovery temperature, the secondary molded product becomes rubbery and recovers the geometric configuration attained by primary molding. This is accomplished by the release of the residual strain energy. When the product in this state is cooled to room temperature, the recovered product solidifies while maintaining its geometric shape, thereby returning to its shape after the primary molding.

It is clearly obvious from this discussion that shape-memory-plastics can memorize the shape corresponding to the completion of the primary molding process. By employing an external stimulus, such as heat, the primary molded product can be easily deformed by secondary molding and its new shape fixed by cooling. By employing external stimuli, it is possible to reversibly transform the product from its primary molded shape to its secondary molded shape, and vice versa.

The shape-memory property involves several processes which enable the material to:

1. Memorize the shape of the primary structure.
2. Fix the shape of the secondary molding.

The processes involved in memorizing the shape of the primary structure include:

a) entanglement of the polymer chains,
b) cross-linking,
c) crystallization,
d) formation of domain structure.

The shape of the secondary molded product is fixed by:

a) heating the material over the glass transition temperature, defined in Figure 6.16, and cooling,
b) melting and re-crystallization of the crystalline phase,
c) melting and reformation of the domain structure.

The secondary molded shape is fixed mainly by physical structural changes, whereas the primary molded shape is chemically and physically memorized.

Shape-memory-plastics (SMP) have several distinct advantages as compared with SMAs. These advantages include the characteristic of being lightweight, freedom from rust, the ability to impact color to products fabricated from SMPs, and the ability to print on these products. Furthermore, SMP products can undergo substantially larger deformations, they are easier to mold, process, and impart memory in addition to lower price. Potential applications of SMPs include:

a) toys, containers, sporting goods, packaging materials,
b) pipe-fitting, electric equipment, automotive parts,
c) orthodontic materials, devices for dilation of blood vessels,

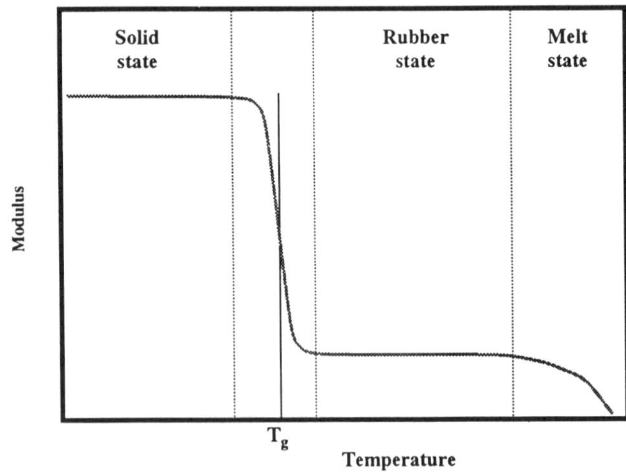

Fig. 6.16 Glass transition of a polymer.

d) clothing, padding material in shoes, eye-glass frames,
e) joining of pipes, auto-chokes for engines, and sensors.

SMPs are currently being investigated by the US Army for stiffening fabric wings, which can be packaged in a standard bag and deployed from a conventional aircraft. The goal of the work is to increase aircraft survivability by providing an offset capability for cargo and personnel. Potential areas that may be stiffened include the leading edge, trailing edge, or other areas where stiffening can be shown to have a significant effect on the performance, such as flattening the wing along its span.

7

Fiber-optic sensors

Smart materials and structures at the most sophisticated level in the foreseeable future will feature a network of embedded sensors and data-links. These sensory and data-transmission systems must typically be small, lightweight, possess geometrical flexibility, operate with low power consumption, feature a wide bandwidth, and be able to withstand the manufacturing environment imposed upon the host structural material without suffering performance degradation. The field of photonics, or optoelectronics, which exploits the ability of light beams to be transmitted through optical fibers of small diameter, has a significant role to play in this endeavor.

Figure 7.1 presents a schematic diagram of a smart structure featuring an embedded fiber-optic sensing system. The light source generates an optical signal which is introduced into the optical network embedded in the structure by an optical interface unit. This signal is subsequently modified as it is transmitted through the structure, and these modifications, which typically include changes of wavelength, phase, frequency, intensity, polarization or modal distribution, are then evaluated to provide information on the state of the structure.

Figure 7.1 provides a glimpse of the embryonic technological field of embedding optical fibers in structural members comprising fiberous composite materials in order to create a smart structural material, and the characteristic diagram provides insight into why fiber-optic sensors and waveguides are strong candidates for satisfying the diverse criteria imposed upon the selection of sensors appropriate for smart materials applications. This notion of embedding optical sensing systems in structures is especially relevant in the context of fiberous polymeric composite materials for aerospace applications where health monitoring capabilities are a major concern, and these concerns have subsequently triggered a major research and development program focused on health management systems.

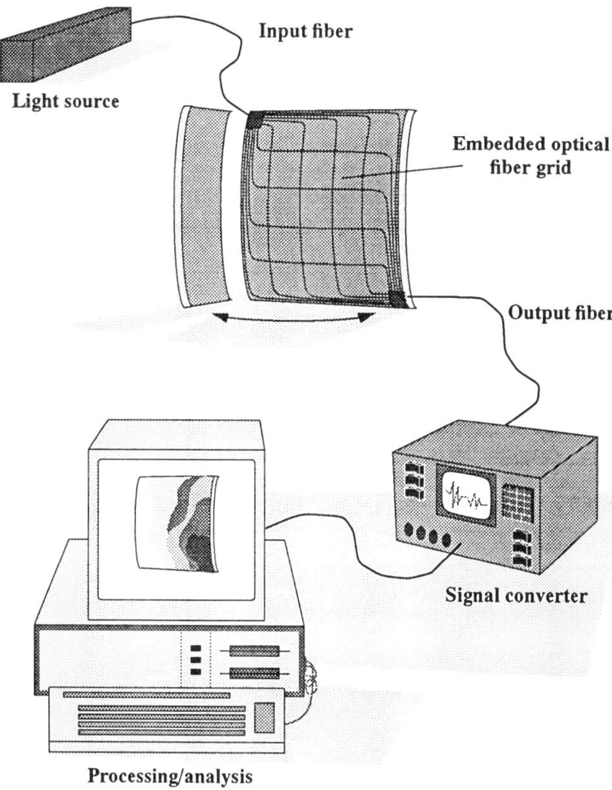

Fig. 7.1 Smart structure featuring an embedded fiber-optic sensing system.

The three principal scenarios for these health management systems are presented in Figure 7.2 from which the biomimetic philosophy is clear. Thus the first phase concerns the birth, or manufacture, of the part. The second phase concerns the life of the part during which the health-monitoring checks are routinely performed on critical performance parameters. Finally, the third phase concerns the death, or failure, of the part relative to prescribed criteria.

For this class of smart structural systems, fiber-optic sensors can make a significant contribution to ensuring that the structural member is manufactured correctly. For example, when processing a fiberous composite part in an autoclave, the pressure and temperature profile are important for the curing process, and in addition, quality control information is required on viscosity, and also the strain, and the moisture content throughout the structure. Furthermore, the embedded fiber-optic sensors can be employed to monitor the dynamic response of the structure under service conditions through the measurement of strain, at a variety of strategic locations

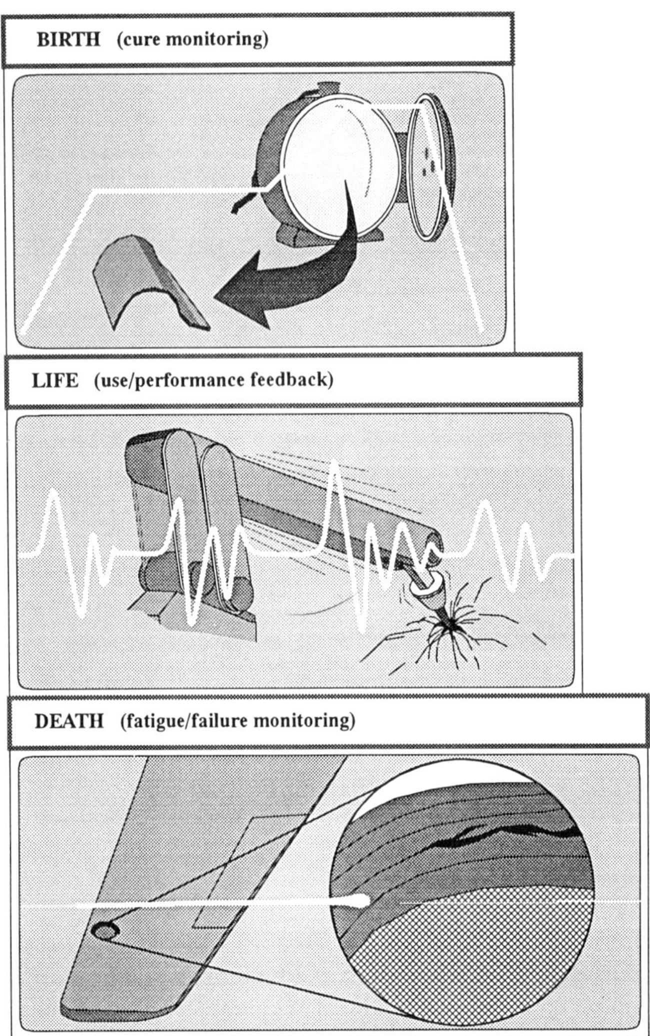

Fig. 7.2 Sensors embedded for feedback throughout the life-cycle of composite components.

throughout the member, for example, both at the surface of the part and also at various depths below the surface. This information can then be employed to actively control the mechanical properties of the smart structure when subjected to excitations from variable service conditions if the part features embedded actuators. Finally, the fiber-optic sensing system can be employed to predict the impending failure of the part, because any crack in the structure will sever the transmission path of the light beams

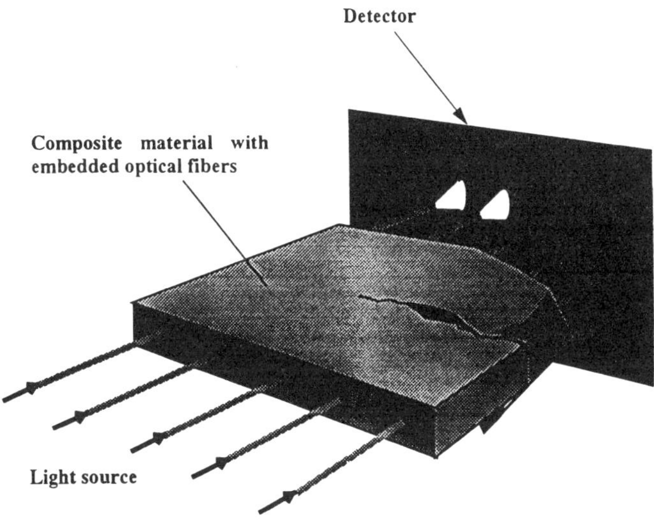

Fig. 7.3 An embedded optic fiber sensing system for damage detection. Embedded optical fibers break at the crack and prevent the light beam from being received at the detector.

travelling in the embedded optical waveguides thereby enabling the extent of structural damage, such as delamination, to be continuously monitored and measured, in order to undertake appropriate corrective action.

Figure 7.3 presents a pictorial representation of such a system at the macromechanical level. A composite plate is shown with five optical fibers embedded in the structure. A crack is assumed to have developed in the structure and is propagating in a direction orthogonal to the longitudinal axis of the fibers. Three optical waveguides have been broken and the optical paths severed. This situation would be detected and appropriate action taken either autonomously or by human intervention depending on the degree of sophistication of the system.

Figure 7.4 presents an illustrative example of how this health monitoring technology would be implemented in practice through the development of self-monitoring smart skins for aircraft. This class of smart aerospace structures would typically feature an instrumented network of sensors, computers and possibly embedded actuators which would simultaneously monitor the flight environment and the structural integrity of the airplane, and where appropriate, would autonomously initiate corrective measures to ensure the survivability of the aircraft. This class of sensing systems is the focus of this chapter.

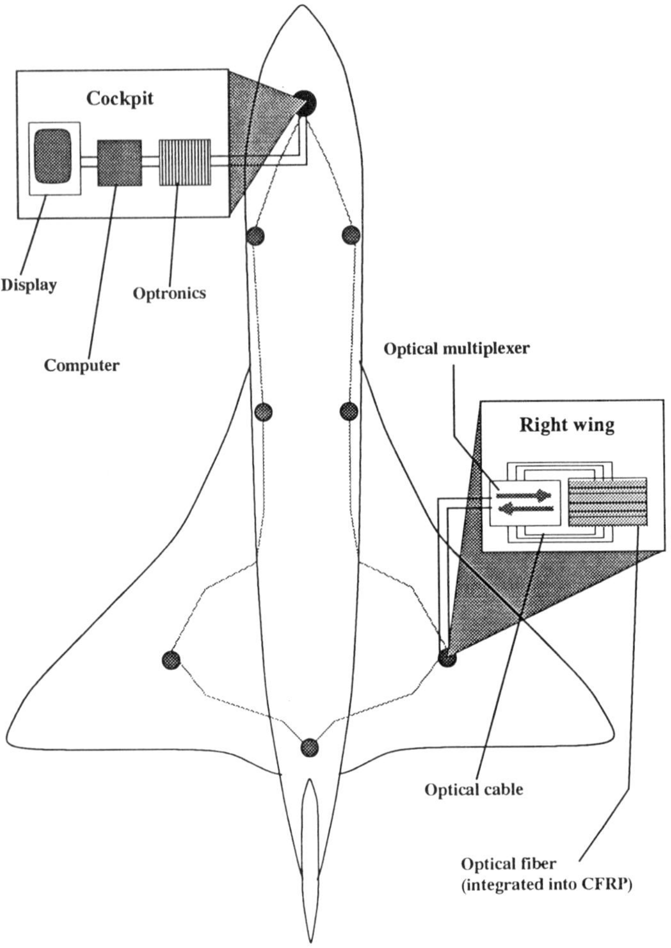

Fig. 7.4 In-flight structure surveillance would require fiber-optic nervous system (FONS) channels in key structural regions of a plane.

7.1 Fiber-optics: an overview

Photonics, opto electronics and fiber-optics technologies involve the transmission of light through transparent fiber waveguides of plastic or glass. The light source is controlled in order to develop a signal representing information which is then transmitted through the optical fiber from the signal source to the receiver. The information is represented by optical signals which feature distinct amplitude, phase, frequency, or polarization characteristics. Thus the combination of the light beam and the optical fiber is analogous to the more conventional information transmission systems involving electrical signals and copper wires. With a conventional fiber-optic system, an optical transmitter system is generally employed to create an appropriate light signal from an electrical signal by employing either a laser diode or a light emitting diode. This light signal is then conducted into the fiber and transmitted to an optical receiver system containing a photodiode which converts the light signal back into an electrical signal prior to amplification and post-processing as shown in Figure 7.5.

While the origins of transmitting information using a light source are not new, as evidenced by the long history of lighthouses which for centuries have warned sailors of maritime dangers, this field has lain dormant until the early 1970s when fiber-optics technologies, or photonics, entered the communications industry. Subsequently, the significant advantages offered by fiber-optic cables were responsible for this technology becoming the preferred choice for telecommunications and data-transmission systems. It was, therefore, only natural for this embryonic technology to penetrate the sensors market as the technology matured, and in the early 1980s fiber-optic

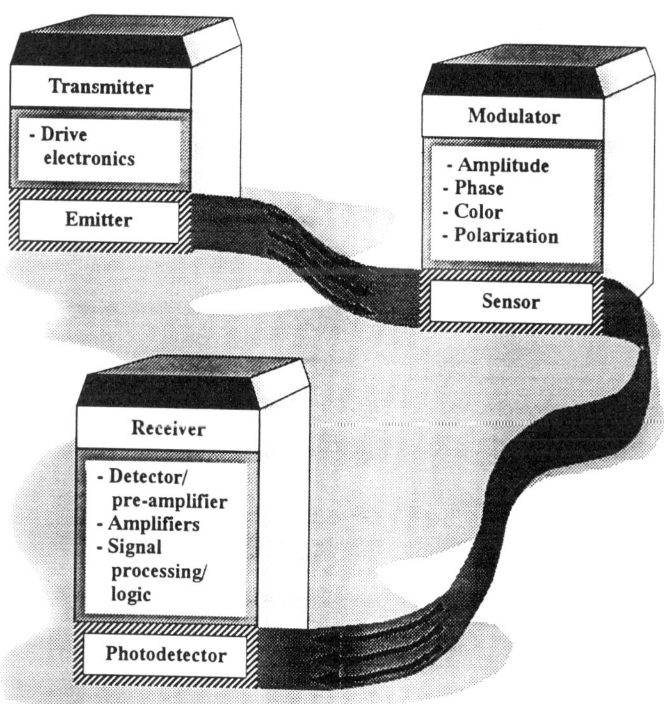

Fig. 7.5 A typical fiber-optic sensing system.

technologies were applied to the development of new types of sensors. This new generation of sensors not only exploits many of the distinct advantages of transmitting information encoded in optical signals, but they also exploit the properties of plastic and silica fibers to generate optoelectronic signals representative of the external stimuli being measured. Thus, in the context of optical sensors, the light signal is transmitted to the region subjected to the external stimulus, such as dynamical loading or temperature for example, and the external stimulus modulates the amplitude, phase, color or polarization characteristics of the light beam. This modulated light is then detected and analyzed by the receiver system to yield information regarding the external stimulus as depicted in Figure 7.6.

The first class of fiber-optic sensors exploited the ability of small diameter optical fibers to transmit electrical signals that had been converted into optical signals. Since the fibers are only employed for signal transmission purposes the fibers are a passive component of the system. This class of sensors is termed extrinsic sensors, in which the sensing function is

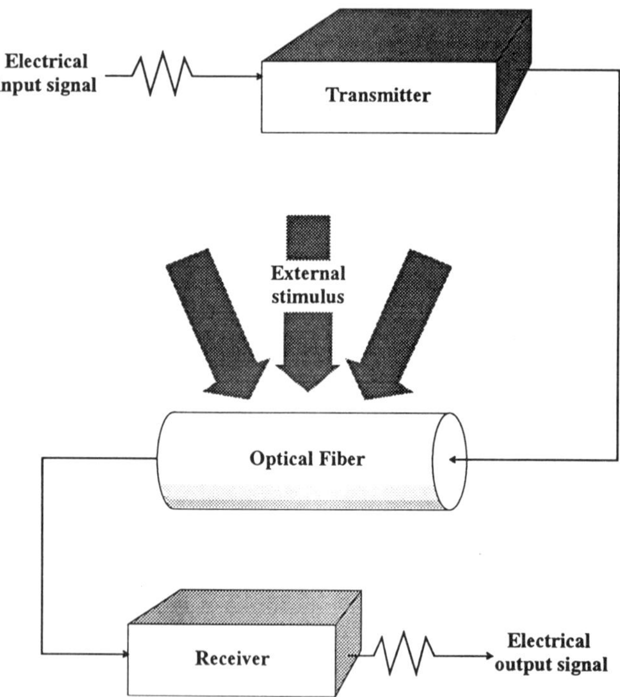

Fig. 7.6 The basis of fiber-optic sensing technologies for smart materials applications.

performed by exploiting a non-fiber-related phenomenon such as the interruption of a light beam or behavior associated with reflected light.

Subsequent developments in fiber-optic technology were responsible for the evolution of a second class of optical sensors, termed intrinsic sensors, in which changes in the fiber properties due to external excitations are the basis for measuring the characteristics of these external stimuli. Since the transmission characteristics of optical fibers are dependent upon excitation from a wide variety of stimuli, this sensitivity has resulted in the phenomenal growth of the field of intrinsic sensing. Indeed it has been predicted that fiber-optic sensors will dominate the sensors market within the next decade for a very broad range of sensing applications.

7.2 Advantages of fiber-optics

Fiber-optic technologies offer the designer of sensors and data-transmission links for smart materials applications, several distinct advantages which include a single common technology-base that enables devices to be developed for sensing numerous physical perturbations of a mechanical, acoustical, magnetic and thermal nature. Figure 7.7 presents a list of several classes of fiber-optic sensors that can be employed to measure a diverse range of variables. The distinct advantages of optical fiber sensors and data-transmission systems over traditional electrical sensors and copper wires has been responsible for significant growth in the photonics industry. In the context of smart structural systems featuring fiberous thermoset composite materials for example; fiber-optic systems can be readily embedded in these composite structures; they do not generate heat which could potentially damage the structure; they are lightweight and small enough to avoid reducing the structural integrity of the composite material; they do not provide a conducting path through the structure; they do not require electrical isolation from the structural material; and fiber-optic systems do not generate electromagnetic interference which can be a crucial advantage in some applications.

Fiber-optic waveguides provide a conduit for the high-density transmission of information. This information-carrying capacity increases with the bandwidth of the transmission medium and the frequency of the carrier. The practical bandwidth of optical fibers far exceeds that of electrical cables since the invention of the laser, which employs light radiation as a carrier, permits frequencies of 100 000 GHz to be attained.

Optical fiber data-transmission systems are characterized by much lower losses than those of systems featuring electrical signals and copper wires. These losses are responsible for the signal losing strength as it propagates along the transmission path. This loss of strength is termed ‘attenuation’, and while the attenuation of a copper cable increases with the modulation

Variable	Methodology
Force	Induced birefringence
Pressure	Piezoelectric Effect
Bending	Piezoabsorption
Density Change	Luminescence
Electric Field	Electro-Optical Effect
Dielectric Polarization	Electrochromatism
Electric Current	Electroluminescence
Magnetic Field	Magneto-Optical Effect Farraday Effect
Magnetic Polarization	Magnetoabsorption
Temperature	Thermal change in refractive index, absorptive properties, or fluorescence, thermoluminescence
Photoelectric Emission	Fiber defects leading to changes in refractive index and absorptive properties
X-rays, Gamma rays	Radiation-induced luminescence

Fig. 7.7 Fiber-optic sensors.

frequency of the signal, the attenuation of a fiber-optic cable is independent of the signal frequency. The significant limitation of electrical systems relative to optical systems can be put in perspective by considering the performance of these competing technologies in trans-Atlantic telephone links. The most advanced coaxial electrical system accommodates approximately 4000 conversations with a repeater spacing every nine kilometers. In sharp contrast to these capabilities, a newly installed fiber-optic link accommodates 39 000 conversations in each direction on two optical fibers with a repeater spacing of 35 kilometers.

Optical fibers are dielectric materials, therefore they do not provide an electrical conduit for the conduction of electricity and consequently they possess electromagnetic immunity. The control of electromagnetic interference in conventional circuitry due to adjacent electrical equipment or high-voltage lines is a major design activity in order to avoid serious problems and inconvenient system malfunctions. These control activities are expensive and they typically involve the deployment of shielded wires or coaxial cables.

Fiber-optic systems are considered to be a highly secure transmission medium since the signals do not radiate energy that can be received by an antenna. It is extremely difficult to tap a fiber surreptitiously and almost impossible to make such a tap undetectable because attempts to access the signal-carrying central section of a fiber-optic cable will disturb the transmission properties of the medium thereby facilitating detection.

Glass fiber cables are lightweight transmission media relative to conventional copper cables. Thus while a typical single-conductor optical cable weighs 9 lb/1000 ft, a comparable electrical copper coaxial cable weighs almost ten times this amount. This situation can be exploited in many smart materials applications especially within the aerospace and transportation industries.

A fiber-optic cable is much smaller than the equivalent copper cable, and furthermore, a single optical fiber can be employed to replace several copper wires. The compact, diminutive characteristics of optical fibers are crucial for the deployment of these systems in smart materials applications. The significant size-reduction features offered by fiber-optics over traditional systems can be illustrated by considering a digital telephony example. A copper coaxial cable 4.5 inches in diameter can accommodate 40 000 conversations. A fiber-optic cable of diameter 0.5 inches containing 150 fibers has the capacity to accommodate 25 000 conversations on each optical fiber, amounting to approximately 1.8 million telephone calls. Thus the volume of traffic accommodated by the fiber-optic cable far exceeds that of the coaxial cable even though it is only one tenth of the size.

Optical fibers are essentially a one-dimensional medium which can be employed for evaluating a measurand. They provide the ability to undertake either line-integrations over a prescribed optical path or else undertake line-differentiations. In the context of line-integrations, high sensitivities may be achieved by prescribing a long optical path in order for the measurand to interact with the optical wave-guide. In the context of line-differentiation, an optical fiber permits the measurand to be determined over any desired path. Consequently, the spatial and the temporal behavior of the measurand field may be obtained at the same time.

Fiber-optic systems are safer than equivalent electrical systems because these dielectric-based systems do not present a spark or fire hazard. Thus fiber-optic sensing technologies may be deployed in potentially explosive or hazardous areas which preclude the use of conventional electrical systems.

The previous paragraphs have synoptically chronicled some of the principal advantages offered by fiber-optic technologies in sensing and data-transmission. Further advantages offered by fiber-optic sensors include the ability to exploit interferometric optical techniques which permit the very high resolution of length measurements in remote locations by non-contacting approaches. Fiber-optic sensors generally feature no moving parts; they are highly compact and mechanically robust; they have high mechanical flexibility which enables their geometry to be adapted to diverse applications; they are of simple construction with low maintenance and high reliability and, while they are generally cheaper than traditional transducers and sensors, they often permit phenomena to be monitored when other more conventional approaches are inadequate. Furthermore, this class of optical sensors is readily employed in conjunction with digital systems and several signal multiplexing methods are available, such as time-division, amplitude-division and digital signal encoding.

The significant advantages offered by the field of photonics is, however, accompanied by a major disadvantage associated with the psychological disposition of *Homo sapiens sapiens*. The challenge, of course, is the lack of understanding and knowledge of fiber-optics technology in the engineering community. Consequently most engineers are comfortable designing with circuitry featuring electrical signals and copper wires, and they prefer to prescribe this technology rather than recommend fiber-optics solutions even though this emerging technology would offer significant advantages. The scenario also applies in practice to other more mature fields of engineering such as composite materials technologies. This situation is not new in the history of humanity as evidenced by an often cited statement made in 1513 by Niccolò Machiavelli that '*it must be remembered that there is nothing more difficult to plan, more doubtful of success, nor more dangerous to manage than the creation of a new system. For the initiator has the enmity of all who would profit by the preservation of the old institution and merely lukewarm defenders in those who would gain by the new one.*'

The reticence to implement this new optics technology for sensing and data-transmission applications will surely be overcome in time with education. This educational process will feature the verification of undeniably cost-competitive fiber-optic solutions, the maturation of the technology through the development of more sophisticated electro-optic electronic components, for example, and also the development of industry standards for both test procedures and also product specifications.

7.3 Light propagation in an optical fiber

Light is a form of electromagnetic energy which is typically transmitted through optical fibers with wavelengths of approximately 1000 nm. The speed of the light transmission is dependent upon the wavelength of the signal and also the medium through which it is being transmitted. When a light beam passes from one medium to a second medium with different

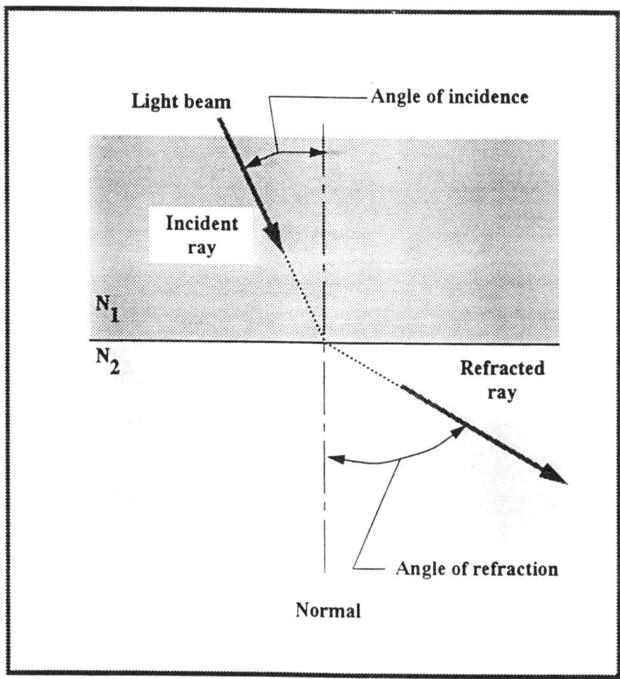

Fig. 7.8 Refraction of a light beam transmitted through media with different refractive indices.

physical properties it changes speed and the direction of the beam also changes at the interface between the two media. This deflection of the beam is termed *refraction*. The refraction of light for a specific medium is characterized by a *refractive index*, N, which is a dimensionless number defined by the quotient $N = c/v$ where c is the velocity of light in a vacuum and v is the velocity of light in the specific medium through which the light beam is being transmitted.

Consider the refraction of a ray of light passing from one medium with a refractive index N_1, to another, with a refractive index N_2, as shown in Figure 7.8. The incident ray in medium N_1 is refracted at the interface between the two media to yield the refracted ray in medium N_2, and the angle of refraction θ_2 is dependent upon the relative magnitudes of the two refractive indices, N_1 and N_2 by Snell's law. Thus

$$N_1 \sin \theta_1 = N_2 \sin \theta_2.$$

Light is refracted away from the normal when the refractive index of the first medium, N_1, is greater than that of the second medium N_2. As the

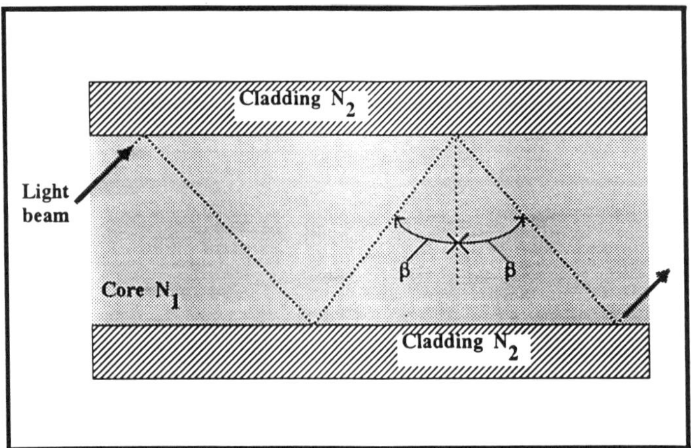

Fig. 7.9 Light transmissions through an optical fiber by total internal reflection.

angle of incidence, θ_1, increases, a critical angle θ_c is defined when the angle of refraction attains a value of 90°. Thus, by Snell's law,

$$\theta_c = \sin^{-1} (N_2/N_1).$$

A further increase in the angle θ_1 results in the light beam not entering the second medium N_2 because it is totally internally reflected. Under these conditions, the angle of incidence, θ_1, equals the angle of reflection θ_2. This condition of total internal reflection is the basic phenomenon exploited by fiber-optics technology, as shown in Figure 7.9, which presents a section of an optical fiber comprising two concentric layers of different refractive indices. The inner layer, or core, has a refractive index greater than the outer layer, or cladding. Thus, a light beam striking the interface between the core and the cladding at an angle greater than the critical angle is reflected back into the core again. The light beam continues this process of intermittent total internal reflection, which ensures that the beam is trapped within the core, as it proceeds along the fiber.

The characteristics of the light propagation through an optical fiber are dependent upon several factors including the characteristics of the light injected into the fiber, the diameter of the fiber, and the chemical composition of the fiber. By controlling these parameters, the number of paths, or modes, of transmission can also be controlled, and this philosophy is responsible for such terms as single-mode fibers and multi-mode fibers. These types of fibers respectively provide a light path for a single light ray, or light paths for over 100 000 light rays. The principle of operation of these optical waveguides is presented in Figure 7.10. The interested reader is referred to the bibliography for further information on these concepts.

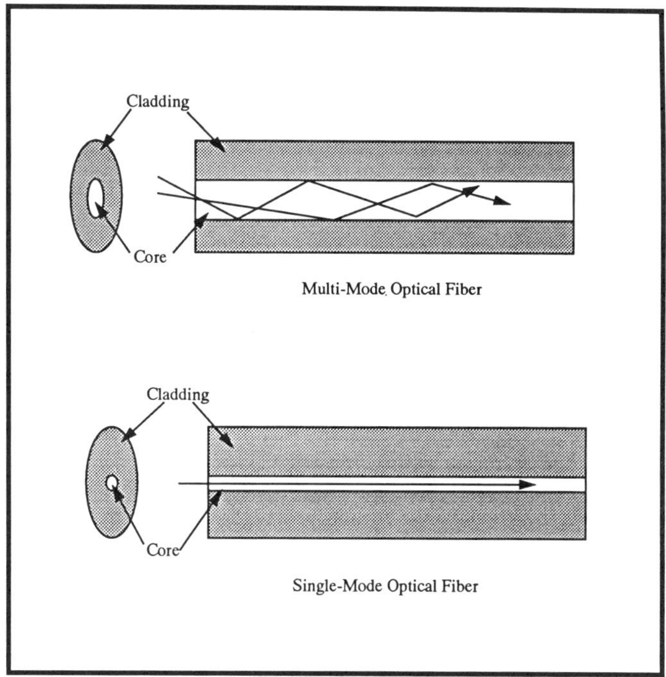

Fig. 7.10 Light transmission through single-mode fibers and multi-mode fibers.

7.4 Embedding optical fibers in fiberous polymeric thermosets

Sensors utilizing fiber-optics technology offer distinct advantages over sensors utilizing electrical technologies in the synthesis of smart materials and structures; consequently, this class of sensor technologies is currently being incorporated in new generations of parts featuring polymeric composite laminate materials. This class of advanced materials is being increasingly employed in a wide range of industrial applications to exploit the high strength-to-weight ratios of these fiberous composite materials, the inherent ability to tailor the stiffnesses and other mechanical properties to comply with design criteria, and also the ease of fabricating these fiberous materials into complex shapes.

While these abilities are indeed laudable, the properties of these advanced materials are, however, dependent upon the cure characteristics of the resin system during manufacture. In addition, in many applications it is desirable to monitor the behavior of a structure and also in particular predict the onset of failure. These diverse requirements have been responsible for optical fibers being embedded into composite structures for health monitoring purposes as shown in Figure 7.2. Thus the birth, or manufacture, of the smart structural composite is monitored, the installation and service life of the smart composite are monitored, and finally the death, or failure, of the smart composite is predicted prior to catastrophic failure. During part manufacture the global cure field and residual stress field can be determined along with the location of flaws. During the service life of the part, the dynamical response to thermal fields and external loads can be monitored. Finally, damage assessment can be undertaken in order to prevent the catastrophic failure of the part.

Damage assessment and the structural integrity of fiberous composite materials is a challenging field of scientific endeavor which has traditionally involved several non-destructive techniques such as ultrasonic methods, visual inspection, and X-ray methods. These examination techniques typically involve specialized equipment and they necessitate a maintenance down-time with the associated costs and inconveniences. The development of viable health monitoring techniques for composite parts involving a network of optical fibers embedded during the manufacture of a part offers significant benefits to the industry.

There have been a number of studies undertaken to measure the structural integrity of fiberous polymeric composite materials featuring embedded optical fibers. Investigations of the behavior of Kevlar-epoxy laminates in which were embedded Corning 1517 multi-mode fibers featuring a 125 μm cladding diameter and a 50 μm core have, for example, shown that the fibers only fracture when they are in the immediate neighborhood of the impact zone. Subsequent work has focused upon increasing the sensitivity of the untreated optical fibers through the segmented etching of the optical fiber. This work has resulted in the development of techniques for measuring the growth of a damage zone and also measuring the internal damage caused by impacts that are barely detectable by visual examination surface of the laminate. The results from this research program indicate that the fracture of optical fibers under these conditions, with the associated profuse bleeding of light, typically results in a 90% loss in transmission capability of the fiber.

The orientation of embedded optical fibers relative to the thickness of the part and relative to the principal fiber directions of adjacent plies are also the subject of evaluation for impact damage detection situations. Current research results suggest that optical fibers should be positioned orthogonally relative to the reinforcing fibers in adjacent plies and if possible embedded between col-linear plies. Other research on optical fibers embedded in fiberous polymeric composite materials has focused on issues of structural integrity. Namely, are the strength, fatigue life, or the fracture properties associated with delamination of a laminate decreased when optical fibers are embedded between plies? This work is still in its infancy due to the diverse properties and characteristics of different thermosets, the location and orientation of the optical fibers, and also the characteristics of the fibers. Consequently, design rules have not yet been established. However, preliminary testing of 8 ply Kevlar-epoxy panels has indicated the static tensile and compressive strengths of the material are unaffected by the embedding of optical fibers in configurations commensurate with the anticipated industrial practice.

Research on smart structural systems featuring composite materials and embedded fiber-optic sensors has, in part, been motivated by the ability to simultaneously integrate the structural fibers of the material with the sensory fibers during the manufacture of the part. Once the macromechanical structure has been synthesized, the fibers and matrix are then subjected to the appropriate processing temperatures and pressures which are dependent upon the type of thermoplastic, thermoset, or metal-matrix composite being produced. The respective processing temperatures for these three principal classes of composites are 600–800°F, 250–650°F, and 1000–2000°F respectively. Consequently, these processing temperatures impose an additional constraint on the selection of appropriate optical fibers for smart materials applications, because the coatings of many fibers are unable to withstand these temperatures without suffering permanent damage.

Research has been undertaken to identify appropriate coatings for fibers that will enable them to withstand the chemical environment and thermal stresses encountered during the processing of composite materials. Typically high temperatures are required for the polymeric systems in order to stimulate a chemical reaction which links together short molecules into long molecular chains. If the required temperature is not attained for the prescribed duration of the cure cycle, the material does not cure properly and inherent flaws are developed which can ultimately initiate failure of the part during service.

Thus the sensing fibers must be able to continuously sense the processing environment within the composite materials during processing in order to permit these critical parameters to be monitored. The bond between the fibers and the adjacent material must ensure that changes in the mechanical or thermal properties of the adjacent host material should not affect the light transmission properties of the optical fibers. The bonding of optical fibers and fiberous composite materials is dependent upon a number of factors including the cleanliness of the optical fibers and the host material; the relative orientation of the structural fibers and the optical fibers; the nature of the composite material, such as a laminate where the optical fibers are located between prepreg plies, or a knitted fabric for a resin transfer molded part where the optical fibers are woven into the cloth; and the properties of the outer coating of the optical fiber. These latter material properties are significant variables that govern the interfacial properties between the optical fiber, which is typically glass rather than plastic, and the composite material.

Research programs have recently been undertaken to investigate the role of fiber coatings on the strength and sensitivity characteristics of fiber-optic sensors embedded in fiberous composite materials. To date, it appears that fibers with high strength coatings comprising thin layers of thermosetting plastics appear to offer the best prospects for successfully interfacing the optical fibers and the host composite structure. Thin high-modulus polyimide coatings of thickness approximately 70 μm on a single-mode fiber with an 8 μm core and an outer diameter of 125 μm prior to coating, have demonstrated high microbend sensitivity in practice which is appropriate for employment in sensing applications where processing temperatures are 350°C or below. The thin coating ensures the deployment of small diameter fibers which will not adversely affect the structural integrity of the part, but the small diameter and low strength fibers are also more susceptible to damage during manufacture.

This sensitivity to damage during manufacture is particularly relevant to the transition region where an optical fiber emerges from the embedded region in the structural member to the unsupported, unprotected domain outside the structural member. Figure 7.11 presents a photograph of this transition region. In order to prevent the catastrophic fracture of the optical transmission medium at the point of emergence from the structure, the

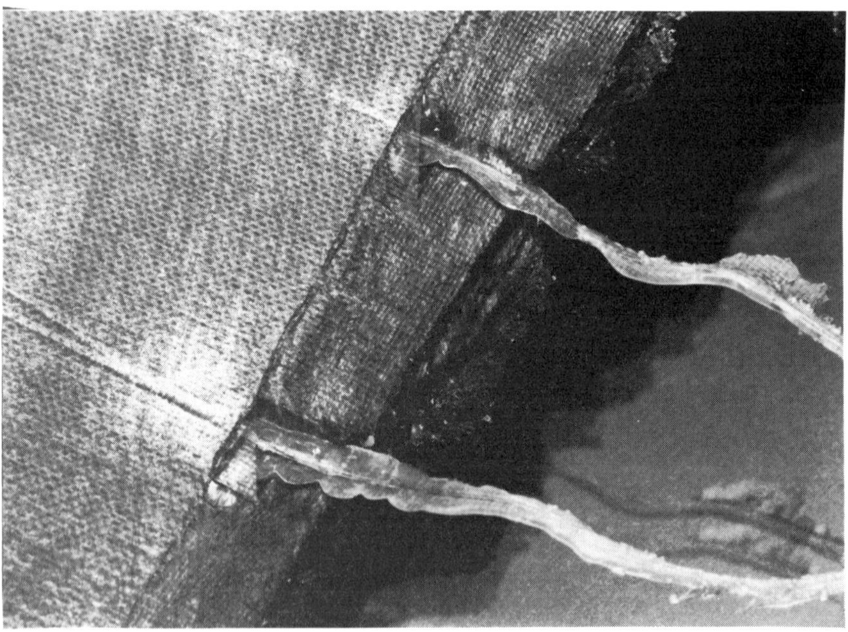

Fig. 7.11 A method for protecting optical fibers as they emerge from an embedded region of a polymeric fiberous composite laminate.

magnitude of the flexural rigidity of the fiber must be gradually transitioned from the rigid framework provided by the structural member to the extremely flexible characteristics of a single unprotected optical fiber. This can be accomplished by using a combination of Kapton film and silicone sealant. The Kapton film sandwiches the fiber as it emerges from the structural member prior to coating the fiber with the silicone sealant, to provide additional transitional protection as illustrated schematically in Figure 7.12. Figure 7.12 indicates how the optical fiber can be configured to emerge from the laminate at the edge of the plate structure. This has the advantage of minimizing potential stress raisers in the structure. Alternatively, the structure can be designed so that the optical fiber emerges from the upper or lower surface of the plate, however, this approach introduces stress spikes in the region where the fiber emerges from the surface ply. An alternative approach is to thread the optical fiber through a small diameter capillary-like flexible Teflon tube embedded in the structural member, and protruding from it, in order to provide the transitional support for the fiber. Figure 7.13 presents a photograph of a plate of smart structural material comprising a graphite-epoxy laminate of AS-4/3501–6 prepreg material with a $[11_4/{-11_4}]_s$ layup in which are embedded two polarimetric fiber-optic sensors that are visible in the lower regions of the plate. The emerging fibers are protected by the technique presented in Figure 7.12.

Implicit in the above discussion on composite plates featuring embedded optical fibers is the fact that fiber-optic sensors can be embedded within a fiberous polymeric structure at numerous different locations throughout the part in order to permit the engineer to develop a *bona fide* three-dimensional measure of the strain field throughout the structural domain.

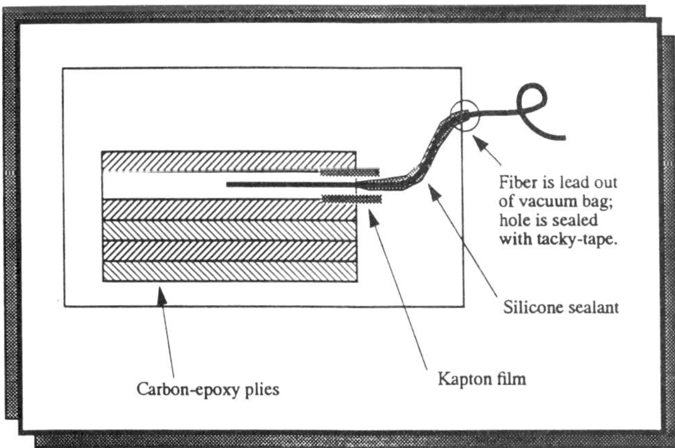

Fig. 7.12 Protection of optical fibers emerging from a fiberous composite laminate by employing a Kapton film sandwich and silicone sealant material.

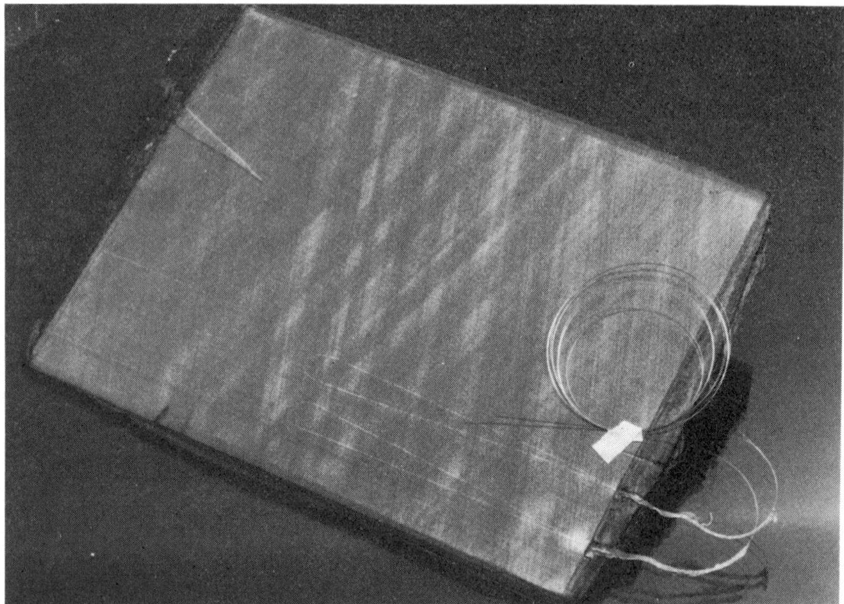

Fig. 7.13 Photograph of a smart structural graphite-epoxy plate in which the two emerging polarimetric fiber-optic sensors are embedded and protected.

Thus, for example, these photonic devices can be positioned below the upper and lower plies at the surfaces of a composite plate in order to monitor the maximum strains in bending, or else they can be positioned at the neutral surface of the plate at the mid-plane to measure the minimum strains. Such a capability is quite unique and cannot easily be obtained using conventional electrical sensors such as resistive strain gauges. This is especially true in the context of graphite-epoxy materials which are electrically conducting. In order to ensure that the sensing system functions correctly under these conditions, a dielectric material must be employed to insulate the graphite structure from those regions of the sensor subjected to an electrical potential difference.

7.5 Fiber-optic strain sensors

Optical fibers serve a dual role in smart structures because they not only provide a sensing capability but they also provide a transmission medium for the light signal. The light signal is transmitted through the fiber to the region which is subjected to the external stimulus and the signal is then modulated by the external stimulus prior to transmitting the modulated light to a receiver system, which then infers information concerning the characteristics of the external stimulus. The design of this arrangement is determined by the type of information sought. Thus, for example, an optical fiber can be embedded along the length of a beam parallel to the longitudinal axis to provide a global, or integrated, measure of the strain in the beam over the domain adjacent to the fiber. While this global information may be relevant in some smart materials applications, generally, local or point sensory information is sought because it has greater relevance in field problems where the magnitude of the quantity being measured varies considerably throughout a spatial domain. Thus, an engineer is generally interested in the magnitude of the strain in a structure at a number of distinct, possibly critical locations, rather than a global measure of the strain in the structure which is difficult to interpret. Consequently there have been vigorous research activities focused on the development of fiber-optic sensors for strain measurement in the local domain or at a point.

These sensors typically exploit the intrinsic characteristics of the fiber-optic waveguide to evaluate the strain field in the smart structure, rather than utilize an additional device to perform this task. Thus the strain field in the structure is coupled directly to the optical fiber by an adhesive film in the case of fiber-optic sensors that are bonded to the surface of the smart structure. Or alternatively, in the case of polymeric composite materials, the strain field in the structural material is coupled directly to the optical fiber by the adhesive properties of the matrix material. Furthermore, as with all sensors, the sensing system must provide a reproducible response under identical conditions, and the sensor should respond only to the strain field while being unaffected by variations in other types of external stimuli such as ultra violet radiation and magnetic fields.

Strain is a quantity whose magnitude typically varies spatially in a complex shaped structure. Thus, in order to measure this quantity, the sensor must be small enough to provide a localized measure of the deformation which requires the utilization of optical connections which are insensitive to strain. Furthermore, the sensor must only respond to the particular measurand which is of interest. While this task is quite specific and well focused, there have been a myriad of ancillary issues associated with this undertaking for installing this class of sensors in polymeric composite materials. For example, in large structures, where it is desirable to form a network of sensors, the sensory system must be amenable to multiplexing where two or more signals are transmitted over a single communications channel by using wavelength-division schemes or time-division multiplexing schemes.

It is generally advantageous for the output signal of any sensor to exhibit a linear response relative to the quantity being measured. A consequence of this criterion in fiber-optics is that interferometric sensors should be maintained at the most phase-sensitive operating condition.

Fiber-optic sensors should be designed for ease of manufacture, and they should also feature commercially available off-the-shelf components in order to ensure minimal cost and facilitate maintainability. The sensor should be mechanically robust to avoid damage during both manufacture of the smart structure and also during the service life of the structure. Furthermore, the sensor should be small enough to ensure that the structural integrity of the material is not compromised. In this regard it is probably advantageous for the sensor to feature a single-ended configuration which exploits a reflective rather than transmissive phenomenon. Such an arrangement would ensure an easier installation procedure with less connections. It would also be advantageous if the sensing system could be easily installed by individuals without specialized knowledge in fiber-optics technologies. Finally, the sensor should possess a dynamic range commensurate with the anticipated properties of the measurand, while possessing adequate resolution and sensitivity for accurately capturing the characteristics of the measurand field.

Fiber-optic strain sensors can be classified as interferometric, polarimetric, and modal interferometric sensors. Each sensing system has distinct advantages and disadvantages which are discussed in the subsequent sections. Some of these systems are presented schematically in Figure 7.14 when they are embedded in a set of beam specimens.

7.5.1 Mach-Zehnder interferometric strain sensors

Interferometric sensors exploit the interference of two light beams. With this class of sensors two coherent beams are created from a single light beam

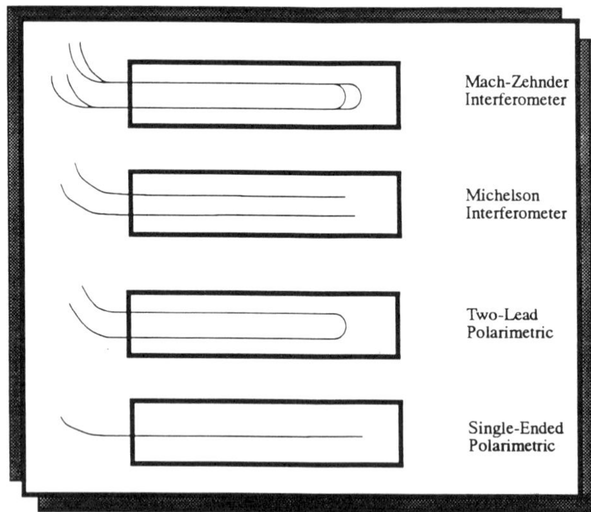

Fig. 7.14 Fiber-optic sensor configurations.

by employing a beam splitting device, and these beams are then coupled into two single mode fibers which are designated the 'reference arm' and the 'sensing arm' in Figure 7.15. The sensing arm is subjected to mechanical deformation while the reference arm is generally protected from the strain field and other external stimuli in the most basic form of this class of sensors, where a measure of the distributed strain over a relatively long optical path length is obtained. A schematic diagram of this system is presented in Figure 7.16. Thus the optical waveguide subjected to mechanical deformation experiences a change in length of the transmission medium, and hence a change in the optical path length of the light beam. This difference in the optical path length of the two beams results in a relative phase shift between them which is detected by observing the shift in the fringe pattern upon recombining the two beams.

The output from this interferometric sensor is an ensemble of fringes which can be transformed into a signal by a slitted photodiode arrangement for example. The fringe pattern is detected by the slitted photodiode and is manifested as a series of impulse-function-like electrical signals in response to the excitation provided by each fringe as it moves across the face of the slit arrangement. The strength of the signal can be amplified by employing a multiple slit array which covers the face of the photodiode.

Since strain is associated with the change in length per unit length, the change in optical path length of the light beam transmitted through the

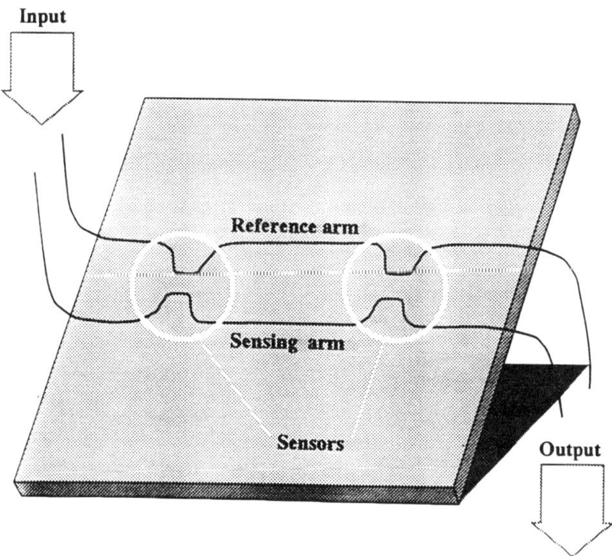

Fig. 7.15 Mach-Zehnder sensing.

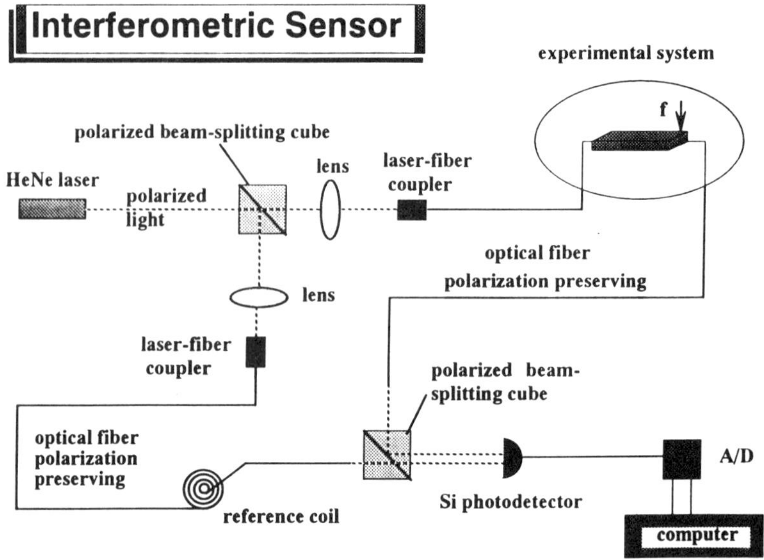

Fig. 7.16 Mach-Zehnder interferometric fiber-optic sensor optical apparatus.

sensing arm with respect to the light beam transmitted through the reference arm, will be governed by the relative length of these two light paths. Therefore, a Mach-Zehnder interferometric sensor can be employed to approximate sensing at a point when the difference in length between the two arms is small, or the sensor can integrate the measurand when the difference in length of the arms is considerably greater, up to the limit imposed by the coherence length of the light source.

With reference to the Mach-Zehnder point sensor shown in Figure 7.15, the shorter of the two optical fibers embedded in the beam is the reference arm of the interferometer. It is positioned immediately adjacent to the sensing arm, which is the longer of the two optical fibers, in order that both optical waveguides are subjected to the same external stimuli prior to entering the sensor gauge length region.

The complexity of the data analysis electronics and the control system for this class of interferometric strain sensors may be simplified by a passive quadrative operation using special couplers. This class of sensors suffers from the necessity to utilize two fibers which involves a physically larger sensor, two connecting leads, and the potential for greater noise sensitivity due to dissimilar optical input paths. Furthermore, this class of sensors is difficult to localize because of the requirement of a reference arm which is isolated from the external stimuli, unless a sensor is required to measure the local deformations.

7.5.2 Michelson interferometric strain sensors

The Michelson strain sensor comprises two optical fibers with mirrored ends to reflect the light. The gauge length of the sensor is the difference in length of the reference arm and the sensing arm of the system, which can be as small as one millimeter in order to furnish information at the local level. This interferometric system and also the Fabry-Perot interferometer provide localized unidirectional sensing capabilities.

One of the fundamental differences between the Michelson interferometer and the Mach-Zehnder interferometer is that while the latter system typically employs two loops, and hence four optical fibers emerge from the structure, the former system features only two fibers emerging from the structure and each fiber accommodates light beams moving in opposite directions. This latter feature is accomplished because the ends of the fibers are mirrored to reflect the incident light path. The light beam subsequently returns along the fiber prior to being decoupled by the same unit that earlier coupled it to the fiber. Clearly this arrangement is more elegant and efficient than the Mach-Zehnder interferometer. Furthermore, the structural integrity of the host structural material is not affected by this arrangement as much as the Mach-Zehnder system, and because the light path of a Michelson interferometer is twice as long as for the Mach-Zehnder interferometer, the Michelson scheme is more sensitive. However, the two optical fibers of the optical sensing system must be carefully bonded to each other, in order to minimize sensor noise attributed to the difference in optical paths of these optical waveguides connected to the sensor region.

7.5.3 Polarimetric strain sensors

This class of fiber-optic strain sensors utilizes the state of polarization of a light beam being transmitted through one of these optical waveguides. Ideally, the characteristics of the light beam entering an optical fiber are the same as the characteristics of the light beam emerging from the fiber. Consequently, the light energy entering each mode is unchanged. However, when the optical fiber is strained, light energy will be distributed differently between the various modes in the fiber, and this change in energy can be detected in order to provide a measure of the strain in the optical fiber and the surrounding medium to which it is bonded.

Light is a form of electromagnetic radiant energy belonging to the same category of energy as television and radio signals, X-rays and radar. Light energy has a higher frequency and shorter wavelength than radio waves, for example, consequently this translates into a high frequency carrier with a larger information-carrying capacity. These electromagnetic signals comprise oscillating electric and magnetic fields that are orthogonal to each other and also orthogonal to the direction of propagation of the light beam. These

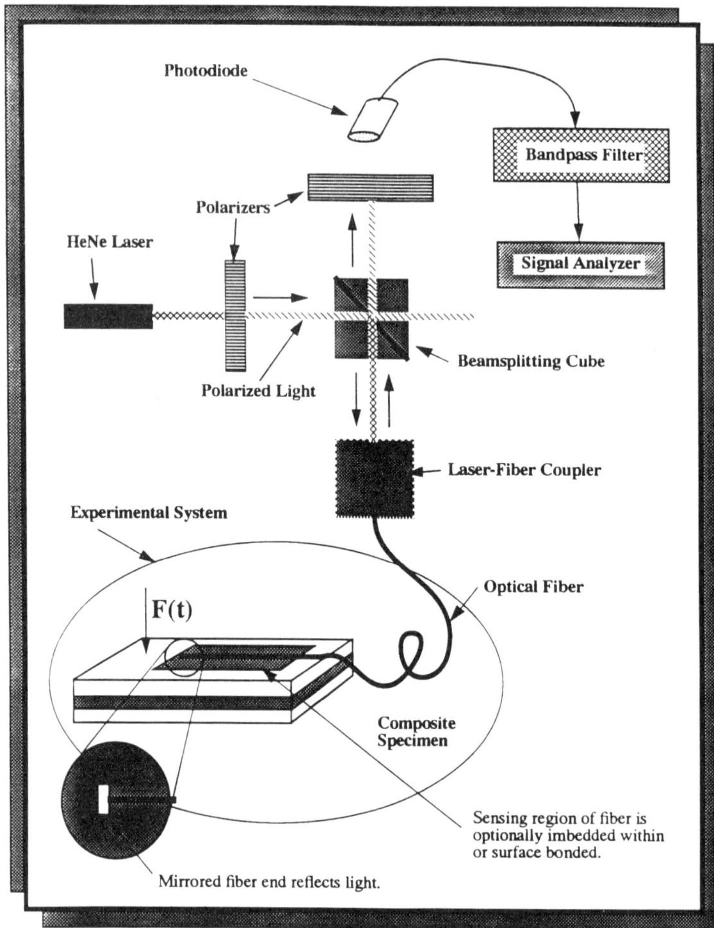

Fig. 7.17 Polarimetric fiber-optic sensor optical apparatus featuring a photonic sensor.

characteristics of light are exploited in polarimetric optical sensors.

Figure 7.17 presents a schematic representation of a polarimetric fiber-optic sensor system for a smart materials application. Light from the Helium-Neon laser source is first polarized and coupled into a single-mode fiber. The polarizer only permits light to pass through it whose electrical field is parallel to the orientation of the polarizer. Consequently, the polarizer can be configured to only permit the passage of all the light energy associated with one of the two orthogonal modes of the single mode fiber, while totally suppressing all the light from the other mode. This capability permits a photodiode and polarizer combination to detect the change in energy in a specific mode caused by the mechanical deformation of the fiber.

This category of optical sensors typically employs a single fiber with a mirrored end to reflect the light beam. Consequently, this class of sensors has a minimal number of fibers emerging from a structural member in which they are embedded. Furthermore, these optical fibers can be designed to be insensitive to external stimuli and this type of sensor can be appropriately configured in order to ensure ease of installation and assembly.

The sensitivity of this class of fiber-optic strain sensors is lower than some of the other classes of highly sensitive optical sensor systems. However, this class of polarimetric distributed strain sensors generates a signal which can

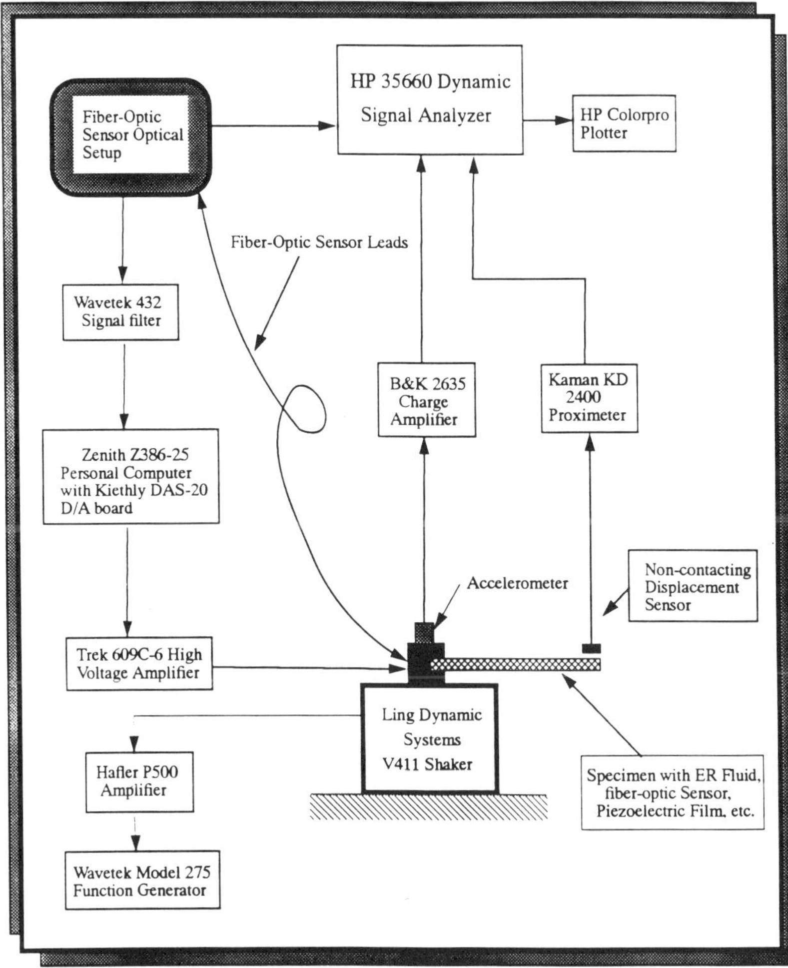

Fig. 7.18 Smart structure experimental apparatus featuring a photonic sensor.

be readily employed for vibration control applications. The integrated strain measured over the length of the sensing arm is responsible for a frequency-dependent phase shift in the polarimetric signal.

In order to evaluate some of the capabilities of a surface-bonded single-ended polarimetric fiber-optic sensor, this class of sensors was configured to be coincident with the longitudinal axis of a graphite-epoxy composite laminate beam as shown in Figure 7.14. Figure 7.18 presents a schematic diagram of the experimental apparatus. The objective of this study was to evaluate the dynamic response and also the frequency response of the beam as measured by the fiber-optic sensor and also a non-contacting displacement sensor at the tip of the cantilever beam which was excited in the transverse direction by a small electro-dynamic shaker.

The comparative experimental results from the two classes of sensors are presented in Figures 7.19 and 7.20. Figure 7.19 presents the elastodynamic response of the beam when it is being excited at its fundamental transverse natural frequency of 41.2 Hz. Figure 7.20 presents the frequency response of the beam based on data furnished by the two types of sensors. These experimental results demonstrate qualitatively the viability of employing this class of fiber-optic sensor in this type of structural application.

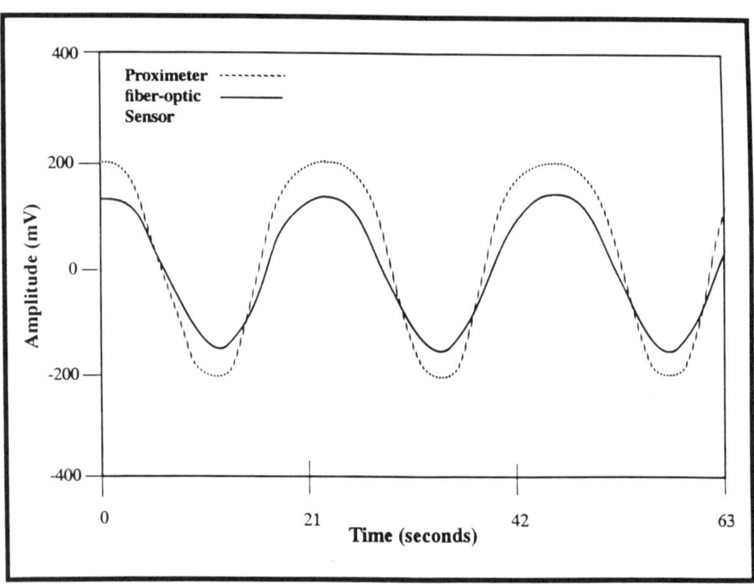

Fig. 7.19 Comparison of the elastodynamic response between a polarimetric fiber-optic sensor signal and a proximeter sensor signal for a smart cantilever beam.

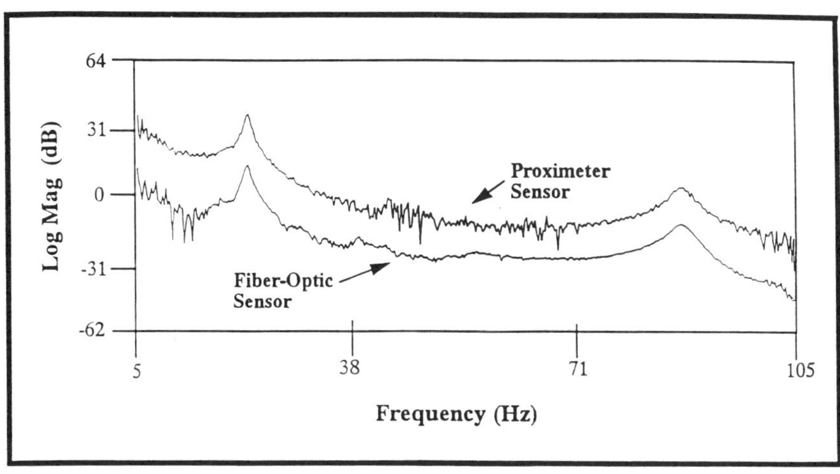

Fig. 7.20 Comparison of the frequency response curves generated utilizing proximeter sensor signals and polarimetric fiber-optic sensor signals.

7.5.4 Fiber-optic strain rosette

The optical strain sensors discussed previously have the ability to either measure mechanical deformation in one direction at a nominal point location, or else provide a measure of the integrated strain throughout a relatively large domain. While these capabilities may indeed be acceptable in many applications, strain is a second-order tensor quantity and the magnitudes of the strain components in several directions at a point are required in order to develop a viable characterization of the deformation field. Furthermore, in the typical irregularly-shaped structural members employed in practical engineering systems that are subjected to diverse boundary conditions and loading regimes, the strain field is of a complex shape which mandates the deployment of multiple gauges in order to ascertain the characteristics of the strain extrema. The principal strains at critical points in the structure are then compared with an appropriate criterion, or else employed to determine the associated principal stresses prior to relating them to the classical failure criteria.

Traditionally this goal of determining the principal strains at a point on the surface of a structure has been accomplished using a resistive strain gauge rosette bonded to the surface of the structure. The rosette typically features three electrical strain gauges whose relative orientation is a known value of 60 degrees or 120 degrees. The magnitude of the principal strains and the orientation of the principal planes are determined by calculation using the strain data provided by the three gauges. Optical strain rosettes have been developed using polarimetric and also Michelson interferometric

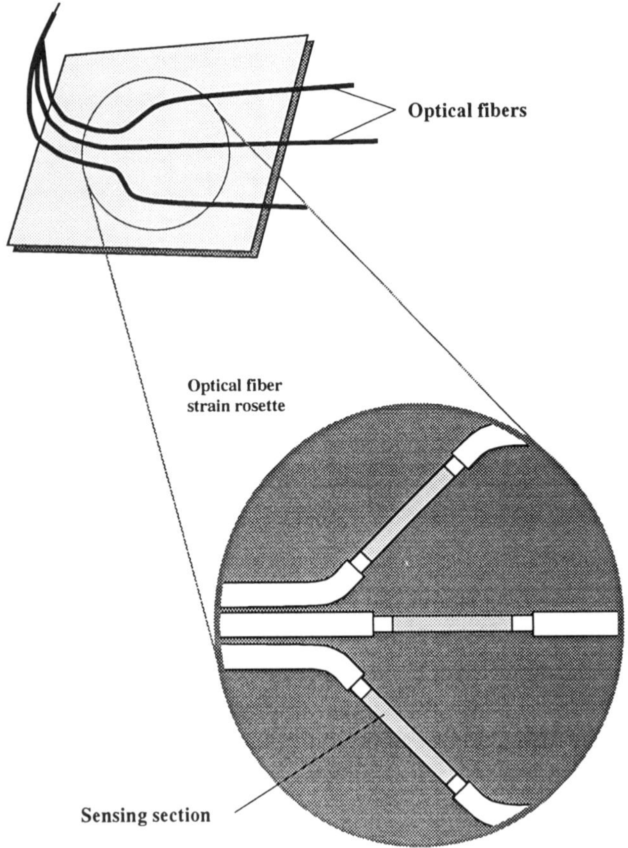

Fig. 7.21 A polarimetric strain rosette bonded to the surface.

techniques to provide an optical analog of the traditional resistive strain rosettes.

Figure 7.21 schematically presents the essential details of a delta rosette configuration featuring three polarimetric fiber-optic strain gauges, which can be embedded at various locations within a smart structure in order to develop a viable three-dimensional measure of the strain field and also the principal strains within the part. The active length of each of these gauges is typically 10 mm and these sensors are isolated by 45° fusion splices from the associated leads. The responses of the three optical gauges readily provide the magnitudes of the two principal strains and the orientation of the principal planes, by utilizing the standard analytical expressions employed in the strain gauge literature for the traditional electrical rosettes. The physical dimensions of this sensor prohibit this class of transducer from attaining the

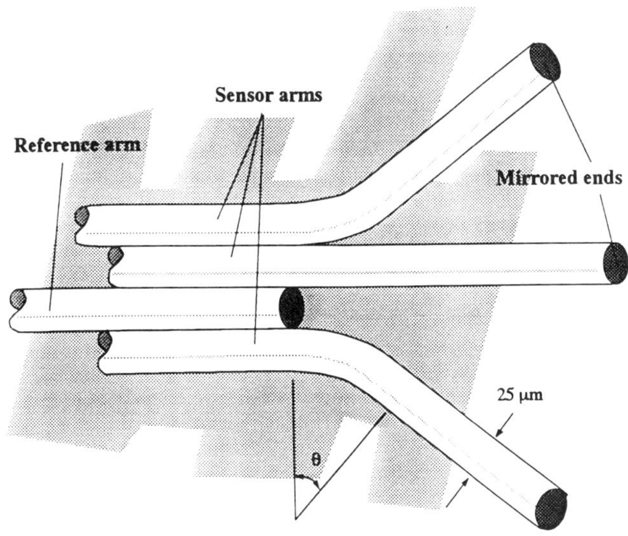

Fig. 7.22 A Michelson fiber-optic strain rosette.

desired sensitivity in a local 'point' region when there are high strain gradients or the magnitude of the strain field is small. Under less demanding conditions, experimental tests have indicated that this class of fiber-optic strain rosette has furnished strain data with an accuracy comparable to data provided by electrical strain gauge rosettes.

Figure 7.22 presents the essential details of a Michelson interferometric strain rosette with optical fibers of 125 μm diameter. It is evident from the diagram that this category of sensors is somewhat smaller and less intrusive than the polarimetric strain rosette design while featuring a superior sensitivity. The configuration of the sensing arms is similar to the classical electrical resistance rosettes. The optical design features a common reference arm located adjacent to the horizontal sensor arms which separate the upper and lower sensing arms. The sensing regions are determined by the mirrored ends of the fibers, and the minimum bend radius of the unbuffed optical fiber is a significant design variable, because the curved region of the sensing arms is a large percentage of the gauge length of the upper and lower sensing regions.

The individual sensor responses are readily transformed by the classical strain rosette formula to yield the magnitudes of the principal strains and the relative orientation of the principal planes. Experimental studies have yielded excellent correlation between this class of optical strain sensor and the classical electrical resistive strain sensors.

8

The epilogue: research issues

The epilogue focuses upon the blue-sky research issues currently confronting researchers in the field of smart materials, or, intelligent materials. Predicting the future is a somewhat risky undertaking, and predicting the evolution of an embryonic field such as smart materials is definitely a hazardous task. Nevertheless projections for the future of this field are discussed in this chapter in an attempt to stimulate further ideas and discussions on this emerging scientific and technological discipline.

The previous chapters have attempted to emphasize the importance of the development of new materials to the evolution of science and technology in the dusk of the twentieth century and beyond. Furthermore, the multidisciplinary nature of the field of smart materials has been expounded and the impact of these materials in many scientific and technologically diverse fields has been enunciated, such as electronics, artificial intelligence, information technology, materials synthesis, physics, chemistry, molecular biology, computer science, the life sciences, control theory, the aerospace industry, materials processing, the defense sciences, transportation and biotechnology. Clearly this embryonic discipline will have a tremendous impact on human civilization and will dramatically change the lifestyles of future generations.

Upon reviewing the history of the science of materials from the beginning of Paleolithic times through the Stone Age, the Bronze Age, and the Iron Age to the current Synthetic Materials Age, it is clearly evident that there has been a distinct evolution from structural materials to functional materials and now smart materials, as the scientific and technological prowess of humanity has matured. This philosophy is presented in Figure 8.1. Thus one million years ago *Homo habilis* fabricated tools and weapons in the mineral flint because this fine-grained, very hard, siliceous rock was the best structural material available in that era. Subsequently this class of structural materials was superseded by implements featuring bronze

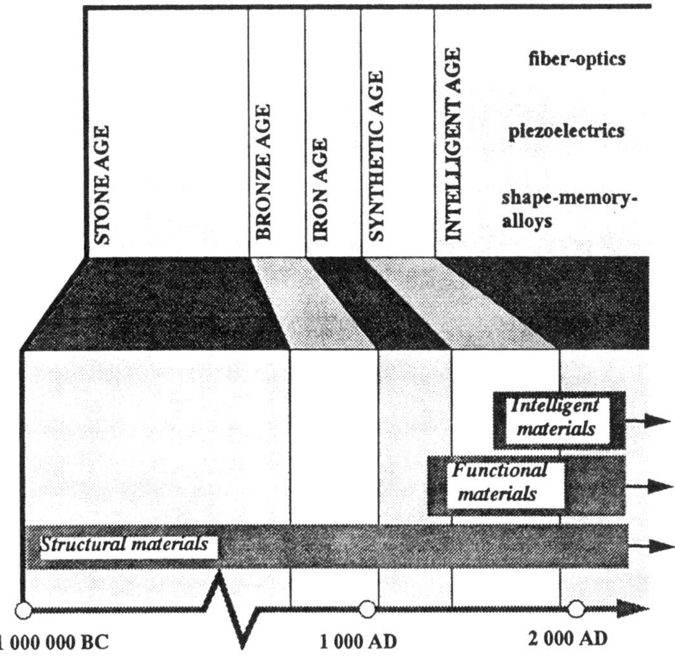

Fig. 8.1 The evolution of materials science.

and iron components, and current industrial practices capitalize on the ability to fabricate members in fiberous polymeric composites, which offer the designer structural properties and engineering capabilities which are significantly superior to those of the traditional monolithic materials.

Thus it is evident how the field of materials science matured as the naturally occurring structural materials of the Stone Age were replaced by technologically-refined metallic implements during the Bronze Age and the Iron Age, prior to being superseded in many technological fields by fiberous composite materials. *Homo sapiens sapiens* has developed distinct design and manufacturing skills in order to synthesize, or optimally-tailor, the macrostructural properties of these man-made materials.

These trends in structural materials have been complemented during the past 200 years or so by the development of a variety of functional materials, such as piezoelectric materials, or magnetostrictive materials for example, in which the functional properties of the material, rather than their structural properties, are exploited in practice. Successes in this field have quite naturally spawned the subsequent development of multi-functional materials in which materials are characterized by several functional properties. The synthesis of synthetic multi-functional materials is a very immature embryonic field, but numerous examples of multi-functional biomaterials exist, for example, bone which ensures structural homeostasis as well as mineral homeostasis. The research efforts focused on multi-functional materials typically exploit the notion of biomimetics, because the molecular units of most biological systems possess multi-functional characteristics which ensure both economy and efficiency. Consequently, the evolution of these multi-functional materials during the millennia has resulted in them being considered to be the ultimate class of materials.

It is clearly evident, therefore, that *Homo sapiens sapiens* has developed the ability to create a diverse range of structural materials, functional materials, and also poly-functional materials by employing theoretical, computational, and manufacturing techniques. These significant scientific and technological capabilities are essential ingredients for the synthesis of the different classes of smart materials. At the most sophisticated level, smart materials will typically feature a combination of sensors, actuators and processors which will permit these materials to respond intelligently and autonomously to dynamically-changing environmental conditions by capitalizing on embedded innovative functions.

These classes of smart materials are currently created by employing commercially available materials such as the various polymeric composite materials, optical fibers, and piezoceramic materials. However, most of these technologies are very mature and the materials have been developed for many decades. Furthermore, the current generation of smart materials are typically hybrid materials where large, relatively homogeneous, material domains are assembled at the macromechanical level to create a material with different properties at the global level. New types of innovative functions are required for the next generation of state-of-the-art materials in order to provide appropriate sensors, processors and actuators, and to overcome the deficiencies of the current materials.

To realize the dream of creating new generations of advanced materials and substances, research must be prosecuted on developing a superior understanding of current materials at the atomic and molecular levels. The primitive functions and molecular structures of these materials must be clearly understood not only to distill guidelines for the design and manufacture of these materials but also to promote thinking on the synthesis of new classes of materials. Thus the fundamental mechanisms associated with intelligence in a material must be clearly elucidated in order to develop viable design rules for creating materials with the desired innovative functions.

8.1 Material intelligence and associated phenomena

Intelligence is evident in a number of existing materials where several innate functions are coordinated, to ensure that a substance will provide the necessary filtering to attenuate a light source of variable intensity for example, or where a self-repair process is undertaken by oxidation. The current generation of materials should be subjected to comprehensive research programs in order to ascertain the underlying mechanisms responsible for these effects. Such programs would ultimately provide a deeper understanding and appreciation of a variety of materials-related phenomena. Thus dynamic reactions can be characterized by pressure generation, for example, optical reactions by fading or luminescence, magnetic reactions by polarization, chemical reactions by reduction or oxidation, and electrical reactions by conduction or exothermal reactions. These reactions are based on a variety of phenomena at the atomic, molecular and mesoscopic levels.

Crystal Structure — Changes occur in the orientation of crystals, in the atomic configuration and in the inter-atomic spacings. These changes are responsible for phase transformations, and in the case of polymeric materials, the molecular chain can be re-configured from a folded state to an extended state, for example.

Molecular Structure — Changes occur in the molecular structure due to the breaking of molecular chains, the re-configuring of intra-molecular bonds and three-dimensional intra-molecular spacings, and anti-genic or enzymatic reactions.

Macroscopic Structure — Global changes occur in the macroscopic structure due to the diffusion and bulk transfer of fluids and ultra-fine powders.

Interfaces — Interfacial changes typically pertain to grain boundary phenomena and also reactions at the surface of a material resulting from interactions with the neighboring environment.

Composition — The interaction between a material and the neighboring environment at the interface may result in a change in the material composition at the surface, due to the attendant chemical reactions.

Energy — The interaction between a material and the neighboring environment at the interface may

	trigger an energy change, or the release of photons or electrons, for example.
Ion Transfer	Ions or atoms may be transformed in materials, and groups or radicals may be transferred along polymer chains.
Charge Transfer	Charges are transferred along polymer chains in organic materials, while charge transfer and subsequent accumulation occurs by conductivity in metals for example.
Electronic Structure	Changes in the orientation of electronic spin vectors can result in a change in the magnetic properties of materials.
Material Constants	The material constants can change in order to manifest mechanisms of intelligence in a material. Thus, for example, changes may occur in the coefficient of thermal expansion, the melting point of a monolithic material, the Curie temperature of a piezoelectric material, and the glass transition temperature of a polymeric material.

It is clearly evident, therefore, that mechanisms manifesting intelligence in a material are typically associated with structural changes in the material. Consequently it is logical that research should focus on these mechanisms in order to develop viable material design methodologies for synthesizing new classes of materials with innovative functions. It is anticipated that the formulation of these design methodologies will be governed by whether the innovative function is dependent upon phenomena at the microscopic level, macroscopic level, or some level in between these upper and lower bounds. Clearly the governing parameters will be dependent upon the level at which the phenomenon is manifested. In the synthesis of certain classes of intelligent materials, it is conceivable that combinations of phenomena at the atomic, molecular, and mesoscopic levels must be coordinated in order that a material manifest a desired innovative function.

Illustrative examples of the relationships between intelligent functions in materials and the associated phenomena are subsequently documented in order to provide the basis for determining appropriate guidelines for prosecuting future research and development in this field.

Self-multiplication or *growth* is manifest as an intelligent function in materials that undergo a change in their crystalline or molecular structure. This class of intelligent materials would be able, for example, to grow by absorbing appropriate substances from within the material or from the neighboring environment, or alternatively breed by developing polymeric chains or expanding collapsible structures. An example of this class of intelligent materials is artificial blood vessels fabricated from collagen and artificial bone (hydroxyapatite).

Self-repair is manifest as an intelligent function in materials that undergo changes in crystalline structure, and in the interfacial conditions at the surface of the material or at the grain boundaries, for example. This class of intelligent materials would be able to autonomously regain their original shape through a phase transformation phenomenon, subsequent to the material suffering permanent deformation from impact for example. These materials may prove to be a candidate for automobile bumpers or the corners of hospital emergency-room trolleys. This class of intelligent self-repair materials would also be able to detect the degradation of the functional properties of a material by monitoring changes at the surface of the material due to contamination and surface impurities. Examples of this class of intelligent materials include shape-memory-materials of non-metallic and metallic composition, and hybrid electrical varistors.

Autolysis is manifest as an intelligent function in materials that change their molecular structure and/or change their macroscopic structure. This class of material, would, for example, be able to vary its adhesive properties by autonomously determining the appropriate time-functional properties relative to the resident conditions. Alternatively, in micro-capsular drug release applications, these materials would determine the appropriate duration and concentration of drug release characteristics, prior to subsequently decomposing to the molecular level of the neighboring biological materials upon completion of the task through the re-configuration of the molecular structure.

Prediction and notification is manifest as an intelligent function in materials that undergo an energy change, a change in molecular structure, or a change in the crystalline structure such as the structural reconfiguration associated with a phase transformation. This class of intelligent materials would, for example, be able to predict the onset of their degenerative state by the transfer of electrons precipitated by a re-configuring of the crystalline structure of the material. Furthermore, these materials would broadcast information pertaining to their state of degradation by changing the color of cholesteric liquid crystals resulting from this process. Other categories of these intelligent materials subjected to fatigue environments would develop stress induced transformations and furnish information by a release of energy, a relaxation of the inherent stresses, or by a change in the color of the materials. A narrow class of copper-zinc-aluminum alloys exhibit some crude prediction-notification properties.

Redundancy is manifest as an intelligent function in materials that change their molecular structure or change their crystalline structure. This class of intelligent materials would be able to relax the stresses in the stress concentration region at the tip of a crack by undergoing a stress-induced transformation. Alternatively, this class of materials could be employed in the fabrication of parts that require high reliability where the material would determine whether the state of stress were attributed to a static load or a shock load and in the latter case the material would undergo a phase transformation in order to develop higher strength properties. Typical examples of this class of materials are the two-phase titanium alloys and aluminum-zinc alloys.

Autonomous diagnosis is manifest as an intelligent function in materials that are either able to change their molecular structure, their crystalline structure or their interfacial properties at the surface, grain boundary and internal interface. These properties are exhibited in materials that are traditionally employed in a non-equilibrium state and change their functional properties when they attain a state of equilibrium. They are also evident in materials that autonomously determine the appropriate time to quickly terminate their functional behavior in response to the ambient environmental conditions, and in substances that detect degradation prior to triggering a stress-induced transformation that reveals this damaged state, by various means such as an energy release or a change in color of the material.

Autonomous assembly is manifest as an intelligent function in materials through changes in molecular structure or crystalline structure. This class of intelligent materials would typically feature assemblages of groups that each manifest their individual functions, alternatively these assembled groups would exist in unison under a prescribed set of operating conditions, but the groups would separate under different conditions and no longer manifest their previous functional form.

Learning is manifest as an intelligent function in materials through changes in the physical constraints, changes in the crystalline structure, or changes in the molecular structure. This class of intelligent materials would be able to facilitate the task of burning a hole in material using a laser beam, by autonomously varying the appropriate heat transfer coefficients of the material as a function of the wavelength of the light, in conjunction with information based upon prior experience. Other classes of these materials would measure the change in their structure in response to environmental stimuli prior to generating an appropriate signal such as an electrical current. Substances with these characteristics include memory and switching devices and optoelectric materials.

Standby is manifest as an intelligent function in materials through energy changes, changes in the molecular structure, and changes in the crystalline structure. An example of this class of intelligent materials includes materials that exploit changes in the molecular structure of thin films to administer appropriate drugs when fevers and other human ailments are detected. Another example of this class of intelligent substances are materials that transform the energy they receive in the form of heat and light, into a form of energy that they can readily store in their microstructure through deformation, for example, prior to releasing this stored energy at the appropriate time.

Feedback is manifest as an intelligent function in materials through a variety of mechanisms including the transfer and accumulation of charges, change in composition, change in molecular structure, the transfer of radicals and ions, and a change in crystalline structure. This class of intelligent materials would, for example, be able to change their characteristics in response to changing environmental conditions. Materials whose transformation temperature is temperature dependent would belong to this class of materials. This class of intelligent materials would be employed in heating and cooling systems, and also new generations of shape-memory metals and plastics.

Information integration is manifest as an intelligent function in materials that undergo a change in structure. This class of intelligent memory materials would accumulate information within themselves by memorizing changes in their structure based upon information gleaned from the neighboring environment. This information would be output as required.

Recognition is manifest as an intelligent function in materials by a variety of phenomena such as the transfer of charges and the accumulation of charges, a change in material composition, a change in molecular structure, a change in the interface properties at the surface, interphases, and grain boundaries, and in a change in crystal structure. This class of intelligent materials would be able to perform a variety of diverse tasks, such as the detection of compositional changes in body fluids by enzyme reaction prior to diagnosing a disease. Other innovative materials would be able to dynamically tailor their porosity, thermal conductivity, and other heat transfer coefficients in response to changing ambient temperature and humidity, in order that clothes attain optimal conditions for the individual wearing them. Materials that can identify viruses, attach to them and gradually release an appropriate drug treatment also belong to this class of innovative materials. Other materials will be able to distinguish between different gases, they will be able to quantify the amount of gas absorbed at levels corresponding to the individual types of gases once the amount of gas absorbed has exceeded a prescribed threshold. Furthermore, materials in this class of intelligent materials will be able to recognize the speed of propagation of stress waves and trigger appropriate dissipative phenomena in order to prevent potential damage.

Homeostasis is manifest as an intelligent function in materials by changes in the physical constraints and also by changes in the crystal structure. An example of this class of materials is a substance that conducts an electrical current by varying the electrical resistivity as a function of the change in temperature.

Time-functional responsiveness is manifest as an intelligent function in materials that undergo changes in their molecular structure, changes in their crystalline structure and changes in their macroscopic structure. These innovative capabilities can be employed to create materials that can be programmed to terminate their functional behavior in response to environmental changes, and also adhesive materials that maintain their adhesive properties until the material itself has determined, possibly upon completion of a prescribed time period, that this function is no longer required, whereupon the adherends easily separate.

Adaptation and adjustment properties are manifest as intelligent functions in materials that undergo a change in physical constants, a change in molecular structure, and a change in crystalline structure. Examples of these materials include films of lubricants which assure fluidic or solid states depending upon the local thermal or dynamic mechanical conditions. Other materials in this classification will be able to control their optical properties in response to changes in external stimuli such as magnetic fields, electrical fields and temperature.

The above list of intelligence functions, the associated phenomena, and the illustrative examples of these different classifications are dependent upon intelligence resident at the most primitive levels in materials. This involves considerations of the three basic classifications of sensors, processors, and actuators.

8.2 Intelligence at the most primitive level in materials

In order to develop viable methodologies for synthesizing and manufacturing commercial intelligent materials in practice, it is of crucial importance to understand and elucidate the intelligent functions at the most primitive levels of the materials. Furthermore, it is also important to understand how these intelligent functions relate to the three generic mechanisms of these classes of materials; namely sensors, processors and actuators.

The sensor function of an intelligent material typically detects and monitors information pertaining to both the external environment in contact with the material, and also the behavior of relevant parameters within the material itself. Examples of sensor functions include photo-chromic effects attributed to the breakdown of molecular bonds, optical effects in ceramics resulting from the re-configuring of the crystalline structure, and changes in the electronic conditions in materials lacking a definite crystalline form. Figure 8.2 presents a synopsis of the inter-relationships between some of the different phenomena associated with intelligent materials and their utility in sensing, relative to different chemical, dynamic, electrical, magnetic, optical and thermal environments.

The processor function of an intelligent material processes and evaluates detected information furnished by the sensor function in conjunction with information already memorized in the material. Examples of processor

Intelligence Phenomena	Sensing Parameters	Sensor Function					
		Chemical	Dynamic	Electrical	Magnetic	Optical	Thermal
Change in molecular structure	Adhesion & Separation	●	●			●	●
	Bond-distance	●	●			●	●
	Three-dimensional structures	●				●	●
Change in crystalline structure	Ceramics		●	●	●	●	●
	Metals		●	●	●		●
	Organic materials		●			●	●
Changes at interfaces	Grain boundaries of ceramics		●	●	●	●	
	Grain boundaries of metals		●	●	●		
	Surface characteristics	●			●	●	
Changes at the macroscopic level			●				
Change of composition		●		●	●	●	
Change in electronic structure					●		
Transfer of ions and radicals					●		●
Charge phenomena				●	●		
Energy absorption and transfer		●	●	●	●	●	●
Change in physical constants		●	●			●	●

Fig. 8.2 Intelligence phenomena and sensing functions.

functions include changes in the molecular structure of polymeric materials, and changes in the interfacial conditions and grain boundaries in ceramic and metallic materials. Examples of memory functions include residual states of distortion in crystalline or molecular structures, and material structures which are unstable. Figure 8.3 presents a synopsis of these inter-relationships.

The triumvirate of functions at this level of a material's structure is completed by the actuator function. This function generates appropriate effects that are dependent upon information from the sensor function and the processor function. Examples of these actuator functions include changes in the absorption properties due to changes in the molecular structure, and changes in color and the external geometry of a material due to phase transformation phenomena. Figure 8.4 synoptically presents the inter-relationships between these phenomena.

Intelligence Phenomena \ Processing Parameters		Processor Function: Chemical	Dynamic	Electrical	Magnetic	Optical	Thermal
Change in molecular structure	Adhesion & Separation	●					
	Bond-distance	●					
	Three-dimensional structures	●					
Change in crystalline structure	Ceramics		●	●	●		
	Metals			●	●		
	Organic materials	●					
Changes at interfaces	Grain boundaries of ceramics		●	●	●		
	Grain boundaries of metals		●	●	●		
	Surface characteristics	●					
Changes at the macroscopic level			●				
Change of composition							
Change in electronic structure					●		
Transfer of ions and radicals		●			●		
Charge phenomena				●	●		
Energy absorption and transfer		●	●	●	●	●	●
Change in physical constants		●	●	●	●	●	

Fig. 8.3 Intelligence phenomena and processing functions.

The transfer of information between sensing functions, processing functions, and actuation functions is accomplished by an information network function. Examples of these functions are similar to those of the processor function category. Energy absorption, conversion, transportation and supply functions are evidenced in intelligent materials through non-linear phenomena and phase transformation phenomena, for example. Examples of these functions include the absorption of energy through the distortion of the crystalline structure and the transportation of energy due to charge transfer.

In order to achieve the broad objectives outlined in the previous paragraphs for the creation of commercially-available intelligent materials, research and development programs have been initiated to achieve these goals. As evidenced from the accomplishments presented in Chapter 3 on smart materials, there has been some significant progress in these endeavors

Intelligence Phenomena	Actuator Parameters	Actuator Function: Chemical	Dynamic	Electrical	Magnetic	Optical	Thermal
Change in molecular structure	Adhesion & Separation	●	●			●	
	Bond-distance	●	●			●	
	Three-dimensional structures	●				●	
Change in crystalline structure	Ceramics		●	●	●	●	●
	Metals		●	●	●	●	
	Organic materials						
Changes at interfaces	Grain boundaries of ceramics	●	●	●	●	●	●
	Grain boundaries of metals		●	●	●		
	Surface characteristics					●	
Changes at the macroscopic level			●	●	●	●	
Change of composition		●	●	●	●	●	
Change in electronic structure				●	●		
Transfer of ions and radicals		●			●		
Charge phenomena				●	●		
Energy absorption and transfer		●	●	●		●	●
Change in physical constants		●	●	●	●	●	●

Fig. 8.4 Intelligence phenomena and actuation functions.

by principally employing much less sophisticated functional materials in structural systems. This current discrete philosophy will ultimately be replaced by coherent design-for-manufacture philosophies at the atomic and molecular levels which will enable a new generation of materials to be synthesized.

This new generation of materials will only be created once a deeper understanding is established of the sensing, processing and actuation functions at the most primitive levels of a material. This must be augmented by a deeper understanding of the relationship between the different phenomena associated with these three classes of functions. Upon achieving these objectives it should be possible to systematically synthesize intelligent materials that feature these different desired attributes, by orchestrating the primitive functions relative to the operating environment.

The synthesis of new generations of materials has traditionally required a comprehensive understanding of the chemical and physical functions and structure of the current generation of materials prior to the field evolving further. At the most primitive levels, the chemical and physical characteristics are governed by the crystalline structure, the interfacial conditions, the molecular structure and composition, and the electronic arrangement and composition. Investigations are being initiated to understand how these characteristics relate to the fundamental sensory, processing, and actuation functions in order to distill theoretical structural-control algorithms at the atomic and molecular levels, which in conjunction with the associated technologies, will facilitate the design and manufacture of intelligent materials.

The scientific method, the traditional cornerstone of all scientific endeavors, mandates that theoretical arguments be evaluated in practice as part of an algorithm involving the distinct phases of hypothesis, the testing of the hypothesis through experimental work prior to a comparison of theory and practice, and then subsequent refinement. Thus experimental techniques, associated with the characterization of materials, must be developed in order to evaluate, enhance, and refine the theoretical tools developed for the creation, design, and manufacture of intelligent materials. Characterization and analysis techniques are important in undertaking a structural or compositional evaluation of an intelligent material. These techniques will typically involve non-destructive strategies and *in situ* evaluations at the atomic level, at the molecular level, and also at the macroscopic level. These experimental techniques will not only be instrumental in stimulating the development of superior theoretical tools but they will be indispensable for discovering new materials through serendipity.

The maturation of the forementioned methodologies and technologies will facilitate the establishment of an intelligent materials data-base. Furthermore, they will ensure that environmentally-compatible materials are synthesized that operate in harmony with humanity.

8.3 Strategies for intelligent materials research

The thesis of this chapter has been the relationship between intelligent functions and the associated structural changes in materials at the atomic, molecular, and macroscopic levels. This domain provides the kernel of knowledge for the synthesis of new classes of intelligent materials by assembling and integrating materials and functions at these different levels. Strategies for prosecuting research on intelligent materials may be considered to belong to the following broad categories: theoretical studies;

design and manufacture; creation of primitive functions; assembly and integration of functions; material characterization; and generic supporting technologies.

8.3.1 Theoretical studies

The development of viable theoretical and computational techniques for elucidating the functions and structural properties of intelligent materials is of crucial importance for this embryonic discipline because it is a foundation for the other research sub-fields. These theoretical techniques should be able to address issues concerned with sensor, processor and actuator functions, in addition to the coordination and orchestration of these functions. The predictive capabilities of these theories will need to be sharpened and honed through the application of the scientific method. Thus, a theoretical model would be formulated to predict the behavior of an intelligent function in a material; experimental investigations would then be undertaken to evaluate the predictive capabilities of the theoretical model under diverse circumstances; experimental data would then be collected and compared with the theoretical predictions prior to modifying the original hypothesis in an iterative fashion, or prior to distilling empirical laws. Subsequently these theoretical or empirical models would be applied to other situations in order to discern their domain of applicability.

8.3.2 Design and manufacture

The development of efficient design and manufacturing strategies are crucial for the commercial exploitation of intelligent material systems. While the design and manufacturing processes associated with many commercial products are typically separate, discrete entities, this somewhat traditional and limited philosophy cannot be adopted and employed in the field of intelligent materials because the two disciplines are inextricably intertwined. Thus, the manufacturing process is responsible for designing, or creating, the structural properties of the material within the distinct constraints and characteristics of that particular manufacturing process. Furthermore, these considerations must be carefully considered in the context of fabricating materials at the atomic, molecular, mesoscopic, and macroscopic levels, depending upon the intelligent functions being embodied in the material. The development of viable design-for-manufacture strategies for the creation of intelligent materials with multi-functional relationships and good quality control measures, will require the accumulation of a significant volume of experimental data in order to establish the governing relationship between material synthesis, structures, and innovative functions.

The design phase of creating an intelligent material must be undertaken within the context of spatially configuring the structure in order to embed the necessary characteristics and primitive functions. Significant progress has already been made towards achieving this goal, since it is now possible to optimize the composition and conditions associated with the phase transformation phenomena of precipitation in some materials using a supercomputer. This computer-aided-design technique permits the movement of individual atoms to be determined which subsequently permits the process to be controlled at the mesoscopic level. Thus, by clarifying the physical image of phenomena associated with intelligence in materials at the atomic level, the basis then exists for imposing control and hence synthesis strategies at the atomic, molecular, or macroscopic levels. This would be an important tool in the design and manufacture of intelligent materials.

8.3.3 Creation of primitive functions

The primitive functions of intelligent materials comprising the sensor function, the processor and memory function, and the actuator function are typically orchestrated to operate harmoniously at the atomic, molecular, mesoscopic, and macroscopic levels. These functions are dependent upon the structural properties of the material at these different levels, and the techniques for structural control are different at the atomic or molecular levels from the techniques that are employed at the mesoscopic or macroscopic levels. At the atomic or molecular levels, systematic structural control typically involves impurities and lattice defects, while at the mesoscopic level the focus of attention is the interfacial conditions or the grain boundary attributes, and at the macroscopic level the focus of attention is typically hetero-dispersion or multi-layer processing.

Structural control at the atomic and molecular levels

Current techniques enable atomic arrangements to be designed and controlled in single atomic layers such as a superlattice. These techniques potentially permit the creation of intelligent materials at the most fundamental levels through the sophisticated control of the structures and compositions. At these levels, hybridization philosophies have an important role to play in the arrangement of different materials with diverse properties wherein the resulting functions are superior to those of the individual substances themselves. Furthermore, these hybridization schemes have the potential to stimulate the evolution of new materials with multiple innovative functions.

In order to create primitive functions it is necessary to develop techniques for controlling individual atoms, individual molecules, and also conditions at the macromechanical level of materials design, such as those involving hetero-dispersion and the creation of materials with numerous strata. At the atomic level, it is anticipated that tools will be developed by capitalizing upon the technologies offered by STM lithography, for example. At the molecular level, sophisticated creation techniques will be employed such as epitaxy techniques, thin film techniques involving lasers and electron beams, surface-control techniques, etching techniques, and techniques involving scanning electron microscopy.

Structural control at the mesoscopic level

Structural control at the mesoscopic level will probably focus on several discrete areas such as those associated with intermediate energy states and also intermediate states of gels or quasi-crystalline structures. Other approaches will involve morphological investigations because some innovative functions are shape-dependent.

Several phenomena provide the basis for controlling the structure of a material at the mesoscopic level. These include the peculiar state of the interactions among groups of ultra-fine particles, where these so-called supermolecules are stabilized by the dynamic interactions between the numerous atomic and molecular groups. Other approaches will focus on a quantum effect for incorporating processing functions and information transfer functions into materials at the mesoscopic level, by creating conductors of only tens of nanometers in length in a material, which are able to generate a vibrational excitation.

It is anticipated that research will also be prosecuted on non-equilibrium states utilizing ion engineering practices to convert thermal energy to kinetic energy, by utilizing pressure, volume and temperature conditions which are augmented by data on stability, composition, and reaction kinematics.

8.4 Assembly and integration functions

In the previous paragraphs on primitive functions and structural control, considerable attention was devoted to concepts for developing materials with numerous orchestrated functions through the assembly and integration of the primitive functions. In other cases, however, it may be more desirable to create materials that feature these different materials.

The fundamentals of this philosophy reside in the structural arrangement of substances, consequently it is of paramount importance to be able to undertake the synthesis of materials and to utilize structural control-algorithms from the microscopic level to the macroscopic level. In order to achieve this objective, it will generally be necessary to coordinate a variety of different functions and also arrange the primitive functions within a spatial domain. Consequently, an appreciation of the assembly and integration processes must be developed within the context of dimensional arguments. Thus a zero-dimensional approach could be relevant to ultra-fine particles; a one-dimensional approach could be relevant to ultra-fine layers; and a three-dimensional approach could be relevant to materials with numerous ultra-thin layers and also materials with multi-dimensional features that are time-dependent.

These different classes of intelligent materials will probably be manufactured by processes that select appropriate species prior to utilizing chemical hyperstacking techniques, to stack and appropriately configure the atomic and molecular structures in ultra-thin layers. Alternatively physical layer stacking techniques may be employed to manufacture ultra-thin layers featuring hybridized atomic or molecular structures with continuous or intermittent characteristics. Other processes that capitalize on clear and zero-strain surface control techniques will be important at the nanometer-order level for interface connectivity, and fine processing.

While the design, control, and integration of primitive functions are indeed a crucial factor in the synthesis of intelligent materials, the orchestration and coordination of these functions in materials are also important factors that must be incorporated in the design and manufacture algorithm. It may, therefore, be appropriate to create the substances that will be performing the processor functions to be a discrete entity, that time-shares the roles of orchestrating the numerous primitive functions with the roles of information transmission and information processing and learning for example.

8.4.1 Analysis and characterization

Accurate analysis and characterization techniques must be developed not only to evaluate the performance of intelligent materials that have been designed and manufactured to satisfy stringent comprehensive material design specifications, but to also characterize innovative materials produced serendipitously and important new phenomenological events not predicted by current theoretical arguments. Furthermore, this class of experimental techniques will be crucial for the maturation of the field of intelligent materials, where progress on the design and development of new generations of materials will undoubtably require the integration and interplay of both analytical and experimental techniques.

Non-destructive and *in situ* measurement techniques are of crucial importance in this endeavor when employed in conjunction with appropriate analysis and characterization techniques. Current approaches typically feature the impingement of an appropriate particulate beam system on the material sample being subjected to evaluation. This beam generally involves electrons, ions, X-rays, or photons, and the reflected beam is subsequently scrutinized in order to characterize the structure of the material sample.

This broad array of analysis and characterization techniques includes scanning tunnelling spectroscopy and microscopy, scanning tunnelling potential and atomic force microscopy which could potentially be employed to identify and analyze individual atoms, Auger electron spectroscopy, X-ray photoelectron spectroscopy, and extended X-ray absorption which could potentially be employed to measure the electronic state at the electron level on surfaces.

Thus these procedures, and the near-term enhancement of these procedures, will enable the structure of a material to be ascertained, the spatial distribution of characteristic domains within the structure to be characterized, and also the process responsible for the creation of the structure to be determined. It is anticipated that some of these techniques will be further enhanced, while others will be integrated in order to develop viable techniques for the analysis and characterization of new generations of multiple-function intelligent materials where the experimental data must be collected at several locations rather than at the single location that is typical in current practice.

8.4.2 Generic supporting technologies

In order to facilitate the prosecution of research and development on intelligent materials, it will be necessary to further develop and enhance a number of technologies that are somewhat generic in nature and common to several of the underlying disciplines in the field. For example, it is anticipated that progress will be accelerated by the development of classes of generic computational techniques for the design and manufacture of intelligent materials. These capabilities will be augmented by the appropriate codification and structuring of user-friendly data-bases on intelligent materials. Such a task will undoubtably involve the accumulation of appropriate data-sets characterizing relevant intelligent functions and properties within the context of standardized methods of data analysis and evaluation.

Techniques for creating ultra-pure materials will be important in order to develop intelligent materials commercially. The techniques will facilitate the purification of numerous exotic materials, such as rare-earth substances, and they will result in the evolution of crystallization processes of high purity.

Other processes will focus on environmental control schemes for creating or measuring non-equilibrium states under extreme conditions. These processes will typically feature both individual environmental conditions or combinations of environments involving extremely low or extremely high temperatures, zero-gravity conditions, and extremely high pressures or extremely low vacuums. Other processes will feature the very rapid change of these environments.

Beam systems are an essential ingredient in the analysis and characterization techniques employed in the research and development of advanced materials. The current generation of techniques must be extended in order to enable them to be applicable to future generations of intelligent materials involving *in situ* measurements, measurements of materials at several points simultaneously and measurements involving several dimensions. These new analysis and characterization systems will feature new types of beam generation devices, multiple beams of different classifications, and feature schemes for controlling the beam characteristics and energy levels.

Finally, in order to further stimulate and promote research and development activities on intelligent materials it is essential to encourage the establishment of appropriate forums for the exchange of ideas, theories, and experimental techniques by the premier groups of international researchers. This is especially important in this particular embryonic field because the science of intelligent materials embraces so many diverse, yet interdisciplinary scientific fields and technologies. The synergistic exchanges between these diverse groups will not only be crucial for the maturation of the individual sub-disciplines but also for the creation of new interdisciplinary concepts and theories unrealizable by individual groups working in isolation. Thus opportunities for the exchange of ideas should be established through conferences, workshops, the establishment of a professional society for intelligent materials practitioners and the launching of new journals where academics, industrialists, and government employees can publish state-of-the-art concepts in the field.

With the establishment of these mechanisms, the science and technologies of intelligent materials and substances will be able to reach fruition and provide an essential ingredient for unforeseen advances in society at the dawn of the twenty-first century. Indeed the impact of these advances will be so profound that future historians will undoubtably characterize this era as the *Age of Intelligent Materials*.

Bibliography

Electro-rheological fluids

Adamson, A.W., *Physical Chemistry of Surfaces*, John Wiley and Sons, New York, 1976.

Adriani, P.M. and A.P. Gast, 'A Microscopic Model of Electro-Rheology,' *Physics of Fluids*, Vol. 31, No. 10, October 1988, pp. 2757–68.

Anaskin, I.F., V.K. Gleb, E.V. Korobko, Khizhinskii, B.P. and Khusid, B.M. 'Effect of External Electric Field on Amplitude-Frequency Characteristics of Electro-Rheological Damper,' *Inzhenerno-Fizicheskii Zhurnal*, Vol. 46, No. 2, 1984, pp. 309–315.

Andrade, E.N. and C. Dodd, 'The effect of an electric field on the viscosity of liquids. Part I,' *Proceedings of the Royal Society*, Vol. 187A, 1946, pp. 296–306.

Andrade, E.N. and C. Dodd, 'The effect of an electric field on the viscosity of liquids. Part II,' *Proceedings of the Royal Society*, Vol. 204A, 1951, pp. 449–464.

Andrade, E.N. and J. Hart, 'The effect of an electric field on the viscosity of liquids. Part III,' *Proceedings of the Royal Society*, Vol. 225A, 1954, pp. 463–472.

Arguelles, J., H.R. Martin, and R. Pick, 'A Theoretical Model for Steady Electroviscous Flow Between Parallel Plates,' *Journal of Mechanical Engineering Science*, Vol. 16, No. 4, 1974, pp. 232–239.

Batchelor, G.K., 'The Stress System in a Suspension of Force-free Particles,' *Journal of Fluid Mechanics*, Vol. 41, April 1970, pp. 545–570.

Bird, R.B., R.C. Armstrong, and O. Hassager, *Dynamics of Polymeric Liquids: Volume I. Fluid Mechanics*, Wiley, New York, 1977.

Block, H. and J.P. Kelly, 'Electro-Rheology,' *Journal Physics D. Applied Physics*, Vol. 21, 1988, pp. 1661–76.

Block, H. and J.P. Kelly, U.K. Patent Application 8,503,581, 1986 and U.S. Patent Specification 4,687,589, 1987.

Booth, F., 'The electroviscous effect for suspensions of solid spherical particles,' *Proceedings of the Royal Society*, Vol. A203, 1950, pp. 533–551.

Brooks, D.A., 'Electro-Rheological Effect Adds Muscle,' *Control & Instrumentation*, October 1982, pp. 57–59.

Brooks, D.A., 'Electro-Rheological Devices,' *Chartered Mechanical Engineer*, London, September 1982, pp. 81–93.

Bullough, W.A. and Peel, D.J. 'Electro-Rheological Oil Hydraulics,' *Proceedings of Japan Society for Hydraulics and Pneumatics*, Vol. 61, No. 17, November 1986, pp. 520–526.

Bullough, W.A. and J.D. Stringer, 'The Utilization of the Electroviscous Effect in a Fluid Power System,' *3rd International Fluid Power Symposium*, Turin, May 1973, Paper # F3–37.

Bullough, W.A., D.J. Peel, and R. Firoozian, 'Electrical Characteristics and other Considerations for Electro-Rheological Fluids,' *Proceedings of the First International Symposium on ER Fluids, 8th Annual Meeting of the Fine Particle Society*, Boston 1987.

Buscall, R., J.W. Goodwin, and R.H. Ottewill, 'The Settling of Particles through Newtonian and Non-Newtonian Media,' *Journal of Colloid and Interface Sciences*, Vol. 85, No. 1, 1982, pp. 78–86.

Carlson, J.D., A.F. Sprecher, and H. Conrad, (eds.), *2nd International Conference on Electro-Rheological Fluids*, Raleigh, NC, August 1989.

Carlson, J.D., A.F. Sprecher, and H. Conrad, 'Electro-Rheology at Small Strains and Strain Rates of Suspensions of Silica Particles in Silicone Oil,' *Materials Science and Engineering*, Vol. 95, 1987, pp. 187–197.

Choi, S.B., B.S. Thompson, and M.V. Gandhi, 'An Experimental Investigation on the Active-Damping Characteristics of a Class of Ultra-advanced Intelligent Composite Materials Featuring Electro-Rheological Fluids,' *Proceedings of the Damping '89 Conference*, West Palm Beach, February 1989, organized by the Flight Dynamics Laboratory of the Air Force Wright Aeronautical Laboratories, Wright Patterson Air Base, Ohio, WRDC-TR-89–3116, Vol. 1, pp. CAC-1 to CAC-14.

Choi, S.B., M.V. Gandhi, and B.S. Thompson, 'Control of Smart Flexible Structures Incorporating Electro-Rheological Fluids: A Proof-of-Concept Investigation,' *1989 Automatic Control Conference, Pittsburgh*, June 1989.

Choi, S.B., M.V. Gandhi, and B.S. Thompson, 'An Experimental Investigation of the Elastodynamic Response of a Slider-Crank Mechanism Featuring a Smart Connecting Rod Incorporating Embedded Electro-Rheological Fluid Domains,' *Proceedings of the 1st National Applied Mechanisms and Robotics Conference*, Cincinnati, Ohio, Vol. II, Nov. 1989, 9B-3.1 to 9B-3.7.

Choi, S.B., B.S. Thompson, and M.V. Gandhi, 'Smart Structures Incorporating Electro-Rheological Fluids for Vibration-Control and Active-Damping Applications: An Experimental Investigation,' 12th Biennial ASME Conference on Mechanical Vibration and Noise, Montreal, Sept. 1989, published in *Machine Dynamics-Applications and Vibration Control Problems*, T.S. Sankar, V. Kamala, P. Kim and D.K. Rao, eds., DE Vol. 18–2, ASME Book H00508B, pp. 229–236.

Choi, S.B., B.S Thompson, and M.V. Gandhi, 'Electro-Rheological Fluids Technology Stimulates a New Generation of Robotic and Machine Systems,' *Proceedings of the Tenth OSU Applied Mechanisms Conference*, New Orleans, Vol. I., December 1987, pp. 3B.2–1 to 3B.2–8.

Choi, Y., A.F. Spencer, and H. Conrad, 'Vibration Characteristics of a Composite Beam Containing an Electro-Rheological Fluid,' *Journal of Intelligent Material Systems and Structures*, Vol. 1, January 1990, pp. 91–104.

DeLang, R.G., C.A. Verbraak, and J.A. Zijderveld, U.S. Patent 3,450,372.

Deinega, Y.F., 'Some Problems of the Electro-Rheology of Dispersed Systems,' *Royal Aircraft Establishment*, Farnborough, England, HC A02/MP A01, 1982.

Dirty Oil: The Discovery and Application of the Winslow Effect, *Hydraulics and Pneumatics*, Vol. 19, 1962, pp. 144–147.

Duclos, T.G., 'Design of Devices Using Electro-Rheological Fluids,' SAE paper #881134, 1988.

Duclos, T.G., D.N. Acker, and J.D. Carlson, 'Fluids that Thicken Electrically,' *Machine Design*, January 1988, pp. 42–46.

Elton, G.A.H. and F.G. Hirschler, 'Electroviscosity: Vol. I and II,' *Proceedings of the Royal Society*, 1948 v.194, pp. 259–288, 275–288.

Elton, G.A.H. and F.G. Hirschler, 'Electroviscosity: Vol. III,' *Proceedings of the Royal Society*, 1949 v.197, pp. 568–572.

Elton, G.A.H. and F.G. Hirschler, 'Electroviscosity: Vol. IV,' *Proceedings of the Royal Society*, 1949 v.198, pp. 581–590.

Filisko, F.E. and W.E. Armstrong, Electric Field Dependent Fluids, U.S. Patent 4,744,914, May 1988.

Freundlich, H., *Colloid & Capillary Chemistry*, Methuen, London, 1926, (First published in German, 1909).

Gast, A.P. and C.F. Zukowski, 'Electro-Rheological Fluids as Colloidal Suspensions,' *Advances in Colloid and Interface Science*, Vol. 30, 1989, pp. 153–202.

Gandhi, M.V., B.S. Thompson, S.B. Choi, and S. Shakir, 'Electro-Rheological-Fluid-Based Articulating Robotic Systems,' *ASME Journal of Mechanisms, Transmissions and Automation in Design*, Vol. 111, 1989, pp. 328–336.

Gandhi, M.V. and B.S. Thompson, 'An Innovative Class of Revolutionary Articulating Mechanism and Robotic Systems Exploiting Smart Materials and Structures: Fiber Optics, Shape-Memory Metals, Piezoelectric Materials and Electro-Rheological Fluids,' *Proceedings of the 1st National Applied Mechanisms and Robotics Conference*, Cincinnati, Ohio, Vol. II, Nov. 1989, 7C-2.1 to 7C-2.7.

Gandhi, M.V., B.S. Thompson, and S.B. Choi, 'A New Generation of Innovative Ultra-Advanced Intelligent Composite Materials Featuring Electro-Rheological Fluids: An Experimental Investigation,' *Journal of Composite Materials*, Vol. 23, 1989, pp. 1232–55.

Gandhi, M.V., B.S. Thompson, and S.B. Choi, 'A Note on an Experimental Investigation of a Slider-Crank Mechanism Featuring a Smart Dynamically-Tunable Connecting-Rod Incorporating an Electro-Rheological Fluid,' *Journal of Sound and Vibration*, Vol. 135, 1989, pp. 511–515.

Gandhi, M.V., B.S. Thompson, and S.B. Choi, 'Smart Ultra-Advanced Composite Materials Incorporating Electro-Rheological Fluids for Military, Aerospace and Automated Manufacturing Applications,' SME/ESD Advanced Composites Conference and Exposition, published in *Advanced Composites III: Expanding the Technology*, Detroit, MI, September 1987, pp. 23–30.

Gandhi, M.V., B.S. Thompson, and S.B. Choi 'Ultra-Advanced Composite Materials Incorporating Electro-Rheological Fluids,' *Proceedings of the 4th Japan–U.S. Conference on Composite Materials*, Washington, DC, June 1988, pp. 875–884.

Gandhi, M.V. and B.S. Thompson, 'A New Generation of Revolutionary Intelligent Composite Materials Featuring Electro-Rheological Fluids,' *Proceedings of the ARO Workshop on Smart Materials, Structures and Mathematical Issues*, Blacksburg, VA, September, 1988, pp. 63–68.

Gandhi, M.V. and B.S. Thompson, 'Dynamically-Tunable Smart Composites Featuring Electro-Rheological Fluids,' *Proceedings of the SPIE 1989 Boston Symposium OE/FIBERS '89 Optoelectronics and Fiber Optic Devices and Applications*, Boston, MA, Vol. 1170, September 1989, pp. 294–304.

Gandhi, M.V. and B.S. Thompson, 'A New Generation of Revolutionary Ultra-Advanced Intelligent Composite Materials Featuring Electro-Rheological Fluids,' *Smart Materials, Structures and Mathematical Issues*, C.A. Rogers, (ed.), Lancaster, PA: Technomic Publishing Company, Inc., 1988, pp. 63–68.

Goodwin, J.W., R.W. Hughes, S.J. Partridge, and C.F. Zukoski, 'The Elasticity of Weakly Flocculated Suspensions,' *Journal of Chemical Physics*, Vol. 85, No. 1, July 1986, pp. 559–566.

Goodwin, J.W., R.H. Ottewill, and A. Parentich, 'Optical Examination of Structured Colloidal Dispersions,' *Journal of Physical Chemistry*, Vol. 84, 1980, pp. 1580–86.

Gorodkin, R.G., Y.V. Kovobko, G.M. Blokh, V.K. Gleb, G.I. Sidorova, and M. Ragother, 'Applications of the Electro-Rheological Effect in Engineering Practice,' *Fluid Mechanics – Soviet Research*, Vol. 8, No. 4, 1979, pp. 48–61.

Gorodkin, R.G., I.V. Bukovich, M.B. Smolskii, and M.M. Ragotner, 'The effect of vibration on structure formation in ER-suspensions,' *Applied Mechanics and Rheophysics*, Minsk, 1983, pp. 75–79.

Hinch, E.J. and J.D. Sherwood, 'The Primary Electroviscous Effect in a Suspension of Spheres with Thin Double Layers,' *Journal of Fluid Mechanics*, Vol. 132, 1983, pp. 337–347.

Honda, T. and T. Sasada, 'Electroviscous Effect in Liquids of Low Permittivity,' *Japanese Journal of Applied Physics*, Vol. 18, No. 8, 1979, pp. 1605–06.

Honda, T. and T. Sasada, 'The Mechanism of Electroviscosity. I. Electrohydro-Dynamic Effect on Polar Liquids,' *Japanese Journal of Applied Physics*, Vol. 16, No. 10, October 1977, pp. 1775–83.

Honda, T. and T. Sasada, 'The Mechanism of Electroviscosity. II. Conductivity Effect of Dielectric Liquids on Electroviscosity,' *Japanese Journal of Applied Physics*, Vol. 16, No. 10, Octover 1977, pp. 1785–91.

Honda, T., K. Kurosawa, and T. Sasada, 'Transient Pressure-Drop Fluctuations in Electroviscous Effect,' *Japanese Journal of Applied Physics*, Vol. 18, No. 11, November 1979, pp. 2059–63.

Honda, T. and T. Sasada, 'The Electroviscous Effect of Water,' *Japanese Journal of Applied Physics*, Vol. 18, No. 3, 1979, pp. 675–676.

Honda, T., K. Kurosawa, and T. Sasada, 'A.C. Characteristics of the Electroviscous Effect,' *Japanese Journal of Applied Physics*, Vol. 19, No. 8, 1980, pp. 1463–66.

Hunter, R.J., *Foundations of Colloid Science, Vol. I and Vol. II*, Oxford Science Publications, Clarendon Press, Oxford, England, 1989.

Klass, D.L. and T.W. Martinek, 'Electroviscous Fluids No. I. Rheological Properties,' *Journal of Applied Physics*, Vol. 38, 1967, pp. 67–74.

Klass, D.L. and T.W. Martinek, 'Electroviscous Fluids No. II. Electrical Properties,' *Journal of Applied Physics*, Vol. 38, 1967, pp. 75–80.

Konig, W., 'Bestimmung einiger Reibungsco-efficienten und Versuche Uber den Einfluss der Magnetisitung und Electrisinung auf die Reibungder Flussigleuten,' *Annals of Physics*, Vol. 25, 1885, pp. 618–624.

Korobko, E.V. and I.A. Chernobal, 'Influence of an External Electric Field on the Propagation of Ultrasound in Electro-Rheological Suspension,' *Inzhenerno-Fizicheskii Zhurnal*, Vol. 48, No. 2, 1985, pp. 219–224.

Lever, D.A., 'Large Distortion of the Electric Double Layer around a Charged Particle by a Shear Flow,' *Journal of Fluid Mechanics*, June 1979, Vol. 92, Part 3, pp. 421–433.

Mandell, M., 'Smart Fluids: Wavelet of the Future?,' *High Technology Business*, July-August 1989, pp. 20–23.

Mullins, J.P., 'Fluid Powerful!,' *Automotive Industries*, September 1988, pp. 68–69.

Peel, D.J., R. Firoozian, and W.A. Bullough, 'Deviation of the Electrical Characteristics of an Electro-Rheological Fluid from Biased Sine Wave Excitation Tests,' *Proceeding of the First International Symposium on ER Fluids, 18th Annual Meeting of the Fine Particle Society*, Boston 1987.

Reiner, M.A. *'Lectures in Theoretical Rheology*,' North Holland, Amsterdam, 1960 [Russian Translation] as Reologiya (Rheology), 1965.

Reiner, M.A. 'Mathematical Theory of Dilatancy,' *American Journal of Mathematics*, Vol. 67, 1945, pp. 350–362.

Rosen, M.J., *Surfactants and Interfacial Phenomena*, John Wiley, Inc.: New York, New York, 1978.

Russel, W.B. and A.P. Gast, 'Non-Equilibrium Statistical Mechanics of Concentrated Colloidal Dispersions: Hard Spheres in Weak Flows,' *Journal Chemical Physics*, Vol. 84, No. 3, February 1986, pp. 1815–26.

Russel, W.B., 'The Rheology of Suspensions of Charged Rigid Spheres,' March 1978 *Journal of Fluid Mechanics*, Vol. 85, Part 2, pp. 209–232.

Saitoh, S., T. Sakai, and H. Kusama, 'Fluidic electro-fluid converter by use of electro-viscous fluid,' *Second International JSME Symposium*, September 1972 (Tokyo).

Schmerling, M.A., M.A. Wilkov, A.E. Sanders, and J.E. Woosley, *Journal of Biomedical Materials Research*, Vol. 10, 1976, pp. 879–902.

Scott, D. and J. Yamaguchi, 'ER fluid devices near commercial stage,' *Automotive Engineering*, Vol. 93, No. 11, November 1985, pp. 75–79.

Sherwood, J.D., 'The primary electroviscous effect in a suspension of spheres,' *Journal of Fluid Mechanics*, Vol. 101, December 1980, pp. 609–630.

Shul'man, Z.P., 'Utilization of Electric and Magnetic Fields for Control of Heat and Mass Transfer in Dispersed Systems (Suspensions),' *Heat Transfer - Soviet Research*, Vol. 14, No. 5, September-October 1982, pp. 1–23.

Shul'man, Z.P. et al, *Elektroreologicheskiy Effekt* (The Electro-Rheological Effect), Nauka i Tekhnika Press, Minsk, 1972.

Shul'man, Z.P., B.M. Khusid, E.V. Korobko, and E.P. Khizhinsky, 'Damping of Mechanical Systems Oscillation by a Non-Newtonian Fluid with Electric-Field Dependent Parameters,' *Journal of Non-Newtonian Fluid Mechanics*, Vol 25, 1987, pp. 329–346.

Shul'man, Z.P., A.D. Matsepuro, and B.M. Khusid, 'Structurization of electro-Rheological suspensions in electric field, Part 1: Qualitative analysis,' Vestsi Akademii Navuk Belaruskai SSR, *Seryya Fizika Energetychnykh Navuk*, No. 3, 1977, pp. 116–121.

Shul'man, Z.P., A.D. Matsepuro, and B.M. Khusid, 'Structurization of electro-rheological suspensions in electric field, Part 2: Quantitative estimates,' Vestsi Akademii Navuk Belarskai SSR, *Seryya Fizika Energetychnykh Navuk*, No. 3, 1977, pp. 122–127.

Shul'man, Z.P., R.G. Gorodkin, V.K. Gleb, and E.V. Korobko, 'The electro-Rheological effect and its possible uses,' *Journal of Non-Newtonian Fluid Mechanics*, No. 8, 1981, pp. 29–41.

Simon, M., R. Kaplow, E. Salzman, and D. Freiman, *Radiology*, Vol. 125, No. 1, 1977, p.92.

Snyder, B., 'Application of Electro-Rheological Fluids to Hydrodynamic Lubrication,' UNR Report # ME-BS-900507, 1990, (proprietary).

Sprecher, A.F., J.D. Carlson, and M. Conrad, 'Electro-Rheology at Small strains and strain rates of suspensions of silicon particles in silicone oil,' *Material Science and Engineering*, Vol. 95, 1987, pp. 187–197.

Stangroom, J.E., UK Patent Specification 1501635, 1978.

Stangroom, J.E., 'Electro-Rheological Fluids,' *Physics in Technology*, Vol. 14, 1983, pp. 290–296.

Stanway, R., J.L. Sproston, and R. Firoozian, 'Identification of the damping law of an electro-Rheological fluid in vibration,' ASME Winter Annual Meeting, Anaheim, California, December 1986, paper 86-WA/DSC-6.

Stevens, N.G., J.L. Sproston, and R. Stanway, 'On the Mechanical Properties of Electro-Rheological Fluids,' *ASME Journal of Applied Mechanics*, Vol. 54, June 1987, pp. 456–458.

Stevens, N.G., J.L. Sproston, and R. Stanway, 'The Influence of Pulsed D.C. Input Signals on Electro-Rheological Fluids,' *Journal of Electrostatics*, Vol. 17, 1985, pp. 181–191.

Strandrud, H.T., 'Electric-Field Valves Inside Cylinder Control Vibrator,' *Hydraulics and Pneumatics*, September 1966, pp. 132–143.

Thompson, B.S. and M.V. Gandhi, 'A Commentary of Flexible Fixturing,' *ASME Applied Mechanics Reviews*, Vol. 39, No.9, 1986, pp. 1365–69.

Thompson, B.S., M.V. Gandhi, and C.Y. Lee, 'A Variational Formulation for Smart Structural Members of Robotic Systems Featuring Embedded Electro-Rheological Fluids,' *Proceedings of the 1st National Applied Mechanisms and Robotics Conference*, Cincinnati, Ohio, Vol. I, Nov. 1989, 2C-2.1 to 2C-2.7.

Treasure, U.Y., F.E. Filisko, and L.H. Radzilowski, 'Polyelectrolytes as Inclusions in Electro-Rheologically Active Materials: Effect of Chemical Characteristics on ER Activity,' *Journal of Rheology*, Vol. 35, No. 6, August 1991, pp. 1051–69.

Uejima, H., 'Dielectric Mechanism and Rheological Properties of Electro-Fluids', *Japanese Journal of Applied Physics*, Vol. 11, No. 3, 1972, pp. 319–325.

Viongradov, G.V., Z.P. Shul'man, Yu. G. Yanovskii, V.V. Barancheeva, E.V. Korobko, and I.V. Bukovich, 'Shear Viscoelastic Behavior of Electro-Rheological Suspensions,' *Inzhenerno-Fizicheskii Zhurnal*, Vol. 50, No. 4, April 1986, pp. 429–432.

Willson, J.R. and K.T. Krueger, U.S. Patent 3,645,443.

Winslow, W.M., 'Induced Vibration of Suspensions,' *Journal of Applied Physics*, Vol. 20, 1949, pp. 1137–40.

Winslow, W.M., 'Method and Means for Translating Electrical Impulses into Mechanical Forces,' *U.S. Patent 2,417,850*, 1947.

Shape-memory alloys

Achenbach, M., T. Atanackovic, and I. Muller, 'A Model for Memory Alloys in Plane Strain,' *International Journal of Solids and Structures*, Vol. 22, No. 2, 1986, pp. 171–219.

Achenbach, M. and I. Muller, 'Creep and Yield in Martensitic Transformation,' *Ingenieur-Archiv*, Vol. 53, 1983, pp. 73–83.

Aifantis, E.C. and J. Gittus, *Phase Transformations*, 1986, Elsevier, Amsterdam.

Baz, A., K. Inman, and J. McCoy, 'Active Vibration Control of Flexible Beams Using Shape Memory Actuators,' *Journal of Sound and Vibration*, Vol. 140, 1990, pp.437–456.

Baz, A., K. Inman, and J. McCoy, 'The Dynamics of Helical Shape Memory Actuators,' *Journal of Intelligent Material Systems and Structures*, Vol. 1, 1990, pp. 105–133.

Banks, R., *Phase Transitions in Shape Memory Alloys and Implications for Thermal Energy Conversion*, Elsevier Applied Science Publishers Ltd.: Essex, UK, 1986, Phase Transformations 77–96.

Buehler, W.J. and R.C. Wiley, 'Nickel-Base Alloys,' U.S. Patent 3,174,851, 1965.

Claus, R.O., 'Optical Fiber Sensors and Single Processing for Smart Materials and Structures Applications,' *Proceedings of the U.S. Army Research Office Workshop*, September 15–16, 1988, Blacksburg, Virginia.

Collins, C. and M. Luskin, 'The Computation of the Austenitic-Martensitic Phase Transition,' *Partial Differential Equations and Continuum Models of Phase Transitions, Lecture Notes in Physics 344*, eds. M. Rascle, D. Serre and M. Slemrod, Springer Verlag, 1989, pp. 34–50.

Cross, W.B., A.H. Kariotis, and F.J. Stimler, 'Nitinol Characterization Study,' NASA CR-1433, September 1969.

Delaey, R.V., H. Tas Krishnan, and H. Warlimont, 'Thermoelasticity, Pseudo-Elasticity, and the Shape-Memory-Effects Associated with Martensitic Transformations,' *Journal of Material Science*, Vol. 9, 1974, pp. 1521–45.

Duerig, T.W., K.N., Melton, D. Stockel, C.M. Wayman, *Engineering Aspects of Shape Memory Alloys*, Butterworth-Heinemann, 1990.

Efsic, E., K. Mukherjee, and C.M. Wayman, 'Observations on the Phase Transition in NH_4Cl Single Crystals at Subzero Temperatures,' *Journal of Metals*, Vol. 16, p. 120.

Funakubo, H. (translated by J.B. Kennedy), *Shape-Memory-Alloys*, 1987, Gordon and Breach Science Publishers, New York.

Goldstein, D., 'A Source Manual for Information on Nitinol and NiTi,' Naval Surface Weapons Center, Silver Spring, Maryland, Report NSWC/WOL TR 78–26, 1978.

Harrison, J.D. and D.E. Hodgson, *Shape-Memory Effects in Alloys*, p. 517, Plenum Press, 1975.

Hashimoto, M., M. Takeda, H. Sagawa, I. Chiba, and K. Sato, 'Application of Shape Memory Alloy to Robotic Actuators,' *Journal of Robotic Systems*, Vol. 2, No. 1, 1985, pp. 3–25.

Hirose, S., K. Ikutra, M. Tsukamoto, K. Sato, Y. Umetani, 'Several Considerations on Design of SMA Actuators,' *Proceedings of the International Conference on Martensitic Transformations*, ICOMAT-86, Nara, Japan, August 1986.

Hoffman, K.H. and M. Niezgodka, 'Mathematical Models of Dynamical Martensitic Transformations in Shape Memory Alloys,' *Journal of Intelligent Material Systems and Structures*, Vol. 1, No. 3, July 1990, pp. 335–374.

Honma, D., Y. Miwa, and N. Iguchi, 'Application of Shape Memory Effect to Digital Control S Actuator,' *Bulletin of Suspense Society of Mechanical Engineers*, Vol. 27, 1984, pp. 1737–42.

Horbogen, E., 'Alloys with Shape Memory – New Materials for Future Technology?', *Metallurgy*, Vol. 41, No. 5, May 1987, pp. 488–493.

Jackson, C.M., H.J. Wagner, and R.J. Wasilewski, '55-Nitinol – The Alloy with a Memory, Its Physical Metallurgy, Properties, and Applications,' NASA-SP-5110, 1972, p. 91.

Likhachev, V.A., S.L. Kuzmin, and Z.L. Kamenceva, *Shape Memory Effect*, Leningrad: Leningrad University Press, (in Russian).

Miwa, Y. 'Shape memory alloy application for sequential operation control,' *System and Control*, Vol. 29, No. 5, May 1985, pp. 303–310.

Muller, I., 'A Model for a Body with Shape-Memory,' *Archives of Rational Mechanics and Analysis*, Vol. 70, 1979, pp. 61–77.

Niezgodka, M. and J. Sprekels, 'Existence of Solutions for a Mathematical Model of Structural Phase Transitions in Shape Memory Alloys,' *Mathematical Methods in the Applied Sciences*, Vol. 10, 1988, pp. 197–223.

Nishiyama, Z., *Martensitic Transformation*, 1978, Academic Press, New York, NY.

Nakai, K., T. Okuda, H. Tanaka, and Sharp, H., 'Heat Conduction in Ni-Ti Shape Memory Alloy in the Presence of Phase Transformation Under Stress,' *Proceedings of the International Conference on Martensitic Transformations*, ICOMAT-86, Nara, Japan, August 1986.

Oonishi, H., *Artificial Organs*, Vol. 12, No. 4, p. 871, 1983, (in Japanese).

Oonishi, H., M. Miyagi, T. Hamada, E. Tsuji, Y. Suzuki, T. Hamaguchi, N. Okabe, and T. Nabeshima, 'Application of the Memory Alloy TiNi as Implant Material,' *Proceedings of the 4th European Conference on Biomaterials*, 1983, p. 149–154.

Oonishi, H., T. Hamaguchi, T. Nabeshima, M. Miyagi, E. Tsuji, T. Hamada, Y. Suzuki, and K. Shikida, 'Studies on the Development of a High Frequency Induction Heating Apparatus to Heat a Shape Memory TiNi Alloy Implanted in a Living Body,' *Proceedings of the 3rd Conference of the Japanese Society for Biomaterials*, 1982, p. 155–161, (in Japanese).

Perkins, J., ed., *Shape Memory Effects in Alloys*, Plenum Press, New York, 1975.

Rogers, C.A., and H.H. Robertshaw, 'Shape Memory Alloy Reinforced Composites,' *Engineering Science Preprints 25*, ESP25.88027, Society of Engineering Sciences, Inc., June 1988.

Rogers, C.A., 'Dynamic Control Concepts Using Shape Memory Alloy Reinforced Plates,' *Proceedings of the U.S. Army Research Office Workshop*, September 15–16, 1988, Blacksburg, Virginia.

Rogers, C.A., C. Liang, and J. Jia, 'Behavior of Shape Memory Alloy Reinforced Composite Plates,' AIAA Paper 89–1389, 1989.

Saburi, T., and C.M. Wayman, 'Crystallographic Similarities in Shape-Memory Martensites,' *Acta Metallurgica*, Vol. 27, 1979, pp. 979–995.

Schetky, L.M., and J. Perkins, 'The Quiet Alloys,' *Machine Design*, Vol. 50, No. 8, 1978, pp. 202–206.

Schetky, L.M., 'Shape Memory Alloys,' *Scientific American*, Vol. 241, 1979, p.74.

Shimizu, K. and K. Otsuka, 'Pseudo-Elasticity and Shape Memory Effects,' *International Metals Review*, Vol. 31, No. 3, 1986, pp. 93–114.

Thumann, M., B. Velten, E. Hornbogen, 'Composites Containing Shape Memory Fibers,' *Proceedings of the International Conference on Martensitic Transformations*, ICOMAT0–86, Nara, Japan, August 1986.

Wayman, C.M., and K. Shimizu, 'The Shape Memory ('Marmem') Effect in Alloys,' *Metal Science Journal*, Vol. 6, 1972, pp. 175–183.

Wayman, C.M., 'Some Applications of Shape-Memory-Alloys,' *Journal of Metals*, June 1980, pp. 129–137.

Wille, J., B. Hansknecht, M.V. Gandhi, and B.S. Thompson, 'A Proof-of-Concept Experimental Investigation on a Four-Bar Linkage with a Smart Rocker Link Featuring a Shape-Memory-Alloy,' *Proceedings 1st National Applied Mechanisms and Robotics Conference*, Cincinnati, Ohio, Vol. II, Nov. 1989, pp. 9B-4.1 to 9B-4.4.

Yeager, J., 'A Practical Shape-Memory Electro-Mechanical Actuator,' *Mechanical Engineering*, Vol. 106, 1984, pp. 52–55.

Piezoelectric actuators

Allik, H. and T.J. Hughes, 'Finite Element Method for Piezoelectric Vibration,' *International Journal of Numerical Methods in Engineering*, Vol. 2, 1979, pp. 151–168.

Bailey, T. and J.E. Hubbard. 'Distributed Piezoelectric-Polymer Active Vibration Control of a Cantilever Beam,' *AIAA Journal of Guidance, Control, Dynamics*, Vol. 8, No. 5, September–October, 1985, pp. 605–611.

Baz, A. and S. Poh, 'Performance of an Active Control System with Piezoelectric Actuators,' *Journal of Sound and Vibration*, Vol. 126, No. 2, 1988, pp. 327–343.

Bleustein, J.L. and H.F. Tiersten, 'Forced Thickness-Shear Vibrations of Discontinuously Plated Piezoelectric Plates,' *The Journal of the Acoustical Society of America*, Vol. 43, No. 6, 1968, pp. 1311–18.

Bondarenko, A.A., N.I. Karas, and A.F. Ulitko, 'Methods for Determining the Vibrational Dissipation Characteristics of Piezoceramic Structural Elements,' *Soviet Applied Mechanics*, 1982, pp. 175–179.

Bryant, M.D. and R.F. Keltie, 'A Characterization of the Linear and Nonlinear Dynamic Performance of a Practical Piezoelectric Actuator,' *Sensors and Actuators*, Vol. 9, 1986, pp. 95–114.

Burke, S.E. and J.E. Hubbard, 'Active Vibration Control of a Simply-Supported Beam Using a Spatially-Distributed Actuator,' *IEEE Control Systems Magazine*, Vol. 7, No. 4, August 1987, pp. 25–30.

Burke, S.E. and J.E. Hubbard, 'Distributed Parameter Control Design for Vibrating Beams Using Generalized Functions,' *1986 IFAC Symposium on Distributed Parameter Systems*, June 1986, Los Angeles, CA.

Burke, S.E. and J.E.Hubbard Jr., 'Performance Measures for Distributed Parameter Control Systems,' *Proceedings of IMACS/IFAC International Symposium on Modeling and Simulation of Distributed Parameter Systems*, October 1987, Hiroshima, Japan.

Cady, W.G., *Piezoelectricity*, 1964, Dover, New York.

Connally, J.A. and J.E. Hubbard, Jr., 'Low Authority Control of a Composite Cantilever Beam in Two Dimensions,' *Proceedings of American Control Conference*, August 1988, pp. 1903–08.

Crawley, E.F. and J. de Luis, 'Use of Piezo-Ceramics as Distributed Actuators in Large Space Structures,' *Proceedings of 26th Structures, Structural Dynamics and Materials Conference*, 1985, pp. 126–133.

Crawley, E.F. and J. de Luis, 'Use of Piezoelectric Actuators as Elements of Intelligent Structures,' *American Institute of Aeronautical Astronautics Journal*, Vol. 25, No. 10, 1987, pp. 1373–85.

Crawley, E.F., J. de Luis, N.W. Hagwood, and E.H. Anderson,

'Development of Piezoelectric Technology for Applications in Control of Intelligent Structures,' *Proceedings of American Control Conference*, 1988, pp. 1890–96.

Crawley, E.F. and E.H. Anderson, 'Detailed Models of Piezoceramic Actuation of Beams,' *Journal of Intelligent Materials and Structures*, Vol. I, 1990, pp. 4–25.

Cudney, H.H., D.J. Inman, and Y. Oshman, 'Distributed Structural Control Using Multilayered Piezoelectric Actuators,' *Proceedings of the 7th VPI & SU/AIAA Symposium on Dynamics and Control of Large Structures*, 1987.

Hagood, N.W., W.H. Chung, and A. von Flotow, 'Modelling of Piezoelectric Actuator Dynamics for Active Structural Control,' *Journal of Intelligent Material Systems and Structures*, Vol. 1, No. 3, July 1990, pp. 327–334.

Hanagud, S., C.C. Won, and M.W. Obal, 'Optimal Placement of Piezoceramic Sensors and Actuators,' *Proceedings of American Control Conference*, August 1988, pp. 1884–89.

Hanagud, S., M.W. Obal, and M. Meyyappa, 'Electronic Damping Techniques and Active Vibration Control,' *Proceedings of 26th Structures, Structural Dynamics and Materials Conference*, 1985, pp. 443–453.

Haun, M.J. and R.E. Newnham, 'Experimental and Theoretical Study of 1–3 and 1–3-0 Piezoelectric PZT-Polymer Composites for Hydrophone Applications,' *Ferroelectrics* 68, pp. 123–139 (1986).

Haun, M.J., P. Moses, T.R. Gururaja, W.A. Schulze, and R.E. Newnhawn, 'Transversely Reinforced 1–3 and 1–3-0 Piezoelectric Composites,' *Ferroelectrics* 49, pp. 259–264 (1983).

Hubbard, Jr., J.E., 'Distributed Sensors and Actuators for Vibration Control in Elastic Components,' *Proceedings of Noise-Conf 87*, 1987, pp. 407–412.

Ikegami, R., D.G. Wilson, J.R. Anderson, and G.J. Julien, 'Active Vibration Control Using NiTiNOL and Piezoelectric Ceramics,' *Journal of Intelligent Materials Systems and Structures*, Vol. 1, April 1990, pp. 189–206.

Im, S. and S.N. Atluri, 'Effects of a Piezo-Actuator on a Finitely Deformed Beam Subjected to General Loading,' *AIAA Journal*, Vol. 27, No. 12, 1989, pp. 1801–07.

Jaffe, B., R. Cook, and H. Jaffe, *Piezoelectric Ceramics*, 1971, Academic Press: New York, NY.

Kawai, H., 'The Piezoelectricity of Poly(vinylidene fluoride),' *Japanese Journal of Applied Physics*, Vol. 8, No. 7, 1969, pp. 975–978.

Kynar Piezo Film Technical Manual, Pennwalt Corporation, Valley Forge, Pennsylvania, 1987.

Lawson, A.W., 'The Vibration of Piezoelectric Plates,' *Physical Review*, Vol. 62, July 1942, pp. 71–76.

Lee, C.Y., S.B. Choi, B.S. Thompson, and M.V. Gandhi, 'A Variational Formulation for the Finite Element Analysis of Linkage and Robotic Systems Featuring Smart Links Incorporating Piezoelectric Materials,' *Proceedings 1st National Applied Mechanisms and Robotics Conference*, Cincinnati, Ohio, Vol. I, Nov. 1989, 2C-1.1 to 2C-1.10.

Liao, C.Y. and C.K. Sung, 'An Elastodynamic Analysis of Flexible Linkages Using Piezoceramic Sensors and Actuators,' *Flexible Mechanisms, Dynamics and Robot Trajectories*, DE-Vol. 24, ASME Book H00623, edited by S. Derby, M. McCarthy, A. Pisano, 1990, pp. 503–510.

Mason, W.P., 'Piezoelectricity its History and Applications,' *Journal of the Acoustical Society of America*, Vol. 70, No. 6, 1981, pp. 1561–66.

Miller, S.E. and J.E. Hubbard, Jr., 'Observability of a Bernoulli-Euler Beam Using PVF2 as a Distributed Sensor,' *Proceedings of 6th VPI & SU/AIAA Symposium on Dynamic and Control of Large Structures*, June 1987.

Miller, S.E. and J.E. Hubbard, Jr., 'Smart Components for Structural Vibration Control,' *Proceedings of American Control Conference*, 1988, pp. 1897–02.

Mindlin, R.D., 'Thickness-Shear and Flexural Vibrations of Crystal Plate,' *Journal of Applied Physics*, Vol. 22, No. 3, 1951, pp. 316–323.

Mindlin, R.D., 'Forced Thickness-Shear and Flexural Vibrations of Piezoelectric Crystal Plates,' *Journal of Applied Physics*, Vol. 23, No. 1, 1952, pp. 83–88.

Mindlin, R.D., 'High Frequency Vibrations of Crystal Plates,' *Quarterly of Applied Mathematics*, Vol. 19, 1961, pp. 51–61.

Mindlin, R.D., 'High Frequency Vibrations of Piezoelectric Crystal Plates,' *International Journal of Solids and Structures*, Vol. 8, 1972, pp. 895–906.

Mindlin, R.D., 'Coupled Piezoelectric Vibrations of Quartz Plates,' *International Journal of Solids and Structures*, Vol. 10, 1974, pp. 453–459.

Mindlin, R.D, 'Equations of High Frequency Vibrations of Thermo Piezoelectric Crystal Plates,' *International Journal of Solids and Structures*, Vol. 10, 1974, pp. 625–637.

Newcomb, C.V. and I. Flinn, 'Improving the Linearity of Piezoelectric Ceramic Actuators,' *Electronics Letters*, Vol. 18, No. 11, 1982, pp. 442–444.

Piezoelectric Ceramics, edited by J. Van Randeraat and R.E. Setterington, N.V. Philips, Eindhoven, 1974.

Piezoelectric Products Company Literature, Piezoelectric Products Inc., Metuchen, NJ, 1984.

Plump, J.M., J.E. Hubbard, Jr., and T. Bailey, 'Nonlinear Control of a Distributed System: Simulation and Experimental Results,' *ASME Journal of Dynamic Systems, Measurement, and Control*, Vol. 109, June 1987, pp. 133–139.

Procopio, G.M. and J.E. Hubbard, Jr., 'Active Damping of a Bernouli-Euler Beam via End Point Independence Control Using Distributed Parameter Techniques,' *Proceedings of 1987 ASME Vibration Conference*.

Seo, I., 'Application of piezoelectric polymer to actuator,' *Journal of the Japanese Society of Precision Engineers*, Vol. 52, No. 5, May 1987, pp. 689–691.

Sessler, G.M., 'Piezoelectricity in Polyvinylidene Fluoride,' *Journal of the Acoustical Society of America*, Vol. 70, No. 6, 1981, pp. 1596–08.

Smith, W.A. and Auld, B.A., 'Modeling 1–3 Composite Piezoelectrics: Thickness-Mode Oscillations', *IEEE Transactions on Ultrasonics, Ferro-Electrics, and Frequency Control*, Vol. 38, pp. 40–47, 1991.

Takahashi, S., 'Piezoelectric actuators and their applications,' *Journal Institute of Electronic Information and Communication Engineering*, Vol. 70, No. 3, March 1987, pp. 295–297.

Tamaru, N., 'Piezoelectric bimorph actuator control using a Kalman filter,'

Transactions of the Society of Instrumentation and Control Engineering, Vol. 23, No. 8, August 1987, pp. 873–875.

Tzou, H.S. and M. Gadre., 'Active Vibration Isolation by Piezoelectric Polymer with Variable Feedback Gain,' *American Institute of Aeronautics and Astronautics Journal*, Vol. 26, No. 8, 1988, pp. 1014–17.

Tzou, H.S. and M. Gadre, 'Theoretical Analysis of a Multi-layered Thin Shell Coupled with Piezoelectric Shell Actuators for Distributed Vibration Controls,' *Journal of Sound and Vibration*, Vol. 132, 1989, pp. 433–450.

Tzou, H.S. and M. Gadre, 'Active Vibration Isolation and Excitation by Piezoelectric Slab with Constant Feedback Gains,' *Journal of Sound and Vibration*, Vol. 136, No. 3, 1989, pp. 477–490.

Tzou, H.S. and C.I. Tseng, 'Distributed Piezoelectric Sensor/Actuator Design for Dynamic Measurement/Control of Distributed Parameter Systems: A Piezoelectric Finite Element Approach,' *Journal of Sound and Vibration*, 1990, Vol. 138, No. 1, pp. 17–34.

Uchino, K., 'Actuator applications of piezoelectric/electrostrictive ceramics,' *Journal of the Japanese Society of Precision Engineers*, Vol. 53, No. 5, May 1987, pp. 686–688.

Piezoelectric sensors

Barsky, M.F., D.K. Linder, and R.O. Claus, 'Robot gripper control system demonstrating PVDF piezoelectric sensors,' *IEEE 1986 Ultrasonics Symposium Proceedings*, pp. 545–548, Vol. 1, November 1986, Williamsburg, VA, USA.

Barsky, M.F., D.K. Lindner, and R.O. Claus, 'Active damping of a robotic gripper PVDF piezoelectric sensor,' *Proceedings of the Eighteenth Southeastern Symposium on System Theory*, April 1986, Knoxville, TN, USA.

Carey, W.P. and B.R. Kowalski,'Chemical piezoelectric sensor and sensor array characterization,' *Analytical Chemistry*, Vol. 58, No. 14, December 1986, pp. 3077–84.

Carlisle, B.H., 'Piezoelectric plastics promise new sensors,' *Machine Design*, Vol. 58, No. 25, October 1986, pp. 105–110.

Crawley, E.F., J. de Luis, H.W. Hagood, and E.H. Anderson, 'Development of Piezoelectric Technology for Applications in Control of Intelligent Structures,' *Proceedings of the 1988 American Control Conference*, Atlanta, GA, Vol. 3, 1988, pp. 1897–02.

Dong Tai-Huo, Pao Cheng-Kang, Kin Dan, 'Accurate measurement of the voltage micro-displacement correlation of piezoelectric sensors with a fiber optic interferometer,' *Proceedings SPIE – International Society Optical Engineers (USA)*, Vol. 798, pp. 304–306, 1987, Fiber Optic Sensors II, March/April 1987, The Hague, The Netherlands.

Hanagud, S., 'Piezoceramic Devices and PVDF Films as Sensors and Actuators for Intelligent Structures,' *Proceedings of the 1988 U.S. Army Research Office Workshop*, Blacksburg, Virginia, September 1988.

Leaver, P., M.J. Cunningham, and B.E. Jones, 'Piezoelectric polymer pressure sensors,' *Sensors and Actuators*, Vol. 12, No. 3, pp. 225–233, October 1987.

Miller, S.E. and J.E. Hubbard Jr., 'Smart Components for Structural Vibration Control,' *Proceedings of the 1988 American Control Conference*, Atlanta, GA, Vol. 3, 1988, pp. 1897–1902.

Measures, R.M., 'Fiber Optic Sensors – the Key to Smart Structures,' *Fiber Optics 89*, SPIE, Vol. 1120, pp. 1120–32.

Measures, R.M., 'Fiber Optic Smart Structures Program at UTIAS,' *Fiber Optic Smart Structures and Skins II*, E. Udd (ed.), SPIE, 1989, Vol. 1170, pp. 92–108.

Motghare, S.J., A.M. Kolte, C.V. Dhuley, and A.G. Karpatal, 'Samariaum doped potassium-niobate – a new piezoelectric sensor,' ISAF '86. *Proceedings of the Sixth IEEE International Symposium on Applications of Ferroelectrics*, June 1986, Bethlehem, PA, USA, pp. 472–475.

Newham, R.E., 'Smart Ceramics,' *Proceedings of the 1988 U.S. Army Research Office Workshop*, Blacksburg, Virginia, September 1988.

Plump, J.M., J.E. Hubbard Jr., and Bailey, T., 'Nonlinear Control of a Distributed System: Simulation and Experimental Results,' *ASME Journal of Dynamic Systems, Measurements, and Control*, Vol. 109, 1987, pp. 133–139.

Safari, A., G. Sa-gong, J. Giniewicz, and R.E. Newnham, 'Composite piezoelectric sensors'; Tressler, R.E., G.L. Mesing, C.G. Pantano, R.E.

Newnham (Editors), 'Tailoring Multiphase and Composite Ceramics', *Proceedings of the Twenty-First University Conference on Ceramic Science,* pp. 445–454, July 1985, University Park, PA, USA.

Wilkes, G.L., 'Materials Issues for Smart Structures,' *Proceedings of the 1988 U.S. Army Research Office Workshop*, Blacksburg, Virginia, September 1988.

Fiber-optic sensors

Allard, F.C., ed., *Fiber Optics Handbook for Engineers and Scientists*, McGraw-Hill Publishing Co.: New York, 1990.

Auch, W., E. Schlemper, and W. Wenzel, 'Fiber optic gyroscope: an advanced rotation rate sensor,' *Electronic Communications*, Vol. 61, No. 4, 1987, pp. 372–378.

Batchellor, C.R., J.P. Dakin, D.A.J. Pearce, 'Fiber optic mechanical sensors for aerospace applications,' *International Journal of Optic Sensors*, Vol. 2, No. 5–6, October-December 1987, pp. 407–411.

Bennett, K.D. and R.O. Claus, 'Analysis of composite structures using optical fiber modal sensing techniques,' *Proceedings IEEE SOUTH-EASTCON '86*, pp. 95–98, March 1986, Richmond, VA, USA.

Bennett, K.D., S.J. Hanna, and R.O. Claus, 'Monitoring of strain in layered media using clad rod acoustic waveguides,' McAvoy, B.R. (Editor), *IEEE 1985 Ultrasonics Symposium Proceedings*, pp. 1064–1067, Vol. 2, October 1985, San Francisco, CA, USA.

Blake, J.N., S.Y. Huang, and B.Y. Kim, 'Elliptical core two-mode fiber strain gauge,' *Proceedings SPIE – International Society Optical Engineers* (*USA*), Vol. 838, pp. 332–339, 1988, Fiber Optic and Laser Sensors V, August 1987, San Diego, CA, USA.

Bucholtz, F., A.D. Kersey, and A. Dandridge. 'DC fiber optic accelerometer with sub-μg sensitivity,' *Electronic Letter*, Vol. 22, No. 9, April 1986, pp. 451–453.

Butler, C.D. and G.B. Hocker, 'Fiber Optics Strain Gauge,' *Applied Optics*, Vol. 17, 1978, pp. 2867–69.

Claus, R.O., B.S. Jackson, and R.G. May, 'Nondestructive evaluation of composites by optical time domain reflectometry in embedded optical fibers,' *Proceedings IEEE SOUTHEASTCON '85*, Raleigh, NC, USA, March 1985, pp. 241–245.

Claus, R.O., B.S. Jackson, and K.D. Bennett, 'Nondestructive testing of composite materials by OTDR in Embedded optical fibers,' *Proceedings SPIE International Society Optical Engineers*, Vol. 566, 1985, pp. 243–248.

Cusworth, D.S. and J.M. Senior, 'Optical fiber sensing using GRIN rod lenses,' *International Journal of Optic Sensors*, Vol. 2, No. 5–6, October-December 1987 pp. 421–436.

Dakin, J.P., 'Optical Fiber Sensors – Principles and Applications,' *Proceedings: Fiber Optics '83*, SPIE, Vol. 374, 1983, pp. 172–182.

DeHart, D.W., 'Air Force Astronautics Laboratory Smart Structures and Skins Program Overview,' SPIE, Vol. 1170, *Fiber Optics Smart Structures and Skins II*, E. Udd, ed., 1989, pp. 11–18.

Farahi, F., T.P. Newson, J.D.C. Jones, and D.A. Jackson, 'Coherence multiplexing of remote fiber optic Fabry-Perot sensing system,' *Optical Communications*, Vol. 65, No. 5, March 1988, pp. 319–321.

Field, S.Y. and K.H.G. Ashbee, 'Weathering of Fiber Reinforced Plastics: Progress of Debonding Detected in Model Systems by Using Fibers as Light Pipes,' *Polymeric Engineering and Science*, Vol. 12, 1972, pp. 30–33.

Flax, A., C. Pennington, and R.O. Claus, 'Single mode optical fiber vibration sensor,' *Proceedings IEEE SOUTHEASTCON '87*, pp. 553–555, Vol. 2,
April 1987, Tampa, FL, USA.

Freidah, J.T., R.E. Wagoner, T.J. Cash, and N.H. Safar, 'Long-lead cable effects on interferometric sensors,' *Proceedings SPIE-International Society Optical Engineers*, Vol. 838, 1988, Fiber Optic and Laser Sensors V, August 1987, San Diego, CA, USA, pp. 372–378.

Hattori, H., 'Development of intelligent optical fiber sensors,' *Instrumentation*, Vol. 30, No. 12, December 1987, pp. 25–28.

Higley, S.E., E. Udd, R.J. Michal, J.P. Theriault, and D.A. Jolin, 'Fiber-Optic Spectrometer,' *Proceedings SPIE-International Society Optical Engineers*, Vol. 838, 1988, pp. 318–324.

Kist, R., 'Line Independence of Optical Fiber Sensors,' *Technica* (Switzerland), Vol. 37, No. 4, March 1988, pp. 13–16.

Kuhlman, R., B. Duncan, and R.O. Claus, 'Fiber optic composite impact monitor,' *Proceedings IEEE SOUTHEASTCON '87*, Vol. 2, April 1987, Tampa, FL, USA, pp. 414–417.

Leka, L.G. and E. Bayo, 'A Close Look at the Embedment of Optical Fibers into Composite Structures,' *Journal of Composites Technology & Research*, Vol. 11, No. 3, 1989, pp. 106–112.

Liu, K., S.M. Ferguson, and R.M. Measures, 'Damage Detection in Composites with Embedded Fiber Optic Interferometric Sensors,' *Proceedings: Fiber Optic Smart Structures and Skins II, SPIE Vol. 1170*, Boston, paper 1170–1122, (1989).

Main, R.P. 'Fiber optic sensors – future light,' *Sensor Review*, Vol. 5, No. 3, July, 1985, pp. 133–139.

Mann, R. 'So what future do you see in fiber optics?,' *Process Engineering (GB)*, Vol. 66, No. 6, June 1985, pp. 79–81.

Martinelli, M. 'Fiber optic sensors,' *Elettron Oggi.*, No. 4, April 1984, pp. 115–124.

Mazus, C.J., G.P. Sendeckyj, and D.M. Stevens, 'Air Force Smart Structures/Skins Program Overview,' SPIE, Vol. 986, *Fiber Optic Smart Structures and Skins*, E. Udd, ed., 1988, pp. 19–29.

Measures, R.M., D. Hogg, R.D. Turner, T. Valis, and M.J. Giliberto, 'Structurally Integrated Fiber Optic Strain Rosette,' *Proceedings: Fiber Optic Smart Structures and Skins I, SPIE Vol. 986*, Boston, paper 986–07, 1988, pp. 32–42.

O'Conner, W.T. and R.O. Claus, 'Optical Fiber Sensor Application in a Simple Robotic Gripper,' *Proceedings of IEEE SOUTHEASTCON '84*, April 1984, Louisville, KY, USA, pp. 295–297.

Rogers, A.J., 'Distributed Optical-Fiber Sensors,' *Journal of Physics D: Applied Physics*, Vol. 19, 1986, pp. 2237–55.

Saleh, B.E.A. and M.C. Teich, *Fundamentals of Photonics*, John Wiley & Sons, 1991.

Shahriari, M.R., G.H. Sigel, and Quan Zhou, 'Porous glass fibers for high sensitivity chemical and biomedical sensors,' *Proceedings SPIE-International Society Optical Engineers*, Vol. 838, 1988, Fiber Optic and Laser Sensors V, August 1987, San Diego, CA, USA, pp. 348–352,

Shankaranarayanan, N.K., K.T. Srinivas, and R.O. Claus, 'Mode-mode interference effects in axially strained few-mode optical fibers,' *Proceedings SPIE-International Society Optical Engineers*, Vol. 838, 1988, pp. 385–388.

Shankaranarayanan, N.K., K.D. Bennett, and R.O. Claus, 'Optical fiber modal domain detection of stress waves,' McAvoy, B.R., editor, *IEEE 1986 Ultra-sonics Symposium Proceedings*, Vol. 2, 1986, November 1986, Williamsburg, VA.

Shih, J.C. and R.O. Claus, 'Concentric Core Optical Fiber Strain Sensor,' *Proceedings IEEE SOUTHEASTCON '86*, March 1986, Richmond, VA, pp. 91–94.

Spillman W.B. and B.R. Kline, 'Fiber Optic Vibration Sensors for Structural Control Applications', *Proceedings of Damping '89*, West Palm Beach, FL, Vol III, pp. ICA-1–ICA-21, 1989, published by Flightdynamics Laboratory, Wright-Patterson Air Force Base, Ohio.

Su, S.F., 'Fiber-optic electric field sensors utilizing electro-absorption,' *Proceedings IEEE SOUTHEASTCON '85*, Raleigh, NC, USA, March/April 1985, pp. 241–245.

Tsutsui, T. and S. Yamamoto, 'Optical fiber sensor,' *Journal of the Japanese Society Precision Engineers*, Vol. 53, No. 12, pp. 1847–51, Dec. 1987.

Turner, R.D., T. Valis, W.D. Hogg, and R.M. Measures, 'Fiber-Optic Strain Sensors for Smart Structures,' *Journal of Intelligent Material Systems and Structures*, Vol. 1, January 1990, pp. 26–49.

Valis, T., E. Tapanes, and R.M. Measures, 'Localized Fiber Optic Strain Sensors Embedded in Composite Materials,' *Proceedings*: *Fiber Optic Smart Structures and Skins II*, *SPIE Vol. 1170*, Boston, paper 1170–53, 1989).

Wade, J.C. and R.O. Claus, 'Interferometric Techniques Using Embedded Optical Fibers for the Quantitative NDE of Composites'; Thompson, D.O., D.E. Chimenti, Editors, *Review of Progress in Quantitative Nondestructive Evaluation*, Vol. 2, August 1982, San Diego, CA, pp. 1731–38.

Waite, S.R., R.P. Tatam, and A. Jackson, 'Use of optical fiber for damage and strain detection in composite materials,' *Composites*, Vol. 19, No. 6, November 1988, pp. 435–442.

Waite, S.R. and G.N. Sage, 'The failure of optical fibers embedded in composite materials,' *Composites*, Vol. 19, No. 4, July 1988, pp. 288–294.

Wiencko, J.A., R.O. Claus, and R.E. Rogers, 'Embedded optical fiber sensors for intelligent aerospace structures,' *Proceedings of SENSORS EXPO, September 1987*, Detroit, MI, USA, pp. 257–262.

Zhipeng Zhand and Qin Zhang and 'Magnetic field sensitivity analysis of an optical fiber with metallic glass strips,' *International Journal of Optic Sensors*, Vol. 2, No. 5–6, pp. 341–347, October-December 1987.

Smart sensors and actuators

'Adjustment of intelligent sensors with a low-cost function tester,' *Elektronik* (West Germany) Vol. 37, No. 6, March 1988, pp. 134–138.

Alba, M., 'Tradeoffs in automotive smart sensors system partitioning,' *Proceedings of SENSORS EXPO*, 401, September 1987, Detroit, MI, USA.

Brignell, J.E. and J.K. Atkinson, 'Sensors, Intelligence, and Networks,' *IEE Colloquium on 'Solid State and Smart Sensors'*, Digest No. 39, March 1988, London, UK.

Caro, J., 'Using numerical processors for intelligent sensor design'; P. Albertos, J.A. De La Puente, Editors, 'Components, Instruments and Techniques for Low Cost Automation and Applications', *IFAC Symposium*, 1988, November 1986, Valencia, Spain, Pergamon, Oxford, UK, pp. 61–65.

De Rossi, D., A. Nannini, and C. Domenici, 'Biomimetic Tactile Sensor With Stress-Component Discrimination Capability,' *Journal of Molecular Electronics*, Vol. 3, 1987, p. 173.

De Rossi, D., A. Nannini, and C. Domenici, 'Artifical Sensing Skin Mimicking Mecho-Electrical Conversion Properties of Human Dermis,' *IEEE Transactions of Biomedical Engineering*, BME-35, 1988, p. 83.

De Rossi, D., L. Lazzeri, C. Domenici, A. Nannini, and P. Basser , 'Tactile Sensing by an Electromechanochemical Skin Analog,' *Sensors and Actuators*, Vol. 17, 1989, p. 107.

Favennec, J.-M., 'Smart sensors in industry,' *Journal of Physics E.*, Vol. 20, No. 9, pp. 1087–90, September 1987.

'Future-Generation Smart Field Sensor and Smart Field Network,' *Instrumentation*, Vol. 30, No. 11, pp. 52–53, November 1987.

Giachino, J.M., 'Smart sensors,' *Sensors & Actuators*, Vol. 10, No. 3–4, November-December 1986, pp. 239–248.

Hanneborg, A. and P. Ohlckers, 'SI's Smart Silicon Sensor,' *Elektro*, Vol. 98, No. 15, September 1985, pp. 8–10.

Hattori, H., 'Development of Intelligent Optical Fiber Sensors,' *Instrumentation*, Vol. 30, No. 12, December 1987, pp. 25–28.

Hester, C.F., H.F. Caulfield, 'Optical AI Architectures for Intelligent Sensors,' *Proceedings SPIE-International Society Optical Engineers, Vol. 754, Optical and Digital Pattern Recognition*, January 1987, Los Angeles, CA, USA, pp. 185–192.

Hockaday, B. and J. Waters, 'Opto-Mechanical Smart Actuator,' *Proceedings of the 1987 American Control Conference*, Vol. 1, June 1987, Minneapolis, MN.

'Intelligent Sensors: One Must First be Convinced,' *Measures*, Vol. 52, No. 7, May 1987, pp. 61–67.

Jayawant, B.V. and J.D.M. Watson, 'Array Sensor for Tactile Sensing in Robotic Applications,' *IEE Colloquium on 'Solid State and Smart Sensors'*, Digest No. 39, March 1988, London, UK.

Legras, J.C., M. Priel, and C. Ranson, 'Application of Multiple Regression Methods and Automatic Testing Equipment to the Characterization of 'Smart' Sensors,' *Sensors and Actuators*, Vol. 12, No. 3, October 1987, 2nd International Meeting on Chemical Sensors, July 1986, Bordaux, France, pp. 235–243.

Middelhoek, S. and A.C. Hoogerwerf, 'Smart Sensors: When and Where?,' *Sensors and Actuators*, Vol. 8, No. 1, September 1985, pp. 39–48.

Montgomery, R.C. and J.P. Williams, 'The Use of Distributed Micro-Computers in the Control of Structural Dynamics Systems,' Hamza, M.H. (Editor), 'Mini and Microcomputers in Control, Filtering and Signal Processing', *Proceedings of the ISMM International Symposium*, December 1984, Las Vegas, NV, USA, pp. 140–144.

Seo, I., 'Development of Intelligent Sensors Using Polymer Materials,' *Instrumentation*, Vol. 30, No. 12, December 1987, pp. 32–35.

Stojilikovic, Z. and J. Clot, 'Integrated behavior of Artifical Skin,' *IEEE Transactions for Biomedical Engineering*, BME-24, 1977, p.396.

Takahashi, K., 'Sensor Materials For the Future: Intelligent Materials,' *Sensors & Actuators*, Vol. 13, No. 1, January 1988.

Tinham, B., 'Micros and Sensors Make Smart Transmitters,' *Control & Instrumentation*, Vol. 18, No. 9, September 1986, pp. 55–57.

Trontelj, J., L. Trontelj, V. Kunc, and M. Stiglic, 'Generalized Smart Sensor Array Electronics,' *Proceedings of the IEEE 1987 Custom Integrated Circuits Conference*, May 1987, Portland, OR, USA, pp. 708–711.

Udd, Eric (ed.), *Fiber Optic Sensors*: *An Introduction for Engineers and Scientists*. New Jersey, John Wiley & Sons, 1991.

Ushikubo, S., 'Field Network of Intelligent Sensors,' *Instrumentation*, Vol. 30, No. 12, December 1987, pp. 40–44.

Magneto-rheological suspensions

Kordonskii, V.I., M.Z. Shul'man, and T.V. Kunevich, 'Investigation of the Magneto-Rheological Effect' in: *Rheology of Polymeric and Disperse Systems and Rheophysics* [in Russian], Part 2, Minsk, 1975, pp. 79–82.

Martinet, A., 'Experimental Evidence of Static and Dynamic Anisotropies of Magnetic Colloids' in: *Thermomechanics of Magnetic Fluids*, Hemisphere, Washington, D.C., 1978, pp. 96–114.

Mikelson, A.E., 'Effect of Magnetic Fields and Thermoelectromotive Force on the Kinetics of the Formation of the Solid Phase,' *Magnitnaya Gidrodinamika*, No. 1, January-March 1981, pp. 66–69.

Sahu, M.D. and V.A. Gadgil, 'Effects of Electromagnetic Fields on Solidification of Some Aluminum Alloys,' *British Foundryman*, No. 3, 1977, pp. 89–92.

Shul'man, Z.P., V.I. Kordonskii, and E.A. Zaltsgendler, 'Measurement of the Magneto-Rheological Characteristics of Ferrosuspensions,' *Magnitnaya Gidrodinamika*, No. 1, January–March 1979, pp. 39–43.

Shul'man, Z.P., V.I. Kordonskii, I.V. Prokhorov, and M.B. Smolskii, 'Experimental Investigation of the Magnetic Characteristics of Magneto-Rheological Suspensions,' *Magnitnaya Gidrodinamika*, No. 3, July–September 1980, pp. 31–37.

Shul'man, Z.P., V.I. Kordonskii, and S.A. Demchuk, 'Electrical Conductivity of Magneto-Rheological Suspensions,' *Kolloidn. Zhurnal*, No. 1, 1979, pp. 193–195.

Uhlmann, D.R., T.P. Seward, and B. Chalmers, 'The Effect of Magnetic Fields on the Structure of Metal Alloy Castings,' *Transactions of the Metallurgical Society AIME*, Vol. 236, 1966, pp. 527–531.

Neural networks

Amit, D.J., *Modelling Brain Function*, Cambridge University Press, 1989.

Grossberg, S., 'Nonlinear Neural Networks, Principles, Mechanisms and Architectures,' *Neural Networks*, Vol. 1, 1988, pp. 17–61.

Hirsch, M., 'Convergent Activation Dynamics in Continuous Time Networks,' *Neural Networks*, Vol. 2, 1989, pp. 341–349.

McCullough, W.S. and W. Pitts, 'A Logical Calculus of the Ideas Imminent in Nervous Activity,' *Bulletin of Mathematical Biophysics*, Vol. 3, 1943, pp. 115–133.

Mead, C.A., *Analog VLSI and Neural Systems*, Addison-Wesley, 1988.

Minsky, M. and S. Pappert, *Perceptrons*, Cambridge, MA. MIT Press, 1986.

Murray, A.F. and L. Tarassenko, *'Programmable Analog Pulse-Firing Neural Networks,'* Oxford University Preprint, 1990.

Rosenblatt, F., *Principles of Neurodynamics*, New York, Spartan, 1962.

Sheperd, G.H., *Neurobiology*, Oxford University Press, 1988.

General publications relevant to smart materials

Abela, G.S., 'Laser Re-Canalization: Preliminary Clinical Experience,' *Cardiovascular Disease Chest Pain*, Vol. 3, 1987, p. 3.

Advanced Materials by Design, U.S. Congress Office of Technological Assessment Report OTA-E-358, June 1988.

Allen, S.M. and J.W. Cahn, 'A Macroscopic Theory for Antiphase Boundary Motion and its Application to Antiphase Domain coarsening,' *Acta Metallurgica*, Vol. 27, 1979, pp. 1085–98.

Ball, J. and R. James, 'Fine Phase Mixtures as Minimizers of Energy,' *Archives Rational Mechanical Analysis*, Vol. 100, 1987, pp. 13–52.

Bassett, C.A.L., 'Biophysical Principles Affecting Bone Structure,' in *The Biochemistry and Physiology of Bone*, (ed. G.H. Bourne) (2nd Edition), Vol. 3, 1971, Academic Press, New York, pp. 1–76.

Bassett, C.A.L., 'Current Concepts of Bone Formation,' *Journal of Bone Jt. Surg.*, Vol. 44, 1962, p. 1217.

Bassett, C.A.L., 'Electrical Effects in Bone,' *Scientific American*, Vol. 213, No. 4, 1965, pp. 18–25.

Bassett, C.A.L. and R.O. Becker, 'Generation of Electric Potentials by Bone in Response to Mechanical Stress,' *Science*, Vol. 137, pp. 1063–64.

Bassett, C.A.L., R.J. Pawluk, and R.O. Becker, 'Effects of Electric Currents on Bone In Vivo,' *Nature*, 1964, Vol. 204, pp. 652–654.

Becker, R.D., C.A.L. Bassett, and C.H. Bachman, 'Bioelectrical Factors Controlling Bone Structure,' in *Bone Biodynamics*, (ed. H.M. Frost), 1964, Little-Brown, Boston, pp. 209–232.

Bendsoe, M. and N. Kikuchi, 'Generating Optimal Topologies in a Structural Design Using a Homogenization Method,' *Computational Methods Applied Mechanical Engineering*, Vol. 71, 1988, pp. 197–224.

Brighton, C.T., J. Black, and S.R. Pollack, *Electrical Properties of Bone and Cartilage: Experimental Effects and Clinical Applications*, 1979, Grune & Stratton, New York.

Bronowicki, A.J., T.L. Mendenhall, and R.M. Manning, *Advanced Composites with Embedded Sensors and Actuators* (*ACESA*), AL-TR-90–022, Astronautics Laboratory, Air Force Space Technology Center, April 1990, Report F04611–88-C-0054.

Brown, W.F., *Magnetostatic Principles in Ferromagnetism*, Vol. 1 of

Selected Topics in Solid State Physics, ed. E.P. Wohlfarth, North-Holland, 1962.

Brown, W.F., *Micromagnetics*, John Wiley and Sons, New York, 1963.

Burstall, A.F., *A History of Mechanical Engineering*, Faber and Faber, London, 1963.

Butler, J.L., *Application Manual for the Design of ETREMA Terfenol-D Magnetostrictive Transducers, Edge Technologies*, Inc., N. Marshfield, 1988.

Calvert, P., 'Biomimetic Processing of Ceramics and Composites,' *Proceedings of the International Workshop on Intelligent Materials, The Society of Non-Traditional Technology,* March 1989, Tsukuba Science City, Japan, pp. 63–72.

Canby, T.Y., 'Reshaping our Lives,' *National Geographic Magazine*, December 1989, pp. 746–781.

Chopra, D., *Quantum Healing*, Bantam Books, New York, 1989.

Clark, A.E., 'Chapter 7 – Magnetostrictive Rare Earth Fe2 Compounds,' *Ferromagnetic Materials*, *Vol. I*, Edited by E.P. Wohlfarth, North-Holland Publishing Co., 1980, pp. 531–589.

Clark, A.E., J.P. Teter, and M. Wun-Fogle, 'Magnetomechanical Coupling in Bridgman-Grown Tb0.3Dy0.7Fe1.9 at High Drive Levels,' presented at the *34th Conference on Magnetism and Magnetic Materials*, Boston, November 1989.

Clark, J.P. and M.C. Flemings, 'Advanced Materials and the Economy,' *Scientific American*, Vol. 255, No. 4, 1986, pp 43–49.

Clark, W.W., H.H. Robertshaw, and T.J. Warrington, 'A Comparison of Actuators for Vibration Control of the Planar Vibrations of a Flexible Cantilevered Beam,' *Journal of Intelligent Material Systems and Structures*, Vol. 1, No. 3, July 1990, pp. 289–308.

Collins, C., D. Kinderlehrer, and M. Luskin, 'Numerical Approximation of the Solution of a Variational Problem with a Double Well Potential' *IMA Preprint 605*, *SIAM Journal of Numerical Analysis* (to appear).

Confino, E., I. Tur-Kaspa, A. DeCherney, R. Corfman, C. Coulam, E. Robinson, G. Haas, E. Katz, M. Vermesh, N. Gleicher, 'Transcervical

Balloon Tuboplasty – A Multicenter Study,' *Journal of the American Medical Association*, Vol. 264, No. 16, October 1990, pp. 2079–82.

Edberg, D.L., 'Control of Flexible Structures by Applied Thermal Gradients,' *AIAA Journal*, Vol. 25, No. 6, June 1987.

The Encyclopaedia Britannica, Encyclopaedia Britannica Inc., Chicago IL, 1989.

Eringen, A.C., *Mechanics of Continua*, R.E. Krieger Publishing Company, Malabar, FL, USA, 1980.

Eringen, A.C. and G.A. Maugin, *Electrodynamics of Continua I: Foundations and Solid Media*, Springer Verlag, New York, 1989.

Eringen, A.C. and G.A. Maugin, *Electrodynamics of Continua II: Fluids and Complex Media*, Springer Verlag, New York, 1989.

Proceedings of the First Joint U.S./Japan Conference on Adaptive Structures, edited by B.K. Wada, J.L. Fanson, and K. Miura, November 13–15 1990, Maui, Hawaii, Technomic Publishing Company.

Fischer, G., 'Strategies Emerge for Advanced Ceramic Business,' *American Ceramic Society Bulletin*, Vol. 65, No. 1,1986, pp. 39–42.

Forester T., ed., *The Materials Revolution*, MIT Press, Cambridge, MA, 1988.

Forward, R.L., 'Electronic Damping of Vibrations in Optical Structures,' *Journal of Applied Optics*, Vol. 18, No. 5, pp. 690–697.

Gandhi, M.V., B.S. Thompson, and F. Fischer, 'Manufacturing-Process-Driven Methodologies for Composites Materials,' *Composite Manufacturing*, Vol. 1, No. 1, March 1990, pp. 32–40.

Gerardi, T., 'Health-Monitoring Aircraft,' *Journal of Intelligent Material Systems and Structures*, Vol. 1, No. 3, July 1990, pp. 375–385.

Gerardi, A.G., 'Health-Monitoring Aircraft,' *Proceedings of ARO/NASA/AFOSR/NASA/ONR, US–Japan Workshop*, Smart Intelligent Materials and Systems, Honolulu, Hawaii, March 1990.

Grippo, P.M., B.S. Thompson, and M.V. Gandhi, 'Flexible Fixturing Systems for CIM Environments,' *International Journal of Computer Integrated Manufacturing*, Vol. 1, No. 2, 1988, pp. 124–135.

Gandhi, M.V. and M. Usman, 'Equilibrium Characterization of Fluid-Saturated Continua, and an Interpretation of the Saturation Boundary Condition Assumption for Solid-Fluid Mixtures,' *International Journal of Engineering Science*, Vol. 27, No. 5, 1989, pp. 539–548.

Gandhi, M.V., M. Usman, A.S. Wineman and K.R. Rajagopal, 'Combined Extension and Torsion of a Swollen Cylinder Within the Context of Mixture Theory,' *Acta Mechanica*, Vol. 79, 1989, pp. 81–95.

Gazes, P.C., *Clinical Cardiology*, 3rd Edition, Lea & Febiger, Philadelphia, 1990, pp. 95–98.

Gurtin, M.E., 'On the Two-Phase Stefan Problem with Interfacial Energy and Entropy,' *Archives Rational Mechanical Analysis*, Vol. 96, 1986, pp. 199–241.

Gurtin, M.E., 'Multi-Phase Thermomechanics with Interfacial Structure. 1. Heat Conduction and the Capillary Balance Law,' *Archive Rational Mechanical Analysis*, Vol. 104, 1988, pp. 195–221.

Hall, D.P. and A.R. Gruentzig, 'Chapter 117: Technique of Percutaneous Transluminal Angioplasty,' *The Heart*, Hurst, J.W., Editor-in-Chief, R.B. Logue, C.E. Rackley, R.C. Schlant, E.H. Sonnenblick, A.G. Wallace, and N.K. Wenger, eds., 6th Edition, McGraw Hill, New York, 1985, pp. 1901–15.

Hiller, M.W., M.D. Bryant, and J. Umegaki, 'Attenuation and Transformation of Vibration through Active Control of Magnetostrictive Terfenol,' *Journal of Sound and Vibration*, Vol. 133, No. 3, 1989, pp. 1–13.

Ikegami, R., D.G. Wilson, and J.H. Laakso, *Advanced Composites with Embedded Sensors and Actuators* (*ACESA*), AL-TR-90–022, Astronautics Laboratory, Air Force Space Technology Center, June 1990, Report F04611–88-C-0053.

Jennings, J.S. and N.H. Macmillan, 'On the Architecture and Function of Cuttle-Fish Bone,' *Journal of Materials Science*, Vol. 18, 1983, pp. 2081–86.

Kohn, R., 'Recent Progress in the Mathematical Modelling of Composite Materials,' *Composite Material Response*: *Constitutive Relations and Damage Mechanisms*, G. Sih *et al.* eds., 1988, pp. 155–177.

Kohn, R. and G. Strang, 'Optimal Design and Relaxation of Variational Problems I-III,' *Communications in Pure Applied Mathematics*, Vol. 39, 1987, pp. 113–137, 139–182, 353–377.

Kranzberg, M. and C.S. Smith, 'Materials Science and Engineering: Its Evolution, Practice and Prospects,' *Material Science and Engineering*, Vol 37, No 1., 1979.

Lemonick M.D., T. McCarroll, J.M. Nash, and D. Wyss, 'Superconductors! The Startling Breakthrough That Could Change Our World,' *Time*, May 11, 1987.

Lubin, G., *Handbook of Composites*, Van Nostrand Reinhold Inc., 1982.

Lurie, K., A. Cherkaev, and A. Fedorov, 'Regularization of Optimal Design Problems for Bars and Plates I, II,' *Journal Optics, Theory, and Applications*, Vol. 37, 1982, pp. 499–522, 523–543.

Mahanti, S.D., and S. Tang, 'Vortices and Strings: Phase Transition in Anisotropic Planar-Rotor Systems,' *Physical Review*, Vol B 33, 1986, pp. 3419.

Mahanti, S.D., S. Tang, and R.K. Kalia, 'Ferroelastic Phase Transitions in Two-Dimensional Molecular Solids,' *Physics Review Letters*, Vol. 56, 1986, p. 484.

Mahanti, S.D., G. Kemeny, and J. Kales, 'Generalized Non-Linear Langevin Equation for a Rotor,' *Physical Review*, Vol. B 33, 1986, p. 3512.

Mahanti, S.D., and S. Tang, 'Phase Transitions in Diatomic Molecular Monolayers,' *Superlattices and Microstructure*, Vol. 1, 1986, p. 517.

Mahanti, S.D., G. Kemeny, and G. Zhang, 'Dynamics of Strongly Coupled Model Rotational-Transitional Systems,' *Physical Review*, Vol. B 35, 1987, p. 8551.

Mahanti, S.D., W. Jin, and S. Tang, 'Dynamics of a Two-Dimensional Diatomic Molecular Monolayer – a Molecular Dynamics Study,' *Solid State Communications*, Vol. 66, 1988, p. 877.

Mahanti, S.D., W. Jin, S. Tang, and R. Kalia, 'Ferroelastic Phase Transition and Phonons in a Diatomic Molecular Monolayer,' *Physical Review*, Vol. B 39, 1989, p. 1989.

Mahanti, S.D., W. Jin, and S. Tang, 'Internal Stress Tensor in Constant-Pressure Molecular Dynamics of Anisotropic Molecular Solids,' *Physical Review*, Vol. B 39, 1989, p. 11928.

Margolis, J.M., *Conductive Polymers and Plastics*, Chapman and Hall, 1989.

Margolis, J.M., 'Composites Challenge Metals for Aircraft/Auto Panel Applications,' *Machine and Tool Blue Book*, February 1987, pp. 45–48.

Moon, F.C., *Magneto-Solid Mechanics*, Wiley Interscence, NY 1984.

Moulson, A.J. and Herbert, J.M., *Electroceramics-Materials, Properties, Applications*, Chapman and Hall, 1990.

Mullins, W.W., 'Two-dimensional motion of idealized grain boundaries,' *Journal of Applied Physics*, Vol. 27, 1956, pp. 900–904.

Murat, F. and L. Tartar, 'Calcul des variations et homogenization,' *Les Methodes de l'Homogenization: Theorie et Applications en Physique*, Collection de la Dir. des Etudes et Recherche d'Electricite de France, Eyrolles, 1985, pp. 319–369.

Negahban, M. and A.S. Wineman, 'Material Symmetry and the Evolution of Anisotropies in a Simple Material: II. The Evolution of Symmetry,' *International Journal of Nonlinear Mechanics*, Vol. 24, No. 6, 1989, pp. 537–549.

Negahban, M. and A.S. Wineman, 'The Evolution of Anisotropies in the Elastic Response of an Elastic-Plastic Material,' *International Journal of Plasticity*, (to appear).

Negahban, M. and A.S. Wineman, 'Following the Mechanical Response in Phase Transitions: Elastic Solid to Elastic Solid Transitions,' *Proceedings of the Twelfth Canadian Congress of Applied Mechanics*, Ottawa, Ontario, 1989, pp. 840–841.

Negahban, M., 'The Effect of Continuous Phase Transition on a Torsional Oscillator,' *62nd Annual meeting of the Society of Rheology*, Santa Fe, New Mexico, October 1990.

Negahban, M., 'Constitutive Modeling of Phase Transition in Smart Materials', *ADPA/AIAA/ASME/SPIE Conference on Active Materials and Adaptive Structures*, Alexandria, Virginia, November 1991.

Olson, D.E. and B. Snyder, 'The Growth of Swirl in Curved Circular Pipes,' *Physics of Fluids*, Vol. 26, No. 2, 1983, pp. 347–349.

Olson, D.E. and B. Snyder, 'The Upstream Scale of Flow Development in

Curved Circular Pipes,' *Journal of Fluid Mechanics*, Vol. 150, 1986, pp. 139–158.

Osada, Y., R. Kishi, K. Umezawa, and H. Yasunaga, 'Mechochemical devices using polymer gels,' *Proceedings of International Workshop on Intelligent Materials*, published by the Society of Non-Traditional Technology, Tokyo 105, Japan, March 1989, pp. 297–302.

Prasad, P.N., 'Multifunctional Molecular and Polymeric Materials for Nonlinear Optics and Photonics,' *Materials Research Society Symposium Proceedings*, Vol. 175, 1990, pp. 79–87.

Prasad, P.N. and D.R. Ulrich, *Nonlinear Optical and Electroactive Polymers*, Plenum Press, New York, 1988.

Purna, S., *The Truth Will Set You Free*, Element Books, New York, 1987.

Rogers, C.A., ed., 'Smart Materials, Structures and Mathematical Issues,' selected papers presented at the *U.S. Army Research Office Workshop on Smart Materials, Structures and Mathematical Issues*, VPI, September 1988, Technomic Publishing, Lancaster, PA, 1989.

Rosato, D., D. Di Mattia, and D. Rosato, *Designing with Plastics and Composites – A Handbook*, Van Nostrand Reinhold Inc., 1991.

Rosenweig, R.E., *Ferrohydrodynamics*, Cambridge University Press, New York, 1985.

Rozvany, G., T. Ong, W. Szeto, R. Sandler, N. Olhoff, and M. Bendsoe, 'Least-weight design of perforated elastic plates I and II,' *International Journal of Solids and Structures*, Vol. 23, 1987, pp. 521–536, 537–550.

Rudd, J.L., 'Air Force Damage Tolerance Design Philosophy,' *ASTM STP 842 on Damage Tolerance of Metallic Structures*: *Analysis Methods and Applications*, pp. 134–141.

Shipman, P., A. Walker, and D. Bichell, *The Human Skeleton*, Harvard University Press, Cambridge, MA, 1985, pp. 53–63.

Snyder, B., 'A New Class of Fluid Instabilities: Vorticity-Induced Waveforms on Falling Parabolic Jets,' *Physics of Fluids*, Vol. 25, No. 5, 1982, pp. 905–907.

Snyder, B., J.R. Hammersley, and D.E. Olson, 'The Axial Skew of Flow in Curved Pipes,' *Journal of Fluid Mechanics*, Vol. 161, 1985, pp. 281–294.

Snyder, B., and D.E. Olson, 'Entrance-Flow Invariance in a Tapering Elliptical Slit,' *Physics of Fluids*, Vol. 29, 1986, pp. 2341–2343.

Snyder, B., D.E. Olson, J.R. Hammersley, C.V. Peterson, and M.J. Jaeger, 'Reversible and Irreversible Components of Central Airway Flow Resistance,' *ASME Journal of Biomechanics Engineering*, Vol. 109, 1987, pp. 154–159.

Snyder, B. and C. Lovely, 'A Computational Study of Developing 2-D Laminar Flow in Curved Channels,' *Physics of Fluids A*, Vol. 2, 1990, pp. 1808–1816.

Snyder, B., K.T. Li, and R.A. Wirtz, 'Secondary Goertler Vortices in a Serpentine Duct,' *Bulletin of American Physical Society*, Series II, Vol. 35, 1990, p. 2331.

Snyder, 'Spatially-Periodic Motion in a Curved Duct,' *Bulletin of American Physical Society*, Series II, Vol. 36, 1991.

Sung, C.K. and B.S. Thompson, 'A Methodology for Synthesizing High-Performance Robots Fabricated in Optimally-Tailored Composite Laminates,' *ASME Journal of Mechanisms, Transmissions, and Automation in Design*, Vol. 109, No. 1, March 1987, pp. 74–86.

Suzuki, A. and T. Tanaka, 'Phase Transition in Polymer Gels Induced by Visible Light,' *Nature*, Vol. 346, No. 6282, July 1990, pp. 345–347.

Takahashi, K., 'Intelligent Materials for Future Electronics,' *Journal of Intelligent Material Systems and Structures*, Vol. 1, No. 2, April 1990 pp. 248–258.

The Concept of Intelligent Materials and Guidelines on their R&D Promotion, Science and Technology Agency in Japan, November 1989, (published in English in January 1990).

Thompson, B.S. and M.V. Gandhi, 'Smart Materials,' in *Encyclopaedia Brittanica Science and the Future 1991 Year Book*, Encyclopaedia Brittanica, Chicago, IL, July 1990, pp. 162–171.

Thompson, B.S. and M.V. Gandhi, 'A Commentary on Flexible Fixturing,' *ASME Applied Mechanics Review*, Vol. 39, No. 9, 1986, pp. 136–69.

Thompson, B.S., 'Composite Laminate Components for Robotic and Machine Systems: Research Issues in Design,' *ASME Applied Mechanical Reviews*, Vol. 40, No. 11, November 1987, pp. 1545–52.

Uchino, K., 'Electrostrictive Actuators: Materials and Applications,' *American Ceramic Society Bulletin*, Vol. 65, No. 4, April 1986, pp. 647–652.

Udd, E., 'Embedded Sensors Make Structures Smart,' *Laser Focus/Electro-Optics*, May 1988, pp. 135–139.

Vaughan, J., *The Physiology of Bone*, 3rd Edition, Clarendon Press, Oxford, 1981, pp. 22–25.

Vincent, J.F.V., *Structural Biomaterials*, Macmillan Press, London, 1982.

Vincent, J.F.V, 'Toughness in Biological Materials,' *Proceedings of International Workshop on Intelligent Materials*, published by the Society of Non-Traditional Technology, Toyko, Japan, March 1989, pp. 131–138.

Vogel, S., 'Smart Skin,' *Discover*, April, 1990, p.26.

Wada, B.K., editor, *Adaptive Structures*, ASME Book AD-Vol. 15, 1989.

Wildey, J.F. II, 'Aging Aircraft,' National Association of Corrosion Engineers, NACE.

Wildey, J.F. II, 'Aloha 737,' *ASTM*, November 1989.

The World Book Encyclopedia, World Book Inc., Chicago, IL, Vol. 1–25, 1988.

Yamauchi, A., H. Okihiko, S. Fujishige, S. Ito, and H. Ichijo, 'Heat Sensitive and Responsible Polymer Gels,' *Proceedings of International Workshop on Intelligent Materials*, published by the Society of Non-Traditional Technology, Tokyo 105, Japan, March 1989, pp. 303–313.

Zhao, M.T., M. Samoc, B.P. Singh, and P.N. Prasad, 'Study of Third-Order Microscopic Optical Nonlinearities in Sequentially Built and Systematically Derivatized Structures,' *Journal of Physical Chemistry*, Vol. 93, 1989, pp. 7916–20.

Index